ELEMENTS
OF ALGEBRA

Prentice-Hall, Inc.
Englewood Cliffs, N.J. 07632

ELEMENTS

Francis J. Mueller

third edition

OF ALGEBRA

Library of Congress Cataloging in Publication Data

Mueller, Francis J
   Elements of algebra.

   Includes index.
   1. Algebra. I. Title.
QA154.2.M83 1981   512.9   80-21273
ISBN 0-13-262469-9

ELEMENTS OF ALGEBRA
Third Edition
Francis J. Mueller

Editorial/Production Supervision: Lynn S. Frankel
Cover Design: Lee Cohen
Manufacturing Buyer: John Hall

Printed in the United States of America

10 9 8 7 6 5 4 3 2 1

Prentice-Hall International, Inc., *London*
Prentice-Hall of Australia Pty. Limited, *Sydney*
Prentice-Hall of Canada, Ltd., *Toronto*
Prentice-Hall of India Private Limited, *New Delhi*
Prentice-Hall of Japan, Inc., *Tokyo*
Prentice-Hall of Southeast Asia Pte. Ltd., *Singapore*
Whitehall Books Limited, *Wellington, New Zealand*

CONTENTS

★Starred sections, throughout this text, indicate material that may be postponed in more basic courses of study, without significant effect upon the remaining parts of the book.

V

SUPPLEMENTARY UNITS

PREFACE

As was the case with the previous editions, this book is intended for the college student whose grasp of algebra is currently weak or marginal. Even those students who come into college without any significant exposure to the subject should find entry into *Elements of Algebra* a relatively easy matter.

Experience with the previous editions has confirmed the importance of a patient and deliberate pace, of taking little or nothing for granted, of offering an abundance of concrete and illustrative examples, of synthesizing in clear and unequivocal terms "what to do," and of supporting the whole fabric of instruction with a great many exercise opportunities and extensive end-of-chapter reviews. All of these emphases have been continued in the present edition.

A prominent feature of this book, as well as its predecessors, is its statements of operational procedures in essentially linear-programmed form. These compact, step-by-step guides are not intended to be memorized blindly, but to serve as models of the ultimate systematic way in which most persons competent in the subject carry out the basic operations of algebraic computation.

Students for whom this book is appropriate often experience difficulty with applications. Consequently, a full chapter has been devoted to the topic. The treatment is extensive and painstaking, the strategy being to keep the student involved long enough to develop a useful degree of competence. Problem solving, like much of the mechanics of mathematics, is a skill that can be acquired through practice.

Consistent attention has been given to keeping the wording of the text clear and simple. The style is direct, almost conversational. That and the heavy emphasis upon programmed procedures and illustrative examples should, for many, make the book practically self-teaching.

While the pedagogy has remained essentially unchanged, there has been some modification of the subject matter in the present edition. The first two chapters of the previous edition have been merged into a more compact and effective opening chapter, Review of Basics. It is intended to be a thorough refresher for all students, and anticipates well the content of the remaining chapters.

The next two chapters, Special Products and Factoring, and Fractions, have been moved forward from their previous positions. In this way, the first three chapters of the book, along with the next, First-Degree Equations and Inequalities, combine to better refresh and clarify the more rudimentary algebraic skills for the student.

Chapter 5, Applications, is the aforementioned in-depth presentation on problem solving by algebraic methods, expanded somewhat from that of the previous edition. New problems of particular interest to business students have been added here and elsewhere throughout the book.

The following chapter, Exponents and Radicals, contains a new section on scientific notation; also its sections on computation with radical expressions have been reorganized and streamlined. Chapter 7, Quadratic Equations and Inequalities, now contains a new and clearer approach to the solution of quadratic inequalities.

There are major changes in the next chapter, Equations and Inequalities in Two Variables. The chapter now includes discussion of distance between points in a plane, slope, the writing of equations for lines, absolute-value equations, and inequalities in two variables (formerly a supplementary unit). Treatment of the concept of function and its notation, previously in this chapter, is now developed in a less formal way in a new supplementary unit.

In this edition a full chapter is devoted to Systems of Equations and Inequalities, including a new section on applications. The section on inequalities was formerly part of a supplementary unit. It has been moved forward to maintain the pattern of treating equations and inequalities in tandem.

In the final chapter, Logarithms, users of earlier editions will note less emphasis upon interpolation, and new sections on exponential equations (previously a supplementary unit) and change of base.

Supplementary units in the present edition are three. A new unit deals with Functions. The other two are holdovers: Progressions and Binomial Theorem. These units may be introduced at various places in the course, depending upon local preferences.

As has been the case with prior editions, there is more subject matter provided in the book than can be conveniently covered in most courses. For the less prepared students, Chapters 1 through 8 and selections from Chapter 9, with deferment of most or all of the starred sections until later courses, will provide a comprehensive basic grounding in algebra. Throughout the book, certain sections, marked with a star, have been identified for possible postponing. The choice to omit any one of them will not handicap development of subsequent topics in the text.

For the better prepared students, a quick review of the first five chapters and concentration on the remaining chapters, plus study of some or all of the supplementary units, will meet the requirements of most courses at the intermediate algebra level.

Variations within these bounds are, of course, both possible and practical.

The evolution of this book to its present state has had, over the years, the helpful guidance of many people—students, teachers, manuscript critics, and editors. By no means are all of them known to the author, unfortunately; but each is due his sincerest thanks. Especially helpful in the prepublication stages of the present edition were the incisive critiques of Professor John W. Hooker, of the Department of Mathematics at Southern Illinois University at Carbondale. His patient comments and those of Professors Terry Czerwinski and Gary Grabner of the University of Illinois at Chicago Circle and Ohio University, respectively, have influenced the content of this book in ways that only the author can recognize and fully appreciate. To all these worthy people, and particularly to Production Editor Lynn Frankel for her dedicated efforts in bringing this work to fruition, I express my deepest gratitude.

Tampa, Florida   **Francis J. Mueller**

LIST OF PROGRAMS

1 REVIEW OF BASICS

1. SETS OF NUMBERS

Algebra has an ancient heritage. The term itself comes from the Arabic, *al-jabr*, the transposition of a quantity from one member of an equation to the other. Historians have traced parts of algebra back to the Egyptians and Greeks, centuries before the time of Christ. The subject as we know it today, however, began to be shaped in the Middle Ages and has continued to evolve. Some of the latest refinements, in emphasis at least, had their beginnings as recently as the last decade or two.

This ever-growing body of knowledge called algebra has its roots in number reckoning. Elementary arithmetic is a rudimentary part of algebra. Principally through the use of letters (variables) to stand for unspecified numbers, algebra has become an enormously more powerful medium for problem solving than arithmetic could ever hope to be.

The natural place, then, to begin a study of algebra is with a discussion of numbers—sets of numbers—and their characteristics.

A **set**, you may recall, is simply a collection of **elements**, called **members**. Sets may be expressed in a variety of ways. The **roster method** is one way, a listing of all of the elements of the set. For example:

$$\{1, 3, 5\}$$

(i.e., "the set whose elements are the numbers 1, 3, 5"). Another way is called the **descriptive method**; for example:

$$\{\text{integers greater than 17}\}$$

(i.e., "the set whose members are the integers greater than 17"). In either case the brace notation, { }, is standard for indicating sets.

Sets may have a few members, such as

$$\{3, 4\}, \quad \{5\}, \quad \{2, 4, 6\}$$

or infinitely many members, such as

$$\{2, 4, 6, 8, 10, \ldots\}$$

(the "..." means "and so on"). It is possible for a set to have no members at all:

$$\{ \ \}$$

Such a set is called the **empty set**, and is usually symbolized by a pair of empty braces, { }, or at times, $\emptyset$.

A partial collection of elements of a given set is called a **subset** of the given set. For example, the set

$$\{1, 2, 3, 4, 5, 6, 7\}$$

has the following sets among its many subsets:

$$\{1, 2, 3\}, \quad \{3, 6\}, \quad \{1, 5, 6\}, \quad \{ \ \}$$

The empty set is a subset of every set.

The set of numbers used primarily in algebra, at least at our level of inquiry, are the **real** numbers. It includes the numbers of arithmetic as well as other types. Each real number corresponds to a unique point on a line. And every point on that line may be thought to correspond to one and only one real number. It would be impossible in a single drawing to label *every* point on the line with its corresponding real number, but *any* point may be properly labeled, as some are in Figure 1.1.

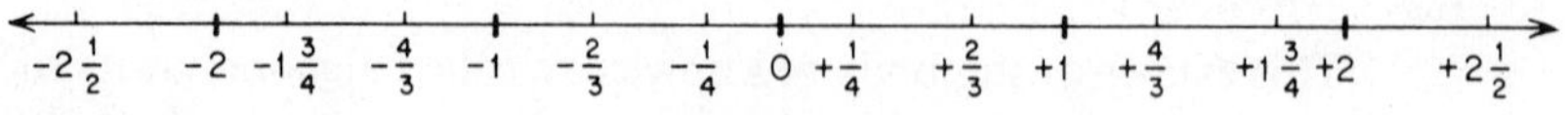

$$-2\tfrac{1}{2} \quad -2 \quad -1\tfrac{3}{4} \quad -\tfrac{4}{3} \quad -1 \quad -\tfrac{2}{3} \quad -\tfrac{1}{4} \quad 0 \quad +\tfrac{1}{4} \quad +\tfrac{2}{3} \quad +1 \quad +\tfrac{4}{3} \quad +1\tfrac{3}{4} \quad +2 \quad +2\tfrac{1}{2}$$

Figure 1.1

Positive real numbers occur to the right of 0 on the real number line, and **negative** real numbers occur to the left of 0.

There are other categories of real numbers—that is, subsets of the set of real numbers—some with overlapping memberships. Among them:

- The **natural numbers**, which are the familiar counting numbers:

$$\{1, 2, 3, 4, 5, 6, 7, \ldots\}$$

- The **whole numbers**, which are the natural numbers, plus 0:

$$\{0, 1, 2, 3, 4, 5, 6, 7, \ldots\}$$

- The **integers**, which are the whole numbers and their respective negatives (0 may be considered to be its own negative):

$$\{0, 1, -1, 2, -2, 3, -3, \ldots 116, -116, \ldots\}$$

- The **rational numbers**, which include the numbers already mentioned, plus the numbers that are the quotients of every possible pair of integers (except divisors of 0), both positive and negative. Some examples:

$$\frac{2}{3} \text{ (quotient of } 2 \div 3); \quad \frac{3}{5}; \quad \frac{-2}{7} \text{ or } -\frac{2}{7}; \quad \frac{8}{5};$$

$$-\frac{9}{5} \text{ or } -1\frac{4}{5}; \quad \frac{6}{2} \text{ or } 3; \quad \frac{4}{1} \text{ or } 4; \quad \text{etc.}$$

Note: Be aware of the distinction between a number and its various numerals or symbols. Each number has many representative numerals. For example, the number seven may be expressed as $7, \frac{7}{1}, \frac{14}{2}$, VII, $3 + 4$, and so on—all numerals, or names, for the same number.

Another way to identify a rational number is to express it in decimal notation. If its decimal version terminates $\left(\text{e.g., as } \frac{3}{5} = 0.6, \frac{7}{8} = 0.875, \frac{5}{4} = 1.25\right)$, or if its decimal equivalent does not terminate but ultimately repeats its digits in some constant pattern $\left(\text{e.g., } \frac{7}{11} = 0.636363\ldots, \text{ or } \frac{4}{3} = 1.3333\ldots\right)$, the number is rational.

The *ratio*nal numbers get their class name from the fact that each may be expressed as a *ratio* (i.e., quotient) of two integers, as we have noted. Those real numbers that *cannot* be so expressed are called **irrational numbers**. Some examples of irrational numbers are $\sqrt{2}, -\sqrt{3}, \pi, \sqrt[3]{5}, \sqrt[5]{-71}$. Because they are real numbers, each irrational number has its own unique location along the real-number line. Their decimal equivalents do not terminate or repeat a constant pattern of digits. For example:

$$\sqrt{2} = 1.414213\ldots \qquad \sqrt[3]{5} = 1.709975\ldots$$

$$-\sqrt{3} = -1.73205\ldots \qquad \sqrt[5]{-71} = -2.345587\ldots$$

$$\pi = 3.14159265\ldots$$

The rational and irrational numbers are said to be **discrete** subsets (i.e., no overlap in membership) of the set of real numbers.

Figure 1.2 summarizes the interrelationship of the various sets of numbers we have been discussing.

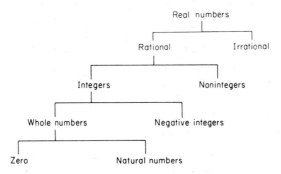

Figure 1.2

Furthermore, the set of real numbers is said to be **ordered**: the numbers can be arranged by size. To put this another way, for any pair of real numbers, one of the two numbers will be greater (and one will be less) than the other.

The usual symbols for such expressions of inequality are

$$<:\quad \text{less than}$$

$$>:\quad \text{greater than}$$

Since each real number has a unique location along the number line, the number line becomes an excellent model for visualizing the relative sizes of two numbers. On a standard number line, as in Figure 1.3:

Figure 1.3

- Numbers located to the *right* of a given number are all *greater than* ($>$) the given number.
- Numbers located to the *left* of a given number are all *less than* ($<$) the given number.

For example,

$$1 > -1 \quad \sqrt{2} > -\sqrt{2} \quad 4 > \sqrt{3} \quad \pi > 3$$

$$-2 < 1 \quad 0 < \sqrt{3} \quad \sqrt{2} < \pi \quad -2 < 0$$

We will study the irrational numbers in greater detail in Chapter 6. Meanwhile, the diagram of Figure 1.4 is intended to show in a slightly different way the interrelationship of the more important sets of numbers that we have been discussing.

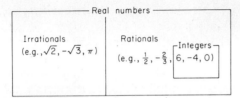

Figure 1.4

Use the code:

Re: real number   In: integer

Ra: rational number   P: positive number

Ir: irrational number   N: negative number

to describe each number. Example: $\sqrt{2}$ Re, Ir, P (i.e., $\sqrt{2}$ is at the same time a real number, an irrational number, and a positive number).

1. 6   **2.** -3 **3.** $-\dfrac{1}{2}$ **4.** $\dfrac{2}{5}$

5. $-3\dfrac{1}{4}$ **6.** $2\dfrac{5}{8}$ **7.** $\sqrt{6}$ **8.** $-\sqrt{5}$

9. 0   **10.** π **11.** $0.626262\ldots$ (repeating)

12. $-1.373737\ldots$ (repeating)   **13.** $-4.62719\ldots$ (nonrepeating)

14. $0.372165\ldots$ (nonrepeating)

Tell in which lettered span, A, ..., E, the following numbers will occur.

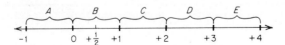

15. $+\dfrac{1}{2}$ **16.** $-\dfrac{1}{2}$ **17.** $3\dfrac{1}{4}$ **18.** $2\dfrac{1}{2}$

19. $-\dfrac{7}{8}$ **20.** $\dfrac{3}{4}$ **21.** $\dfrac{8}{5}$ **22.** -3.1

23. $1.62626262\ldots$ **24.** $-0.00031323\ldots$

Refer to the number line of Figure 1.3. Tell whether the statement is true or false.

25. $-\sqrt{2} < 0$ **26.** $\sqrt{2} < 1$ **27.** $2 > \pi$

28. $4 > -2$ **29.** $-2 < 4$ **30.** $\sqrt{3} > \sqrt{2}$

31. $\sqrt{2} > -2$ **32.** $\pi > 3$ **33.** $0 < -\sqrt{2}$

34. $-\sqrt{2} > \sqrt{3}$ **35.** $0 > -2$ **36.** $-2 > \sqrt{3}$

2. SUMS AND DIFFERENCES OF REAL NUMBERS

The operations of arithmetic—addition, subtraction, multiplication, and division—also hold for the set of real numbers. We review them in this section and the next in program form.

Recall that the **absolute value** of a number may be thought of as the "distance" that a number lies from 0 on the number line, irrespective of direction. Note in Figure 1.3, in the previous section, that both $+2$ and -2 lie at a distance of 2 units from 0.

Using the symbol $|\ \ |$ to indicate absolute value, we have

$$|+2| = |-2| = 2$$

which is read "the absolute value of $+2$ equals the absolute value of -2 equals 2."

Similarly,

$$\left| -1\frac{1}{2} \right| = \left| +1\frac{1}{2} \right| = 1\frac{1}{2}$$

$$|+1000| = |-1000| = 1000$$

Thus a number and its negative have the same absolute value which is always positive. In practice, the absolute value of a nonzero number can readily be established by simply ignoring the sign of its numeral.

When the operation is addition, the numbers that go into the operation are called addends, and the result is called their sum:

$$\text{addend} + \text{addend} = \text{sum}$$

PROGRAM 1.1

To add two or more real numbers that agree in sign (i.e., all terms positive or all negative):

Step 1: Compute the sum of the absolute values of the numbers.

Step 2: Prefix the sum of Step 1 with the common sign.

EXAMPLES

1. Find the sum of -2 and -4 according to Program **1.1**.

SOLUTION *Step 1:* The absolute value of -2 is 2.
The absolute value of -4 is 4.
Sum of the absolute values: $2 + 4 = 6$.

Step 2: Common sign of the addends: $-$(minus).
Sum of -2 and -4 is -6.

2.   Add $+3$ and $+4$.

Step 1: $|+3| + |+4| = 3 + 4 = 7$

Step 2: Common sign of the addends: $+$(plus), thus:

$$(+3) + (+4) = +7$$

3.   Add $(-5) + (-2) + (-4)$.

Step 1: $|-5| + |-2| + |-4| = 5 + 2 + 4 = 11$

Step 2: The sum is given the common sign of the addends; thus:

$$(-5) + (-2) + (-4) = -11$$

4.   Add $(+6) + (+2) + (+5.7)$.

Step 1: $|+6| + |+2| + |+5.7| = 6 + 2 + 5.7 = 13.7$

Step 2: $(+6) + (+2) + (+5.7) = +13.7$

5.   Add $\left(-\dfrac{1}{5}\right) + \left(-\dfrac{3}{5}\right)$.

Step 1: $\left|-\dfrac{1}{5}\right| + \left|-\dfrac{3}{5}\right| = \dfrac{1}{5} + \dfrac{3}{5} = \dfrac{1+3}{5} = \dfrac{4}{5}$

Step 2: $\left(-\dfrac{1}{5}\right) + \left(-\dfrac{3}{5}\right) = -\dfrac{4}{5}$

PROGRAM 1.2

To add two real numbers that differ in sign:

Step 1:   Subtract the smaller absolute value of the two numbers from the greater.

Step 2:   Prefix the result of Step 1 with the sign of the number having the greater absolute value.

EXAMPLES

1.   Find the sum of $+2$ and -6 according to Program **1.2**.

Step 1: The absolute value of $+2$ is 2.

The absolute value of -6 is 6, and $6 > 2$.

Difference of absolute values: $6 - 2 = 4$.

Step 2: The sign of the number having the greater absolute value is $-$; so $(+2) + (-6) = -4$.

2.   Add $(+6) + (-5)$.

SOLUTION *Step 1:* $|+6| = 6$ (greater absolute value)
$|-5| = 5$
$6 - 5 = 1$

Step 2: The number having the greater absolute value has $+$ for its sign; that sign is given to the difference:

$$(+6) + (-5) = +1$$

3.   Add -120 and $+85$.

SOLUTION *Step 1:* $|-120| = 120$
$|+85| = 85$
-120 has the greater absolute value.

$$(-)\,\frac{\begin{array}{r} 120 \\ 85 \end{array}}{35}$$

Step 2: The sum of -120 and $+85$ is -35.

PROGRAM 1.3

To add three or more real numbers that differ in sign:

Step 1:   Compute the sum of the positive numbers.
Step 2:   Compute the sum of the negative numbers.
Step 3:   Add the sums of Steps 1 and 2, using Program **1.2**, for the desired sum.

EXAMPLES

1.   Add $(+3) + (-2) + (+5) + (+6) + (-14)$.

SOLUTION *Step 1:* Compute the sum of the positive numbers:

$$(+3) + (+5) + (+6) = +14$$

Step 2: Compute the sum of the negative numbers:

$$(-2) + (-14) = -16$$

Step 3: Add the sums of Steps 1 and 2:

$$(+14) + (-16) = -2$$

Thus:

$$(+3) + (-2) + (+5) + (+6) + (-14) = -2$$

2. Add $(+2) + (-3) + (-4) + (-13) + (+16)$.

Step 1:   *Step 2:*   *Step 3:*

$$
\begin{array}{r}
+\ 2 \\
+16 \\
(+)\ \overline{+18}
\end{array}
\qquad
\begin{array}{r}
-\ 3 \\
-\ 4 \\
-13 \\
(+)\ \overline{-20}
\end{array}
\qquad
\begin{array}{r}
+18 \\
-20 \\
(+)\ \overline{-2} \quad \text{(sum)}
\end{array}
$$

3. Compute the sum of $62,\ -35,\ -87,\ 210,$ and 34.

Step 1: Sum of the positive numbers:

$$62 + 210 + 34 = 306$$

Step 2: Sum of the negative numbers:

$$(-35) + (-87) = -122$$

Step 3: $(+306) + (-122) = +184$

There is an important distinction between a negative number and the negative *of* a number. A negative number is one that is less than zero. The *negative of* a number is one that has a sign "opposite" that of the given number. A number and its negative always add to zero. For example,

The negative of $+3$ is -3.
The negative of -3 is $+3$.
The negative of $-\dfrac{1}{2}$ is $+\dfrac{1}{2}$.

This distinction is important to Program **1.4**, which follows, and elsewhere in mathematics.

The negative of a number is also known as the **additive inverse** of the given number. The sum of a number and its additive inverse is always 0.

PROGRAM 1.4

To subtract one real number from another:

Add the negative of the number to be subtracted to the other number.

Note: An alternative statement would be: Reverse the sign of the number to be subtracted and add.

EXAMPLES

1. Subtract $(+8) - (+5)$.

SOLUTION
The negative of $+5$ is -5:
$$(+8) - (+5) = (+8) + (-5) = +3$$

2. Subtract $(-3) - (-4)$.

SOLUTION
The negative of -4 is $+4$:
$$(-3) - (-4) = (-3) + (+4) = +1$$

3. Subtract $(-6) - (+3)$.

SOLUTION
The negative of $+3$ is -3:
$$(-6) - (+3) = (-6) + (-3) = -9$$

4. Subtract the lower number from the upper.

SOLUTION

(a)
$$\begin{array}{r} -15 \\ (-)\underline{ -\ 3} \end{array} \Rightarrow \begin{array}{r} -15 \\ (+)\underline{+\ 3} \\ -12 \end{array}$$

(b)
$$\begin{array}{r} +36 \\ (-)\underline{-57} \end{array} \Rightarrow \begin{array}{r} +36 \\ (+)\underline{+57} \\ +93 \end{array}$$

(c)
$$\begin{array}{r} -\dfrac{3}{8} \\ (+)\underline{+\dfrac{1}{8}} \end{array} \Rightarrow \begin{array}{r} -\dfrac{3}{8} \\ (+)\underline{-\dfrac{1}{8}} \\ -\dfrac{4}{8} = -\dfrac{1}{2} \end{array}$$

(d)
$$\begin{array}{r} +71 \\ (-)\underline{+90} \end{array} \Rightarrow \begin{array}{r} +71 \\ (+)\underline{-90} \\ -19 \end{array}$$

EXERCISE 1-2

Write the absolute value of each.

1. $+5$

2. $+8$

3. -6

4. -12

5. $+\dfrac{1}{4}$

6. $-\dfrac{3}{4}$

7. 0

8. $-1\dfrac{1}{2}$

9. $-7\dfrac{2}{5}$

10. $+\dfrac{8}{3}$

Add.

11.
$$\begin{array}{r} +13 \\ (+)\underline{+\ 6} \end{array}$$

12.
$$\begin{array}{r} -86 \\ (+)\underline{-13} \\ -73 \end{array}$$

13.
$$\begin{array}{r} +27 \\ (+)\underline{-15} \\ 12 \end{array}$$

10 Chapter 1 REVIEW OF BASICS

14. $+31$
            -14
        $(+)$ ‾‾‾‾‾

15. -62
            -27
        $(+)$ ‾‾‾‾‾

16. $+39$
            $+47$
        $(+)$ ‾‾‾‾‾

17. $+42$
            -62
        $(+)$ ‾‾‾‾‾

18. $+63$
            -15
        $(+)$ ‾‾‾‾‾

19. $+128$
            $-\ 67$
        $(+)$ ‾‾‾‾‾

20. -384
            $+126$
        $(+)$ ‾‾‾‾‾

21. $+376$
            -954
        $(+)$ ‾‾‾‾‾

22. -183
            $+274$
        $(+)$ ‾‾‾‾‾

23. $(+176) + (-82)$

24. $(-863) + (-175)$

25. $(+185) + (+76)$

26. $(-162) + (+81)$

27. $\left(+\dfrac{1}{7}\right) + \left(+\dfrac{3}{7}\right)$

28. $\left(-\dfrac{17}{23}\right) + \left(-\dfrac{4}{23}\right)$

29. $(-8) + (-3) + (+2)$

30. $(+16) + (-9) + (-7)$

31. $(-14) + (-6) + (-8)$

32. $(-21) + (+62) + (-14)$

33. $(+13) + (-15) + (-19)$

34. $(-112) + (-104) + (+63) + (+52)$

35. $(+416) + (-18) + (-16) + (-198)$

36. $(+216) + (-173) + (-412) + (+63)$

Subtract the lower number from the upper.

37. $+14$
            $+\ 8$
        $(-)$ ‾‾‾‾‾

38. $+43$
            $+28$
        $(-)$ ‾‾‾‾‾

39. -18
            $+\ 6$
        $(-)$ ‾‾‾‾‾

40. -27
            -13
        $(-)$ ‾‾‾‾‾

41. $+13$
            $+\ 8$
        $(-)$ ‾‾‾‾‾

42. $+32$
            $+18$
        $(-)$ ‾‾‾‾‾

43. -29
            -17
        $(-)$ ‾‾‾‾‾

44. -84
            -21
        $(-)$ ‾‾‾‾‾

45. $+13$
            $-\ 6$
        $(-)$ ‾‾‾‾‾

46. $+27$
            $-\ 5$
        $(-)$ ‾‾‾‾‾

47. -43
            $+12$
        $(-)$ ‾‾‾‾‾

48. -68
            $+19$
        $(-)$ ‾‾‾‾‾

Subtract.

49. $(-34) - (+38)$

50. $(-67) - (+84)$

51. $(+53) - (-61)$

52. $(+94) - (-98)$

53. $(+47) - (+50)$

54. $(+67) - (+79)$

55. $(-38) - (-41)$

56. $(-76) - (-89)$

57. $(+114) - (+216)$

58. $(+306) - (+169)$

59. $(-200) - (-200)$

60. $(-182) - (-282)$

Supply the missing addend in the sum.

61. $(+6) + (\ \ ) = +2$

62. $(+3) + (\ \ ) = 0$

63. $(\ \ ) + (-3) = +4$

64. $(\ \ ) + (+3) = -7$

65. $(-9) + (+2) + (\quad) = -16$ **66.** $(-4) + (\quad) + (-17) = +2$
67. $(\quad) + (-13) + (-19) = -5$ **68.** $(\quad) + (+16) + (-19) = -22$

3. PRODUCTS AND QUOTIENTS OF REAL NUMBERS

In this section we review the procedures for multiplying and dividing real numbers. Recall the terminology of multiplication:

$$\text{factor} \times \text{factor} = \text{product}$$

PROGRAM 1.5

To multiply two or more real numbers:

Step 1:   Compute the product of the absolute values of the numbers.

Step 2:   Prefix the product of Step 1 with a plus sign if the number of negative factors in the product is even (i.e., none, two, four, etc.) or with a minus sign if the number of negative factors in the product is odd (i.e., one, three, five, etc.).

EXAMPLES

1.   Multiply $(-6) \times (-3) \times (+10)$.

SOLUTION   *Step 1:* Compute the product of the absolute values of the factors:

$$|-6| \times |-3| \times |+10| = 6 \times 3 \times 10 = 180$$

Step 2: The number of negative factors (two) is even; therefore, the product is positive:

$$(-6) \times (-3) \times (+10) = +180$$

2.   Multiply $(-3) \times (-2) \times (-1) \times (-7) \times (-3)$.

SOLUTION   *Step 1:* The product of the absolute values of the factors:

$$3 \times 2 \times 1 \times 7 \times 3 = 126$$

Step 2: The number of negative factors is odd (five); therefore, the product is negative:

$$(-3) \times (-2) \times (-1) \times (-7) \times (-3) = -126$$

3. Multiply $\left(-\dfrac{1}{2}\right) \times \left(+\dfrac{3}{5}\right).$

SOLUTION *Step 1:* $\left| -\dfrac{1}{2} \right| \times \left| +\dfrac{3}{5} \right| = \dfrac{1}{2} \times \dfrac{3}{5} = \dfrac{1 \times 3}{2 \times 5} = \dfrac{3}{10}$

Step 2: The number of negative factors is odd (one); so the product is negative:

$$\left(-\dfrac{1}{2}\right) \times \left(+\dfrac{3}{5}\right) = -\dfrac{3}{10}$$

Note: When we deal with numbers exclusively, as we do in arithmetic, the "times" sign $\times$ is appropriate. However, in algebra the factors of a product are usually written without the $\times$ sign, which can be confused with the variable x. Sometimes a dot is used—for instance, $6 \cdot 4 = 6 \times 4$. More generally, multiplication is indicated by enclosing the factors in parentheses, with no operation symbol between the factors—for instance, $(6)(4) = 6 \times 4$.

We turn now to the division of real numbers. Recall the terminology of division:

$$\text{dividend} \div \text{divisor} = \text{quotient}$$

or

$$\dfrac{\text{dividend}}{\text{divisor}} = \text{quotient}$$

PROGRAM 1.6

To divide one real number by another:

Step 1:   Compute the quotient of the absolute values of the two numbers.

Step 2:   Prefix the quotient of Step 1 with a plus sign if the two numbers have the same sign and with a minus sign if the two numbers have opposite signs.

EXAMPLES

1.   Divide $(-36) \div (-9).$

SOLUTION *Step 1:* Compute the quotient of the absolute values of dividend and divisor:

$$|-36| \div |-9| = 36 \div 9 = 4$$

Step 2: (-36) and (-9) are both negative; therefore, the quotient is positive:

$$(-36) \div (-9) = +4$$

2. Divide $(-42) \div (+7)$.

Step 1: Compute the quotient, using absolute values:
$$|-42| \div |+7| = 42 \div 7 = 6$$

Step 2: (-42) and $(+7)$ are negative and positive, respectively; therefore, the quotient is negative:
$$(-42) \div (+7) = -6$$

3. Simplify $\dfrac{(-3) \times (+8)}{-2}$.

$$\frac{(-3) \times (+8)}{-2} = \frac{-24}{-2} = (-24) \div (-2) = +12$$

EXERCISE 1-3

Compute the products.

1. $(-5) \times (+3) \times (-2)$ **2.** $(-4) \times (-2) \times (-6)$
3. $(+4) \cdot (+10) \cdot (+3)$ **4.** $(-3) \cdot (+2) \cdot (-7)$
5. $(-5)(-2)(-7)$ **6.** $(+8)(+2)(+4)$
7. $(-3)(-4)(+2)(-5)$ **8.** $(-6)(+3)(+4)(+2)$
9. $(-2)(-3)(-5)(+7)(+5)(-2)$
10. $(+2)(+6)(-5)(-3)(+4)(+3)$
11. $(-1)(-1)(+3)(+2)(-1)(-3)(+6)$
12. $(-3)(+6)(-9)(+4)(0)(-8)(+5)$
13. $(+2) \times (-6 + 3)$ **14.** $(-2) \times (+10 + 2)$
15. $(-5) \times (+2 - 5 - 3)$ **16.** $(+3) \times (-4 - 2 + 8)$
17. $(-6)(-5 + 7 - 2)$ **18.** $(+5)(-6 + 3 - 1)$
19. $(-2 + 3)(+4 - 6)$ **20.** $(+7 - 8)(-6 + 5)$

Compute the quotients.

21. $(+4) \div (-2)$ **22.** $(-6) \div (+3)$ **23.** $(-20) \div (-5)$
24. $(+9) \div (+3)$ **25.** $(-12) \div (+4)$ **26.** $(-36) \div (-6)$
27. $(+56) \div (+8)$ **28.** $(+30) \div (-10)$ **29.** $(-1) \div (-5)$
30. $(+12) \div (-12)$ **31.** $(+7) \div (+2)$ **32.** $(-10) \div (-3)$

Simplify.

33. $\dfrac{(-3) \times (+2)}{(+6)}$ **34.** $\dfrac{(-3) \times (+4)}{(+2)}$ **35.** $\dfrac{(-5) \times (+4)}{(-10)}$

36. $\dfrac{(+6) \times (-3)}{(-9)}$ **37.** $\dfrac{(-8) \div (-2)}{(+2)}$ **38.** $\dfrac{(+49) \div (-7)}{(+3)}$

39. $\dfrac{(-4) \div (+9)}{(-3) \times (+2)}$ **40.** $\dfrac{(+3) \div (-2)}{(+5) \times (-3)}$

4. SOME PROPERTIES OF OPERATIONS

As we perform various operations on numbers, we may or may not be aware of certain basic properties of the operations. Our general number sense is usually enough to keep our thinking straight. In algebra, however, we deal with more abstract concepts and an awareness of certain properties is often the key to understanding.

For instance, it is well known that addition and subtraction "undo" one another:

$$6 + 2 = 8 \quad \text{(addend + addend = sum)}$$

$$8 - 2 = 6 \quad \text{(sum } - \text{ addend = other addend)}$$

So do multiplication and division "undo" one another:

$$3 \times 4 = 12 \quad \text{(factor} \times \text{factor = product)}$$

$$12 \div 3 = 4 \quad \text{(product} \div \text{factor = other factor)}$$

Operations that "undo" one another are called **inverse operations** of one another. Thus addition and subtraction are inverse operations of each other, and multiplication and division are inverse operations of each other. You have probably used inverse operations in arithmetic to check your work. That is, you have checked a subtraction by addition, a multiplication by division, or a division by multiplication.

Here are a few more important properties of operations that you should know about.

commutative property

The commutative property has to do with the order, or sequence, in which the operations are performed. The real numbers are commutative under the operations of addition and multiplication. That is, the order or sequence in which we add real numbers, or multiply them, makes no difference as far as the result is concerned. To illustrate,

$$\text{addition:} \qquad 2 + 3 = 5 \qquad 3 + 2 = 5$$

$$\text{multiplication:} \qquad \frac{1}{2} \times 6 = 3 \qquad 6 \times \frac{1}{2} = 3$$

The commutative property does not hold for the inverse operations of subtraction and division, as demonstrated by these examples.

$$\text{subtraction:} \quad 6 - 7 = -1 \qquad 7 - 6 = +1$$

$$\text{division:} \qquad 2 \div 4 = \frac{1}{2} \qquad 4 \div 2 = 2$$

associative property

The associative property concerns *grouping*. The real numbers are associative under the operations of addition and multiplication. That is, how we group the addends in an addition, or the factors in a multiplication, has no bearing on the result. For example,

$$3 + 2 + 6 = (3 + 2) + 6 = 3 + (2 + 6) = 11$$
$$5 \quad + 6 = 3 + \quad 8$$

$$4 \times 2 \times 3 = (4 \times 2) \times 3 = 4 \times (2 \times 3) = 24$$
$$8 \quad \times 3 = 4 \times \quad 6$$

distributive property

The real numbers are said to be distributive for multiplication over addition. What that means is, if we are to multiply a number and the sum of several other numbers, we may:

   A. Multiply each of the several numbers by the first number, then add the products; or
   B. Add the several numbers, then multiply that sum by the first number.

EXAMPLES

1. $3 \times (2 + 5) = ?$

SOLUTION

Method **A**:   Method **B**:

$3 \times 2 = \qquad 6 \qquad 2 + 5 = \qquad 7$

$3 \times 5 = \qquad (+)\dfrac{15}{21} \qquad\qquad (\times)\dfrac{3}{21}$

2. $-4 \times (6 + 2) = ?$

SOLUTION

$$\text{A:} \quad -4 + (6 + 2) = -24 - 8 = -32$$

$$\text{B:} \quad -4 \times (6 + 2) = -4 \times 8 = -32$$

3. When the numbers in the parentheses imply a subtraction, the distributive property still holds true.

$$(-5)(6 - 2) = ?$$

SOLUTION

$$\text{A:} \quad (-5)(6 - 2) = (-5)(6) + (-5)(-2) = -30 + 10 = -20$$

$$\text{B:} \quad (-5)(6 - 2) = (-5)(4) = -20$$

Thus it may be said that "The real numbers are distributive for multiplication over subtraction."

The use of parentheses in the Examples above brings up a basic convention. It is generally accepted that operations within parentheses are to be carried out first. For example:

$$4 + (2 - 5) = 4 + (-3) = 1$$

Further, when there are choices, the order of operations is commonly accepted to be:

- Multiplication and Division first, in order, left to right.
- Then Addition and Subtraction.

For example:

$$6 - 3 \times 2 = 6 - (3 \times 2) = 6 - 6 = 0$$
$$6 - 3 \times 2 \neq (6 - 3) \times 2 = 3 \times 2 = 6$$

That is, multiplication is performed before subtraction.

In the case of:

$$18 \div 6 \times 3 = (18 \div 6) \times 3 = 3 \times 3 = 9$$
$$18 \div 6 \times 3 \neq 18 \div (6 \times 3) = 18 \div 18 = 1$$

Here division is performed first, then the multiplication, because that is their order, left to right.

When there are several symbols of grouping, the practice is to begin with the innermost, and work outward.

EXAMPLES

1. Simplify $3 + [2 - (6 - 5)]$.

SOLUTION

Start within the parentheses ( ), then the brackets [ ].

$$3 + [2 - (6 - 5)] = 3 + [2 - (1)]$$
$$= 3 + [1]$$
$$= 4$$

2. Simplify $6\{[2 - (4 + 1)] - 3 + [2 - (6 - 8)]\}$

SOLUTION

Start with the numbers within parentheses ( ), then brackets [ ], and finally the outmost symbol of grouping, the braces { }.

$$6\{[2 - (4 + 1)] - 3 + [2 - (6 - 8)]\}$$
$$= 6\{[2 - (5)] - 3 + [2 - (-2)]\}$$
$$= 6\{[-3] - 3 + [4]\}$$
$$= 6\{-3 - 3 + 4\}$$
$$= 6\{-2\}$$
$$= -12$$

5. DIVISION BY ZERO

Multiplication and division, we have seen, are inverse operations of one another. Each operation has its particular reverse operation. For instance,

Operation	Reverse
$8 \times 3 = 24$	$24 \div 3 = 8$ $24 \div 8 = 3$
$6 \div 2 = 3$	$3 \times 2 = 6$ $2 \times 3 = 6$

When the divisor is 0, however, we meet with either a contradiction or an indeterminate result. Let us consider the former, the contradiction.

Suppose that $8 \div 0$ did have a real number quotient, which we will call q. In this case,

$$\text{If } 8 \div 0 = q, \text{ then } 0 \times q = 8.$$

We know that anytime 0 is a factor in a multiplication, the product must be 0 no matter what the other factor may be. So to admit that $8 \div 0$ has some real number q for a quotient leads us to a contradiction. The same contradiction would develop for any other nonzero number divided by 0, not just 8, as we have used in our example.

What about $0 \div 0$? Instead of no possible quotient as before, we have an infinite number of them! We could say $0 \div 0 = 23$, since $23 \times 0 = 0$. Or we could say $0 \div 0 = -\frac{1}{2}$, since $0 \times \left(-\frac{1}{2}\right) = 0$; and so on. In fact, we might say that $0 \div 0$ equals any number and confirm it by citing the fact that "any number times 0 is 0."

Therefore, because division by 0 leads either to a contradiction or a non-unique (indeterminate) result, we exclude it. In mathematical terms, we say that *division by zero is undefined.*

EXERCISE 1-4

Identify the basic property reflected in each case.

1. $5 \times 4 = 4 \times 5$
2. $(4 + 2) + 8 = 4 + (2 + 8)$
3. $(3 \times 6) \times 5 = 3 \times (6 \times 5)$
4. $3 \times (2 + 1) = (3 \times 2) + (3 \times 1)$
5. $9 \times (1 + 17) = (9 \times 1) + (9 \times 17)$
6. $(42 \times 31) \times 0 = 42 \times (31 \times 0)$
7. $(18 + 6) \times 4 = (18 \times 4) + (6 \times 4)$
8. $* + \$ = \$ + *$
9. $(\# \times ?) \times * = \# \times (? \times *)$
10. $\# \times \& = \& \times \#$
11. $(\$ + ?) + \% = \$ + (? + \%)$
12. $\$ \times (\# + *) = (\$ \times \#) + (\$ \times *)$

By inspection alone, tell whether the number sentence is true or false.

13. $687 + 876 = 876 + 687$

14. $32(84 - 16) = (32 \times 84) - (32 \times 16)$

15. $248 - 37 = 37 - 248$

16. $422 \times 36 = (400 \times 36) + (22 \times 36)$

17. $98 \times 64 = (100 \times 64) - (2 \times 64)$

18. $67 + (21 + 32) = (67 + 21) + 32$

19. $3(36 + 57) > (3)(36) + (3)(57)$

20. $(6)(8)(10)(12) > (12)(10)(8)(6)$

21. $(3)(3)(7)(9) < (3)(3)(7)(10)$

22. $67 + 95 < 95 + 67$

23. $89(100 + 16) = (100)(89) + (100)(16)$

24. $42 - 36 - 17 = 17 - 36 - 42$

Simplify.

25. $3 \times 2 - 6$ **26.** $5 + 6 \div 2$ **27.** $2 - 6 \times 5$

28. $9 \div 3 - 2$ **29.** $6 \div 4 \times 2$ **30.** $4 \times 3 \div 2$

31. $2 \times 3 - 5 + 7 \times 2$ **32.** $5 + 2 \times 3 \div 3 - 1$

33. $3 - 2 \times 9 \div 3$ **34.** $4 + 3 \div 2 \times 6$

Remove all signs of grouping and simplify.

35. $6 + (3 + 4) - 5$ **36.** $8 - (3 - 2) + (4 - 6)$

37. $5(4 - 6) - 2(3 + 1)$ **38.** $5 - [6(3 - 5) + 2]$

39. $-[3(2 - 5) - (-6 + 2)]$ **40.** $4\{[6 - 7(3 - 2 - 4) + 8] + 3\}$

41. $20 - \{3[4(6 - 7 + 15) - 18]\}$

42. $\{2 - 3\}\{6 - [5 + (2 - 1)] + 4\}\{-7[3 - (6 - 6)]\}$

Write a number equivalent to each if the expression is defined.

43. $(3)(0)$ **44.** $0 \div 3$ **45.** $3 \div 0$

46. $\dfrac{9}{0}$ **47.** $\dfrac{0}{6}$ **48.** $(0)(0)$

49. $\dfrac{0}{0}$ **50.** $\dfrac{-8}{0}$ **51.** $\dfrac{0}{-1}$

52. $\dfrac{3 - 7}{2 - 2}$ **53.** $\dfrac{9 - 9}{2 - 3}$ **54.** $3 \div (3 - 3)$

6. SOME ALGEBRAIC TERMINOLOGY

A symbol that represents any of the numbers in some specified set is called a **variable**. Usually, variables are letters of the alphabet. The specified set is the **domain of the variable**, often called the **replacement set**.

Variables make generalizations much easier to express. For example, if x (variable) represents any number in the set of real numbers (domain of the variable), then it is true that:

$$\text{(a)} \quad (2)(x) = x + x \quad \text{and} \quad \text{(b)} \quad x - x = 0$$

In (a) we express the generalization that twice any real number is equal to the sum of that number and itself; (b) is the generalization that any real number subtracted from itself is 0.

Each generalization can be verified for specific numbers if we replace the variable, wherever it occurs, with a number from the domain of the variable. For instance, the real numbers 7, $-\frac{1}{4}$, $2\frac{1}{2}$, and -3 are in the domain of the variable, and

$$2 \times 7 = 7 + 7 \qquad\qquad\qquad 2\frac{1}{2} - 2\frac{1}{2} = 0$$

$$2 \times \left(-\frac{1}{4}\right) = \left(-\frac{1}{4}\right) + \left(-\frac{1}{4}\right) \qquad (-3) - (-3) = 0$$

When a number, say 3, precedes a variable, say n, multiplication is implied: $3n = (3)(n)$ or $3 \cdot n$. Moreover, a succession of variables in algebra implies multiplication; for example, $ab = a \cdot b$, and $2psm = 2 \cdot p \cdot s \cdot m$ or $(2)(p)(s)(m)$.

Algebraic expressions such as $3n$ or $5x$ are called **monomials**. Usually, monomials consist of a numeral called the **numerical coefficient** and one or more variables:

$$23\ n$$

numerical coefficient ⸺⸺⸺ variable(s)

$$6\ xy$$

When no numerical coefficient appears in a monomial, the numerical coefficient is understood to be 1:

$$s = 1s = 1 \cdot s$$

$$mn = 1mn = 1 \cdot m \cdot n$$

In the case of negative expressions, such as $-y$, the numerical coefficient is understood to be -1:

$$-y = -1y = -1 \cdot y$$

Technically, numbers alone, such as 6, -3, $+14$, called **constants**, are also classified as monomials. Monomials which differ by, at most, their numerical coefficients are called **like monomials**. For example, $4x$, $5x$, and x are like monomials because they agree in variable. On the other hand, $2x$ and $2y$ are not like monomials even though they agree in numerical coefficients.

Sums and differences of two monomials are called **binomials** (e.g., $m + n$, $3b + 2$, and $c - 5d$). Sums and differences of three monomials are called **trinomials** (e.g., $5x + 2y - 5$). Collectively, monomials, binomials, trinomials, and the sums and differences of four or more monomials are called **polynomials**.

7. SUMS AND DIFFERENCES OF MONOMIALS

The basic operations of algebra are the same as those of arithmetic: addition, subtraction, multiplication, and division. The most elementary apply to monomials.

The usual procedure for computing sums and differences of like monomials is based upon the distributive property. For example, if the like monomials are $8n$ and $5n$, their sum may be expressed as a binomial, $8n + 5n$, or more simply as a monomial, $13n$:

$$8n + 5n = \underbrace{(8 \cdot n) + (5 \cdot n) = (8 + 5) \cdot n}_{\text{distributive property}} = (13) \cdot (n) = 13n$$

Similarly for their difference:

$$8n - 5n = 8n + (-5n) = [(8) + (-5)] \cdot n = (3) \cdot (n) = 3n$$

We may program the procedure for computing sums and differences of like monomials as follows.

PROGRAM 1.7

To add (subtract) like monomials:

Step 1:   Add (subtract) the numerical coefficients of the monomials.

Step 2:   Combine the sum (difference) of Step 1 and the common variable of the terms added (subtracted).

EXAMPLES

1.   Add $4x + x + 7x$.

SOLUTION   *Step 1:*   Add the numerical coefficients: $4 + 1 + 7 = 12$.

Step 2:   Combine the common variable (x) and the sum of Step 1: $12x$ is the sum of $4x + x + 7x$.

2. Add $3y + (-2y) + 4y + \frac{1}{2}y$.

SOLUTION *Step 1:* Add: $3 - 2 + 4 + \frac{1}{2} = 5\frac{1}{2}$.

Step 2: The common variable is y: $5\frac{1}{2}y$ is the desired sum.

3. Subtract $(6y) - (2y)$.

SOLUTION *Step 1:* Subtract the numerical coefficients: $6 - 2 = 4$.
Step 2: The common variable is y: $4y$ is the desired difference.

4. Subtract $(-3p) - (2p)$.

SOLUTION *Step 1:* $(-3) - (2) = (-3) + (-2) = -5$.
Step 2: The common variable is p: $-5p$ is the desired difference.

5. Subtract $(-6ab) - (-8ab)$.

SOLUTION *Step 1:* $(-6) - (-8) = (-6) + (+8) = 2$.
Step 2: The common variable is ab: $2ab$ is the desired difference.

6. Add $(4p) + (3q)$.

SOLUTION The monomials are unlike; the simplest expression of their sum is the binomial: $4p + 3q$.

7. Subtract $(6a) - (5b)$.

SOLUTION The monomials are unlike; the simplest expression of their difference is the binomial: $6a - 5b$.

EXERCISE 1-5

Add.

1. $(+14x) + (+7x)$
2. $(-18y) + (-6y)$
3. $(+13p) + (-6p)$
4. $(+17d) + (-14d)$
5. $(+31s) + (-65s)$
6. $(-42k) + (-57k)$
7. $(-56x) + (+18x)$
8. $(-3z) + (+2y)$
9. $(-16m) + (+94n)$
10. $(+18r) + (-22r)$
11. $(+14f) + (+6f) + (-8f)$
12. $(-27x) + (-18x) + (+6x)$
13. $(+18m) + (-16x) + (-18m)$
14. $(-37y) + (+14y) + (+19y)$
15. $(-42a) + (-36a) + (-51a)$
16. $(+14x) + (+13y) + (+17x)$
17. $(-37x) + (+14c) + (+23x)$
18. $(+15s) + (+12t) + (-20s)$
19. $(+16m) + (-17k) + (-19m)$
20. $(-12b) + (-14b) + (+107a)$

Subtract.

21. $(+14r) - (+7r)$ 22. $(-18m) - (-6m)$

23. $(+12k) - (+14k)$ 24. $(+16a) - (-37b)$

25. $(+17c) - (-12b)$ 26. $(-21t) - (+13t)$

27. $(-27d) - (-34e)$ 28. $(+26d) - (-18d)$

29. $\left(+\dfrac{1}{2}x\right) - \left(-\dfrac{3}{4}x\right)$ 30. $(-44p) - (-44p)$

Simplify.

31. $4xy + 3xy$ 32. $3ab - 2ab$

33. $5m + 3m - 8m$ 34. $3xy - 5xy - 7xy$

35. $3a - 5a + 4k$ 36. $2m - m - \dfrac{1}{2}m$

37. $32s - 56s - 7s$ 38. $3d - 8d + 5d$

39. $5xy + 2xy - 7xy$ 40. $-3q - 6p + 9q$

41. Check the result of Exercise 33 by letting $m = 2; 3; -1$.

42. Check the result of Exercise 35 by letting $a = 4$ and $k = 5$.

43. If $x = 2$ and $y = 3$, what would be the value of the result of Exercise 39?

44. If $p = -1$ and $q = 0$, what would be the value of the result of Exercise 40?

8. POWERS

Since we agree, in algebra, to write $3 \cdot n$ as $3n$, and $m \cdot n$ as mn, then to be consistent, $n \cdot n$ is nn. In this last case, we have a special type of product, one in which both factors are the same. A product in which all factors are identical is called a **power** of the repeated factor.

Instead of listing all of the factors (as is necessary when they are all different), a highly useful shorthand notation may be employed: a numeral written above and to the right of a factor indicates the number of times that factor is used in the product. Thus:

$$5 \times 5 \times 5 \times 5 = 5^4$$

$$a \cdot a \cdot a = a^3$$

$$b \cdot b \cdot b \cdot b = b^4$$

$$ab \cdot ab \cdot ab = a \cdot b \cdot a \cdot b \cdot a \cdot b$$

$$= a \cdot a \cdot a \cdot b \cdot b \cdot b$$

$$= a^3 b^3$$

In such expressions, the repeated factor is called the **base**; the upper-level numeral that indicates the number of times the factor is used is called the **exponent**.

As with numerical coefficients, when no exponent is attached to an expression, whether numerical or literal, the exponent is understood to be 1:

$$a = a^1 \qquad 3 = 3^1 \qquad ab = a^1b^1 \qquad 4d = 4^1d^1$$

If, as we have noted above,

$$a^4 = a \cdot a \cdot a \cdot a \quad \text{and} \quad a^3 = a \cdot a \cdot a$$

then

$$a^4 \cdot a^3 = (a \cdot a \cdot a \cdot a) \cdot (a \cdot a \cdot a)$$

$$= a \cdot a \cdot a \cdot a \cdot a \cdot a \cdot a = a^7$$

Thus we could have obtained the same resulting exponent (7) by simply adding the exponents of the two like-based factors:

$$a^4 \cdot a^3 = a^{4+3} = a^7$$

PROGRAM 1.8

To multiply powers having the same base:

Step 1:   Add the exponents of the like-based factors.
Step 2:   Write as the desired product the common base with an exponent equal to the sum of the exponents found in Step 1.

EXAMPLES

1.   Find the product of $a^4 \cdot a^3 \cdot a^7$.

SOLUTION   *Step 1:* Add the exponents: $4 + 3 + 7 = 14$.
Step 2: The common base is a; the product is a^{14}.

2.   Find the product of $a^3b^2 \cdot a^2b^2 \cdot a^4b$.

Note: When all factors are not of the same base, powers of the same base are multiplied together independently.

$$a^3b^2 \cdot a^2b^2 \cdot a^4b^1 = a^3 \cdot a^2 \cdot a^4 \cdot b^2 \cdot b^2 \cdot b^1$$

$$= a^{3+2+4}b^{2+2+1}$$

$$= a^9b^5$$

9. PRODUCTS OF MONOMIALS

We observed previously that a monomial may be considered to be a product in itself, for example, $8pq = 8 \cdot p \cdot q$. Computing the product of several monomials is essentially a matter of reorganizing and collecting factors according to the associative and commutative properties. For instance, to multiply $(6ab)(2ab^2c)$:

$$(6ab)(2ab^2c) = (6 \cdot a \cdot b) \cdot (2 \cdot a \cdot b \cdot b \cdot c)$$
$$= 6 \cdot a \cdot b \cdot 2 \cdot a \cdot b \cdot b \cdot c$$
$$= 6 \cdot 2 \cdot a \cdot a \cdot b \cdot b \cdot b \cdot c$$
$$= 12 \cdot a^2 \cdot b^3 \cdot c$$
$$= 12a^2b^3c$$

PROGRAM 1.9

To multiply several monomials:

Step 1:   Multiply the numerical coefficients of the several monomials for the numerical coefficient of the product.

Step 2:   Multiply the variables of the several monomials for the variable of the product.

EXAMPLES

1.   Multiply $(4a^2b)(-2ac^2)(bc^3)$.

SOLUTION   *Step 1:* Numerical coefficients: $(4)(-2)(1) = -8$.
Step 2: Variables: $(a^2b)(ac^2)(bc^3) = a^2 \cdot b^1 \cdot a^1 \cdot c^2 \cdot b^1 \cdot c^3$
$$= a^{2+1}b^{1+1}c^{2+3} = a^3b^2c^5.$$

The product is $-8a^3b^2c^5$.

2.   Multiply $(-3a^2b)(-2a^3b)(4)$.

SOLUTION   *Step 1:* $(-3)(-2)(+4) = +24$
Step 2: $(a^2b)(a^3b) = a^2 \cdot b^1 \cdot a^3 \cdot b^1 = a^5b^2$
The product is $24a^5b^2$.

3.   Multiply $(6xy^3)(-3xy)(-x^3y)$.

SOLUTION   *Step 1:* $(+6)(-3)(-1) = +18$

Note: $-x^3y$ may be thought of as $-1x^3y$.

Step 2: $(xy^3)(xy)(x^3y) = x^1 \cdot y^3 \cdot x^1 \cdot y^1 \cdot x^3 \cdot y^1 = x^5y^5$
The product is $18x^5y^5$.

Express each product as a monomial.

1. $a^3 \cdot a^2$
2. $b^4 \cdot b^3 \cdot b$
3. $d^2 \cdot d \cdot d^{10}$
4. $x^3 \cdot x^5 \cdot x^7$
5. $m^2 \cdot m^3 \cdot m^5$
6. $c^2 \cdot c^2 \cdot d$
7. $a^2b^2 \cdot a^4b^2 \cdot ab^3$
8. $a^3b^2 \cdot a^5 \cdot b^7 \cdot a$
9. $x^3y \cdot x^2y^3 \cdot x^4$
10. $m^2n \cdot mn^2 \cdot m^3 \cdot n^3$
11. $(-2)(5x)$
12. $(4)(-3x)$
13. $(2n)(5n)$
14. $(-3x)(-7x)$
15. $(-2x)(4y)$
16. $(6x)(5xy)$
17. $(-2m)(-3mn)$
18. $(-4a)(12bc)$
19. $(-3ab^2)(20a^2b)$
20. $(4xy)(4yz)$
21. $\left(-\dfrac{1}{2}a^2\right)(6b^2)$
22. $(16x^3)\left(-\dfrac{1}{4}x\right)$
23. $(-5a)(2b)(-3a)$
24. $(-4x)(-3y)(2x)$
25. $(-3x)(-2b)(3x)$
26. $(4m)(-3m)(-7m^2)$
27. $(3st)(3st)(3st)$
28. $(4ab)(3bc)(2ac)$
29. $(-2mn)(-3mp)(-4np)$
30. $(-3xy)(-2xy^2)\left(\dfrac{1}{2}x^2y^2\right)$
31. $(6ab)(-8)(-19b)(0)$
32. $(-x^3)(-y^5)(x^2)(m)$

Give the numerical value of the mathematical expression in:

33. Exercise 15 if $x = 2$ and $y = 5$.
34. Exercise 17 if $m = 20$ and $n = 10$.
35. Exercise 27 if $s = -1$ and $t = -3$.
36. Exercise 29 if $m = -1$, $n = 2$, and $p = 0$.

10. QUOTIENTS OF MONOMIALS

Since

$$a^5 = a \cdot a \cdot a \cdot a \cdot a \quad \text{and} \quad a^3 = a \cdot a \cdot a$$

then

$$a^5 \div a^3 = \frac{a^5}{a^3} = \frac{\cancel{a} \cdot \cancel{a} \cdot \cancel{a} \cdot a \cdot a}{\cancel{a} \cdot \cancel{a} \cdot \cancel{a}} = a^2 \quad (a \neq 0)$$

and

$$a^3 \div a^5 = \frac{a^3}{a^5} = \frac{\cancel{a} \cdot \cancel{a} \cdot \cancel{a}}{\cancel{a} \cdot \cancel{a} \cdot \cancel{a} \cdot a \cdot a} = \frac{1}{a^2} \quad (a \neq 0)$$

In both of these examples we could have obtained the exponent of the quotient by subtracting the smaller exponent from the larger exponent of the two original terms. Although later, in Chapter 6, we shall find it useful to define negative exponents, for now we handle problems such as those above with a two-part program for dividing powers having the same base.

PROGRAM 1.10 (a)

To divide powers having the same base when the exponent of the dividend is greater than the exponent of the divisor:

Step 1: Subtract the exponent of the divisor from the exponent of the dividend.

Step 2: Write as the desired quotient the common base (zero excepted) with an exponent equal to the difference of the exponents found in Step 1.

EXAMPLES

1. Divide b^4 by b^3 $(b \neq 0)$.

SOLUTION *Step 1:* Subtract exponents: $4 - 3 = 1$.
Step 2: The quotient is b^1 (or b).

2. Divide $a^9 \div a^3$ $(a \neq 0)$.

SOLUTION *Step 1:* $9 - 3 = 6$
Step 2: The quotient is a^6.

PROGRAM 1.10 (b)

To divide powers having the same base when the exponent of the divisor is greater than the exponent of the dividend:

Step 1: Subtract the exponent of the dividend from the exponent of the divisor.

Step 2: Write as the desired quotient a fraction whose numerator is 1 and denominator is the common base (zero excepted) with an exponent equal to the difference of the exponents found in Step 1.

EXAMPLES

1. Divide k^5 by k^8 $(k \neq 0)$.

SOLUTION *Step 1:* Subtract exponents: $8 - 5 = 3$

Step 2: The quotient is $\dfrac{1}{k^3}$.

2. Divide $x^3 \div x^7 \ (x \neq 0)$.

SOLUTION
$$x^3 \div x^7 = \frac{1}{x^{7-3}} = \frac{1}{x^4}$$

Note: Later, in Chapter 6, with the development of negative exponents, Programs **1.10 (a)** and **1.10(b)** combine into a single general procedure.

Computing the quotient of two monomials essentially involves (a) expressing the dividend monomial as the numerator of a fraction and the divisor as the denominator of that fraction, and (b) reducing that fraction to lowest terms by dividing out common factors. In arithmetic, this is a familiar process:

$$60 \div 84 = \frac{60}{84} = \frac{\cancel{2} \times \cancel{2} \times \cancel{3} \times 5}{\cancel{2} \times \cancel{2} \times \cancel{3} \times 7} = \frac{5}{7}$$

PROGRAM 1.11

To divide one monomial by another:

Step 1:   Express dividend and divisor as the numerator and denominator of a fraction, respectively.

Step 2:   Simplify the fraction by dividing out common factors.

EXAMPLES

1.   Divide $36a^3b^2c$ by $9ab^2$.

SOLUTION   *Step 1:* $\dfrac{36a^3b^2c}{9ab^2}$

Step 2: $\dfrac{2 \cdot 2 \cdot \cancel{3} \cdot \cancel{3} \cdot \cancel{a} \cdot a \cdot a \cdot \cancel{b} \cdot \cancel{b} \cdot c}{\cancel{3} \cdot \cancel{3} \cdot \cancel{a} \cdot \cancel{b} \cdot \cancel{b}} = 4a^2c$

2.   Divide $-6a^3bc^2$ by $2ab^4$.

SOLUTION
$$\frac{\overset{-3a^2}{\cancel{-6a^3bc^2}}}{\underset{b^3}{\cancel{2ab^4}}} = \frac{-3a^2c^2}{b^3}$$

3.   Divide $(-25x^3) \div (-5x^2yz)$.

SOLUTION
$$\frac{\overset{+5\ x^1}{\cancel{-25x^3}}}{\cancel{-5x^2yz}} = \frac{5x}{yz}$$

28 Chapter 1 REVIEW OF BASICS

Compute the quotient.

1. $2^5 \div 2^3$ 2. $2^2 \div 2^3$ 3. $2^4 \div 2^7$

4. $x^8 \div x^4$ 5. $x^8 \div x^2$ 6. $x^7 \div x^9$

7. $(xy)^3 \div (xy)^5$ 8. $(ab)^2 \div (ab)$ 9. $(-a)^7 \div (-\dot{a})^5$

10. $(-2)^3 \div (-2)^6$ 11. $14a \div (-2)$ 12. $-15x \div 5$

13. $-24a^2 \div 6a$ 14. $18x^4 \div 2x$ 15. $25p^2q \div (-5pq)$

16. $(-84a^3b^2) \div (7a^2b^2)$ 17. $(26a^3b) \div (-13a^3b)$

18. $40x^2y^3z \div 16x^2y$ 19. $32a^2y \div 4ay^3$

20. $8a^2b \div 4ab^2$ 21. $9ab^3 \div 3ac$

22. $(-87a^2b^3c) \div (3bc^2)$ 23. $16a^2bx \div 5ax^2$

24. $(-4ab^3) \div (-ab^4)$ 25. $(-2xy^2) \div (3xy^3)$

26. $(-9t^2v^3) \div (21t^2vx^2)$ 27. $(-13a^2b^5) \div (-26a^6b)$

28. $35x^2y^7 \div 7xy^6$ 29. $(-8x^2yz) \div (12xz^2)$

30. $9m^2n \div 4mq$ 31. $7x^2y^7 \div 7xy^5$

11. SUMS AND DIFFERENCES OF POLYNOMIALS

Computing the sum of two or more polynomials is essentially a matter of adding the like monomial parts of the polynomials:

$$(3a + 2b - 4c) + (2a - b + c) = 3a + 2b - 4c + 2a - b + c$$
$$= 3a + 2a + 2b - b - 4c + c$$
$$= (3a + 2a) + (2b - b) + (-4c + c)$$
$$= (5a) + (b) + (-3c)$$
$$= 5a + b - 3c$$

Program **1.12** suggests a more practical way to arrive at the sum of several polynomials.

PROGRAM 1.12

To add polynomials:

Step 1:   Arrange each polynomial addend so that like monomials are in the same column.

Step 2:   Add each column separately.

EXAMPLES

1. Add $(3a + b - c) + (4a - 3b + 6c) + (2a + b + c)$.

SOLUTION

Step 1: Arrange addends:

$$\begin{aligned} 3a + b - c \\ 4a - 3b + 6c \end{aligned}$$

Step 2: Add by columns: $(+)\dfrac{2a + b + c}{9a - b + 6c}$ (sum)

2. Add $(4p + 6q + 11r) + (3p - 2r) + (3q - 4r)$.

SOLUTION

Step 1: Arrange addends:

$$\begin{aligned} 4p + 6q + 11r \\ 3p - 2r \end{aligned}$$

Step 2: Add by columns: $(+)\dfrac{ + 3q - 4r}{7p + 9q + 5r}$ (sum)

As with monomials, the difference of two polynomials is readily found by adding the negative of the polynomial being subtracted to the other polynomial, since

$$a - b = a + (-b)$$

(Recall Program **1.4.**)

The **negative of a polynomial** is a polynomial in which the signs of the terms of the given polynomial have been reversed. For instance, the negative of the polynomial $2a - b + c$ is the polynomial $-2a + b - c$. We can verify that one is the negative of the other (i.e., additive inverse) by adding the two polynomials together, using Program **1.12**, and noting that the sum is 0.

Thus to subtract $2a - b + c$ from $3a + 2b - 4c$, say, we have

$$(3a + 2b - 4c) - (2a - b + c) = (3a + 2b - 4c) + (-2a + b - c)$$

$$= 3a + 2b - 4c - 2a + b - c$$

$$= a + 3b - 5c$$

PROGRAM 1.13

To subtract one polynomial from another:

Add the negative of the polynomial being subtracted to the other polynomial.

Note: In effect, "Reverse the signs of each term of the polynomial being subtracted and then add.

1. Subtract $(6x - 4y) - (3x + 8y)$.

SOLUTION

The negative of $(3x + 8y)$ is $(-3x - 8y)$; so

$$(6x - 4y) - (3x + 8y) = (6x - 4y) + (-3x - 8y)$$
$$= 6x - 4y - 3x - 8y$$
$$= 3x - 12y$$

2. Subtract $(4a - 2b + c)$ from $(3a - 4b + 8c)$.

SOLUTION

$$(3a - 4b + 8c) - (4a - 2b + c) = (3a - 4b + 8c) + (-4a + 2b - c)$$
$$= 3a - 4b + 8c - 4a + 2b - c$$
$$= -a - 2b + 7c$$

or in vertical form

$$
\begin{array}{r}
+3a - 4b + 8c \\
+4a - 2b + c \\
\hline
\end{array}
(-)
\quad\Rightarrow\quad
\begin{array}{r}
+3a - 4b + 8c \\
-4a + 2b - c \\
\hline
-a - 2b + 7c
\end{array}
(+)
$$

3. Subtract from $4x - 3y$ the binomial $3x - 2z$.

SOLUTION

$$
\begin{array}{r}
+4x - 3y \\
+3x - 2z \\
\hline
\end{array}
(-)
\quad\Rightarrow\quad
\begin{array}{r}
+4x - 3y \\
-3x + 2z \\
\hline
+x - 3y + 2z
\end{array}
(+)
$$

Add.

1.
$$
\begin{array}{r}
6xy - 2xy^2 \\
-3xy - 17xy^2 \\
4xy + 12xy^2 \\
-xy - xy^2 \\
\hline
\end{array}
(+)
$$

2.
$$
\begin{array}{r}
3x - 4p \\
-2x \\
3x - 3p \\
- 4p \\
\hline
\end{array}
(+)
$$

3.
$$
\begin{array}{r}
7xy - 3y + 2z \\
- 4y + 7z \\
3xy + 5y + 2z \\
-2xy - 8z \\
\hline
\end{array}
(+)
$$

4.
$$
\begin{array}{r}
x + 3y - 12p^2 \\
-x + 5y \\
- 5y + 17p^2 \\
3x - 4p^2 \\
\hline
\end{array}
(+)
$$

5. $(4x - 3y) + (3d + 4x) + (3y + 2z) + (5z - 8x)$

6. $(3x - 2y) + (4x - 3y + 2z) + (4z - 6x) + (2y + 5z)$
7. $(3x - 2y) + (6x + 2y) + (-5x - 4y) + (2y)$
8. $(4x - 3y) + (2x - 3y) + (-6x + 6y)$

Subtract.

9. $(3x - 2y) - (6x - 7y)$ 10. $(3x - 2y) - (2y - 3x)$
11. $(3x - 4y) - (6y + 12)$ 12. $(x + 3p) - (3p + 2)$
13. $(x - 4y) - (-3y)$ 14. $(6a - 5b) - (-2b + 5a)$
15. $(3 + 4a) - (-5 + 6b)$ 16. $(17a - 4b) - (-12c + 3d)$
17. $(3x + 4y - 7) - (2x - 2y + 5)$
18. $(3x - 7y) - (2p + 3x - 15y)$
19. $(a - 34b - 27c) - (6d - 32a - 5c)$

Simplify.

20. $3x - 2y + 3x - 5x + 3y - x - 3y$
21. $5a - 3b + 2a - 6c - 7a - 15b - 4c$
22. $6ab - 4b + 5ab - 8b + 4ab - 3b$
23. $9x - 3y + 3x - y + 2x$
24. $2g - 5h + 4g + 8h + 8g - 3h$
25. $6y + (6y - 2x)$ 26. $2x + (5x - 3y)$
27. $(4m + 2n) - 3m$ 28. $(3a - 2b) + 2b$
29. $2y - (6y + 2x)$ 30. $5b + (3x - 5x)$
31. $-3p + (2p + q) - p^2$ 32. $2x - (3x + y) - 7$
33. $2x^2 - (3x + 6x^2) - x^2$ 34. $3a + (2b - 3a^2) + b^2$
35. $-(3a - 5b + 3c) - (a + 7b - 4c)$
36. $(2x - y) + (4x + 2y) - (6x + 3y)$
37. $(5x + 3y) - [(2y - 6x) - (2x + 8y)]$
38. $-[(2a - b) + (5b + 6a)] - (8a + 3b)$
39. $[(4m - s) - (2m + 5s)] + (m - 4s)$
40. $[5 - (2a - 4)] - [a - (6a - 3)]$
41. $6x - \{2x - [5x - (3x - 8) - 8] + 4\}$
42. $7f - \{3f - [5 - (4f - 2) + 6f] - 5\}$

Rewrite the expressions below, enclosing the last three terms in parentheses, preceded by a minus sign.

43. $2x + 6y - 3t - 8$ 44. $4x - 6y + 3z + 5$
45. $ab - c + gk - h$ 46. $-st + q - a - c - d$
47. Add $(3x - 2y)$ to itself and then subtract the sum from $(6y - 7x)$.
48. Subtract $6x - 3y$ from 1.   49. Subtract $4x - 3y$ from 0.
50. What must be added to $(3x - 2y)$ to make the sum $(4y + z)$?

12. PRODUCTS OF POLYNOMIALS

The usual method for computing the product of a monomial and a polynomial is based upon the distributive property:

$$a \cdot (b + c) = (a \cdot b) + (a \cdot c)$$

as can be noted in the following program and examples.

PROGRAM 1.14

To multiply a polynomial by a monomial.

Step 1: Multiply each term of the polynomial by the monomial.

Step 2: Add the products of Step 1 for the desired product.

EXAMPLES

1. Multiply $(6a)(3b - 2a)$, or $6a(3b - 2a)$.

SOLUTION *Step 1:* Multiply each term of the binomial, $3b - 2a$, by $6a$:

$$(6a)(3b) = 18ab$$
$$(6a)(-2a) = -12a^2$$

Step 2: Add the products of Step 1: $18ab - 12a^2$.

2. Multiply $5a^3b(3a^2 - 2b + 4)$.

SOLUTION *Step 1:* Multiply each term of the trinomial by $5a^3b$:

$$(5a^3b)(3a^2) = 15a^5b$$
$$(5a^3b)(-2b) = -10a^3b^2$$
$$(5a^3b)(+4) = 20a^3b$$

Step 2: Add products: $15a^5b - 10a^3b^2 + 20a^3b$.

3. Multiply $-3a^3(4c - 2a^4 - 6ab^2)$.

SOLUTION *Step 1:*

$$(-3a^3)(4c) = -12a^3c$$
$$(-3a^3)(-2a^4) = 6a^7$$
$$(-3a^3)(-6ab^2) = 18a^4b^2$$

Step 2: Add products: $-12a^3c + 6a^7 + 18a^4b^2$.

4. Multiply $-c(a^3 - 2b^2 + d)$.

SOLUTION

$$-c(a^3 - 2b^2 + d) = (-c)(a^3) + (-c)(-2b^2) + (-c)(d)$$
$$= (-ca^3) + (2cb^2) + (-cd)$$
$$= -ca^3 + 2cb^2 - cd$$

Computing the product of two polynomials involves repeated application of the distributive property. For instance, in $(a + b)(c + d + e)$, we consider the second factor—for the moment—as a single term:

$$(a + b)(c + d + e) = [a(c + d + e)] + [b(c + d + e)]$$

Then applying the distributive property a second time, within the bracketed expressions,

$$[a(c + d + e)] + [b(c + d + e)] = [(ac) + (ad) + (ae)] + [(bc) + (bd) + (be)]$$
$$= [ac + ad + ae] + [bc + bd + be]$$
$$= ac + ad + ae + bc + bd + be$$

A shortened version of the foregoing is given in the next program.

PROGRAM 1.15

To multiply two polynomials:

Step 1: Multiply each term of one polynomial by every term of the other polynomial.

Step 2: Simplify by collecting like terms.

EXAMPLES

1. Give the product of $(2x + 1)(3x + 2)$.

SOLUTION *Step 1:* Multiply each term of the second polynomial, $3x + 2$, by every term of the first polynomial, $2x + 1$:

$$(2x + 1)(3x + 2) = (2x)(3x) + (2x)(2) + (1)(3x) + (1)(2)$$
$$= 6x^2 + 4x + 3x + 2$$

Step 2: Collect terms:

$$= 6x^2 + 7x + 2$$

2. Multiply $(3a - 4c)(2a + 3y)$.

SOLUTION Multiply the terms of one polynomial by those of the other:

$$(3a - 4c)(2a + 3y) = (3a)(2a) + (3a)(3y) + (-4c)(2a) + (-4c)(3y)$$
$$= 6a^2 + 9ay - 8ac - 12cy$$

3. Multiply $(4x^2 - 3x)(2x + 7)$.

SOLUTION
$$(4x^2 - 3x)(2x + 7) = (4x^2)(2x) + (4x^2)(7) + (-3x)(2x) + (-3x)(7)$$
$$= 8x^3 + 28x^2 - 6x^2 - 21x$$
$$= 8x^3 + 22x^2 - 21x$$

4. Multiply $(4a - 2b + 3c)(2a - b - 4c)$.

SOLUTION When the polynomials are extensive, a vertical arrangement like the one below is useful.

$$
\begin{array}{r}
4a - 2b + 3c \\
(\times)\ \underline{2a - \ b - 4c} \\
8a^2 - 4ab + \ 6ac \\
-\ 4ab \qquad\quad + 2b^2 - 3bc \\
-\ 16ac \qquad\quad + 8bc - 12c^2 \\
\hline
8a^2 - 8ab - 10ac + 2b^2 + 5bc - 12c^2
\end{array}
$$

EXERCISE 1-9

Multiply.

1. $(3x)(4x + 2y - z)$
2. $(-x)(4a - 2x + 3y)$
3. $(3ab)(-2a + 3b - ab)$
4. $(2x - 3b + 4c)(8a)$
5. $(-3ax)(-2a + 5x - 6)$
6. $(2a)(a - 2 + a + 3)$
7. $(-2a^2c)(3a - 4c - 2ac)$
8. $(-8a^2)(3b^2 - 4a - 2b)$
9. $(-pqr)(3p + 2qr + 7)$
10. $(-6ab^2c^3)(3a^2bc - 3ab + 3ac)$
11. $(2x - 7)(5x + 2)$
12. $(5a - 2x)(a + 3x)$
13. $(m - n)(2p - m)$
14. $(x - y)(3x - 2y + 6)$
15. $(3x - 5)(3x^2 - 2x + 7)$
16. $(x - 2y)(6x^2 - 3xy + 5y^2)$
17. $(a^2 - 2a)(5a - 2a^2 - 3a^3)$
18. $(-d + 2c)(-8d^2 + 4cd - 6c^2)$
19. $(3ab - 4ab^2)(a^5b - 6a^4b^2 + b^3)$
20. $(3x^2 - 2x + 1)(5x^2 - 3x + 7)$
21. $(6m^2 - 5mn + n^2)(2m^2 + 3mn - 6n^2)$
22. $(3a - 3b + c)(4a^2 - 2b + c)$
23. What is the area of a rectangle $(3x + 2)$ ft by $(4x - 7)$ ft?
24. What would be the sum if $(s - 3t)$ were added repeatedly $(2s - 3t)$ times?

25. What polynomial, when divided by $(x - y)$, will yield a quotient $x^2 + xy + y^2$?

26. What is the product if one factor is a increased by b and the other is a decreased by b?

13. QUOTIENTS OF POLYNOMIALS

Division is said to be *distributive from the right*:

$$(6 + 4) \div 2 \stackrel{?}{=} (6 \div 2) + (4 \div 2)$$

$$10 \quad \div 2 \stackrel{?}{=} \quad 3 \quad + \quad 2$$

$$5 = 5$$

But division is *not distributive from the left*:

$$2 \div (6 + 4) \stackrel{?}{=} (2 \div 6) + (2 \div 4)$$

$$2 \div \quad 10 \quad \stackrel{?}{=} \quad \frac{1}{3} \quad + \quad \frac{1}{2}$$

$$\frac{1}{5} \quad \neq \quad \frac{5}{6}$$

We take advantage of the property of right distributivity to develop the following program.

PROGRAM 1.16

To divide a polynomial by a monomial:

Step 1:   Divide each term of the polynomial by the monomial.

Step 2:   Add the quotients of Step 1 for the desired quotient.

EXAMPLES

1.   Divide $6a^4b^3 - 3a^2b^3$ by $3ab$.

SOLUTION   *Step 1:* Divide each term of the polynomial, $6a^4b^3 - 3a^2b^3$, by the divisor, $3ab$:

$$\frac{\overset{2}{\cancel{6}}a^{\overset{3}{\cancel{4}}}b^{\overset{2}{\cancel{3}}}}{\cancel{3}\cancel{a}\cancel{b}} = 2a^3b^2$$

$$\frac{\overset{-1}{-\cancel{3}}a^{\overset{1}{\cancel{2}}}b^{\overset{2}{\cancel{3}}}}{\cancel{3}\cancel{a}\cancel{b}} = (-1)(ab^2) = -ab^2$$

Step 2: The desired quotient is the sum of the quotients of Step 1: $2a^3b^2 - ab^2$.

2. Divide $(18x^4y - 15x^2y^3) \div 3x^3y^2$.

SOLUTION *Step 1:* Divide $18x^4y$ and $-15x^2y^3$ by $3x^3y^2$:

$$\frac{\overset{6}{\cancel{18}}x^4\overset{1}{\cancel{y}}}{\cancel{3}x^3y^{\cancel{2}}^1} = \frac{6x}{y}$$

$$\frac{\overset{-5}{\cancel{-15}}x^2\overset{1}{\cancel{y^3}}}{\cancel{3}x^{3}\cancel{y^2}} = \frac{-5y}{x} = -\frac{5y}{x}$$

Step 2: The desired quotient is $\dfrac{6x}{y} - \dfrac{5y}{x}$.

3. Compute $(8x^3 + 4x^2y + 16xy^2) \div 4x$.

SOLUTION Divide each term of the trinomial by $4x$:

$$(8x^3 + 4x^2y + 16xy^2) \div 4x = (8x^3 \div 4x) + (4x^2y \div 4x) + (16xy^2 \div 4x)$$

$$= \frac{\overset{2x^2}{\cancel{8x^3}}}{\cancel{4x}} + \frac{\overset{x}{\cancel{4x^2y}}}{\cancel{4x}} + \frac{\overset{4}{\cancel{16xy^2}}}{\cancel{4x}}$$

$$= 2x^2 + xy + 4y^2$$

Division of a polynomial by a polynomial is usually a tedious process. It is similar to long division in arithmetic in that the quotient is assembled piecemeal.

PROGRAM 1.17

To divide one polynomial by another:

Step 1: Arrange both polynomials in descending powers of the same variable.

Step 2: Find the first term of the quotient by dividing the first term of the dividend by the first term of the divisor.

Step 3: Multiply the divisor by the quotient term of Step 2.

Step 4: Subtract the product of Step 3 from the dividend; bring down the next term of the original dividend to form the new dividend.

Step 5: Repeat the loop or sequence of Steps 2, 3, and 4 on the new dividend; keep repeating this loop of steps until the exponent of the first term of any remainder is less than the exponent of the first term of the divisor.

EXAMPLES

1. Compute the quotient $(3x + 9x^2 - 2) \div (2 + 3x)$.

Step 1: Arrange both polynomials in descending powers of the same variable:

$$3x + 9x^2 - 2 \rightarrow 9x^2 + 3x - 2$$
$$2 + 3x \rightarrow 3x + 2$$

$$3x + 2 \overline{)9x^2 + 3x - 2}$$

divisor ⏌   dividend ⏌

Step 2: Divide the first term of the dividend, $9x^2$, by the first term of the divisor, $3x$, to obtain the first term of the quotient, $3x$:

$$\frac{9x^2}{3x} = 3x$$

Step 3: Multiply the divisor, $3x + 2$, by the result of Step 2, $3x$:

$$(3x)(3x + 2) = 9x^2 + 6x$$

Step 4: Subtract the product of Step 3 from the dividend; then bring down the next term in the dividend, -2, to form a new dividend:

$$
\begin{array}{r}
3x \phantom{{}+9x^2+3x-2} \\
3x + 2 \overline{)9x^2 + 3x - 2} \\
(-)\underline{9x^2 + 6x \phantom{{}- 2}} \\
- 3x - 2 \quad \text{(new dividend)}
\end{array}
$$

Step 5: Repeat Steps 2, 3, and 4:

Step 2: $\dfrac{-3x}{3x} = -1$

Step 3: $(-1)(3x + 2) = -3x - 2$

Step 4: Subtract.

$$
\begin{array}{r}
3x - 1 \\
3x + 2 \overline{)9x^2 + 3x - 2} \\
(-)\underline{9x^2 + 6x \phantom{{}-2}} \\
- 3x - 2 \\
(-)\underline{- 3x - 2} \\
0
\end{array}
$$

2. Divide $8x + 6x^2 + 2$ by $2x + 2$.

Step 1: Arrange the terms in descending powers of the same variable:

$$8x + 6x^2 + 2 \rightarrow 6x^2 + 8x + 2$$
$$2x + 2 \rightarrow 2x + 2$$

$$2x + 2 \overline{)6x^2 + 8x + 2}$$

Step 2: Divide the first term of the dividend, $6x^2$, by the first term of the divisor, $2x$, to obtain the first term of the quotient:

$$\frac{6x^2}{2x} = 3x$$

$$\begin{array}{r} 3x + 1 \\ 2x + 2\overline{)6x^2 + 8x + 2} \end{array}$$

Step 3: $(3x)(2x + 2) \longrightarrow 6x^2 + 6x$

Step 4: Subtract and $\left.\begin{array}{l}\\ \text{bring down } +2\end{array}\right\}$ $\quad\longrightarrow 2x + 2$
$\quad\longrightarrow 2x + 2$

Step 5: $\left\{\begin{array}{l} \dfrac{2x}{2x} = 1 \\[2mm] (1)(2x + 2) = \overset{\frown}{2x + 2} \\[2mm] (2x + 2) - (2x + 2) = 0 \end{array}\right.$

3. Compute $(2x + 6x^2 - 20) \div (7 + 3x)$.

SOLUTION *Step 1:* $(6x^2 + 2x - 20) \div (3x + 7)$

Step 2: $\dfrac{6x^2}{3x} = 2x$

Step 3:
$$\begin{array}{r} 2x - 4 \\ 3x + 7\overline{)6x^2 + 2x - 20} \\ 6x^2 + 14x \end{array}$$

Step 4: Subtract and $\longrightarrow$
bring down -20
$$\begin{array}{r} -12x - 20 \\ -12x - 28 \\ \hline + 8 \end{array}$$

Step 5: $\left\{\begin{array}{l} \dfrac{-12x}{3x} = -4 \\[2mm] (-4)(3x + 7) = \overset{\frown}{-12x - 28} \\[2mm] (-12x - 20) - (-12x - 28) = +8 \quad \text{(remainder)} \end{array}\right.$

As in arithmetic, when a remainder occurs, it can be treated simply as a remainder or added to the quotient as a fraction whose denominator is the divisor. Thus

$$(6x^2 + 2x - 20) \div (3x + 7) = \left\{\begin{array}{c} 2x - 4, \text{ R}(+8) \\ \text{or} \\ 2x - 4 + \dfrac{8}{3x + 7} \end{array}\right.$$

4. Divide $x^4 - y^4$ by $x - y$.

SOLUTION It usually helps to write missing terms of the dividend with zero coefficients:

$$
\require{enclose}
\begin{array}{r}
x^3 + x^2y + xy^2 + y^3 \\
x - y \overline{)\,x^4 + 0x^3y + 0x^2y^2 + 0xy^3 - y^4} \\
\underline{x^4 - x^3y} \\
x^3y + 0x^2y^2 \\
\underline{x^3y - x^2y^2} \\
x^2y^2 + 0xy^3 \\
\underline{x^2y^2 - xy^3} \\
xy^3 - y^4 \\
\underline{xy^3 - y^4}
\end{array}
$$

5. Divide $3a^3 + a^2 - 2a + 7$ by $a^2 - 4$.

SOLUTION

$$
\begin{array}{r}
3a + 1 \quad \text{(quotient)} \\
a^2 - 4 \overline{)\,3a^3 + a^2 - 2a + 7} \\
\underline{3a^3 - 12a} \\
a^2 + 10a + 7 \\
\underline{a^2 - 4} \\
10a + 11 \quad \text{(remainder)}
\end{array}
$$

(Review Step 5 of Program **1.17**.)

EXERCISE 1-10

Divide; express quotients in simplest terms.

1. $(12a^2 - 18a^4) \div (6a^2)$ **2.** $(t^3 - t^2) \div (-t^2)$

3. $(x - 6x^2y^2) \div (x^2)$ **4.** $(5xyz - 10x^2z) \div (-5xz)$

5. $(12k^2n^3 - 28k^3n^2) \div (4k^2n^2)$ **6.** $(-12a^2b^3 + 48a^4b^4) \div (-6a^2b^3)$

7. $(-21a^3b - 12a^2b^2 + 3ab) \div (-3ab)$

8. $(2x^4 - 6x^3 + 10ax) \div (-4x^3)$ **9.** $\dfrac{-18x^3y + 33x^2y - 27xy^3}{3x^2}$

10. $\dfrac{28x^2yz^3 - 21x^3y + 14x^3y^2z}{-7x^2y}$ **11.** $\dfrac{25a^3b^2c^2 - 20a^2b^3c^2 - 15a^2b^2c^4}{-5a^2b^2c^2}$

12. $\dfrac{30x^4y - 5x^2y^2 + 12xy^2}{-5x^2y^2}$ **13.** $\dfrac{a^3 - 3b^2 - 6a^5}{-3a^2}$

14. $\dfrac{12ab^2 - 9a^3b - 15b^3}{3b}$ **15.** $\dfrac{m^2 - \dfrac{1}{2}mn - \dfrac{3}{4}m^2n^2}{2m}$

16. $\dfrac{a^3 - \dfrac{2}{3}ab + a^2b^2}{2a}$ **17.** $(3x^2 - 5x + 2) \div (x - 1)$

18. $(4x^2 - x - 5) \div (x + 1)$ 19. $(2x^2 + x - 15) \div (x + 3)$
20. $(4x^2 - 5x - 6) \div (x - 2)$ 21. $(3a^3 - 6a^2 - 7a - 6) \div (a - 3)$
22. $(a^3 - 5a^2 - 4a + 2) \div (a + 1)$
23. $(m^3 + 2m^2 - 7m + 4) \div (m - 2)$
24. $(6n^3 - 19n^2 + n + 8) \div (n - 3)$
25. $(x^3 - 3x + 8) \div (x - 2)$ 26. $(3x^3 + 5x^2 - 2) \div (x + 1)$
27. $(a^4 - 3a^3 - a^2 - 11a - 4) \div (a - 4)$
28. $(x^4 - 2x^3 + x^2 - 6x - 1) \div (x - 3)$
• 29. $(x^4 + x - 2) \div (x - 1)$ 30. $(2a^4 - 3a^2 + 1) \div (a - 1)$
31. $(6a^2 + 11a + 3) \div (3a + 1)$ 32. $(12x^2 + 7xy - 12y^2) \div (4x - 3y)$
33. $(8x - 8x^2 + 6) \div (2x - 3)$ 34. $(2y^2 - 17yz + 35z^2) \div (y - 5z)$
35. $(53a^2 + 15a^3 - 8 - 30a) \div (3a - 2)$
36. $(4x^3 + 1 - 3x) \div (2x - 1)$ 37. $(a^3 - 3a^2 + 2a - 6) \div (a^2 + 2)$
38. $(m^3 + 2m^2 - m - 2) \div (m^2 - 1)$
39. $(x^3 - x^2 + x + 1) \div (x^2 + 1)$
40. $(x^3 + 2x^2 - 5x) \div (x^2 - 5)$
41. $(2a^3 + 2a^2 - a - 1) \div (2a^2 - 1)$
42. $(6w^3 - 18w^2 + w - 3) \div (6w^2 + 1)$
43. $(2x^5 + 4x^3 + x^2) \div (2x^3 + 1)$
44. $(3a^4 + 2a^3 - 3a^2 - 2a + 3) \div (3a^2 + 2a)$
45. $(12x^7 - 6x^5 - 6x^2 + 2) \div (6x^2 - 3)$

| REVIEW

PART A

Answer True or False.

1. $4, \dfrac{8}{2}$, and IV are three different numerals for the same number.

2. The decimal equivalent for every rational number terminates.

3. $0 \div 0 = 0$

4. The commutative property relates to the order or sequence of numbers under an operation; the associative property relates to the grouping of the numbers.

5. The product of thirty-four factors, seventeen of which are positive numbers and the rest negative numbers, will be a negative number.

6. The numerical coefficient for such monomials as t, xy, and q is zero.

7. In the expression, m^2n, the exponent for m is 2 and the exponent for n is 1.

8. When we have a product of identical factors, we have what is called a power of the repeated factor.

9. $a^2 \cdot a^5 = a^7$ would be true for any real number that might replace a.

10. $b - a$ is the negative of $a - b$.

PART B

Given the set of numbers $\left\{6, \dfrac{2}{3}, -5, \sqrt{2}, 3.3333\ldots, -0.123, 0, \pi\right\}$

1. Name the integers in the set.
2. Name the negative numbers in the set.
3. Name the rational numbers in the set.
4. Name the irrational numbers in the set.

Replace the □ *with the appropriate symbol:* = *(equals),* < *(is less than),* > *(is greater than).*

5. 42 □ 24

6. $\dfrac{5}{8}$ □ 1

7. 18 × 5 □ 250 − 3

8. $\dfrac{1}{2}$ × 64 □ 50 − 26

9. +2 □ −4
10. −4 □ +1
11. −8 □ −12
12. −9 □ −15
13. |−8| □ |+8|
14. 0 □ −5

Write (a) the negative and (b) the absolute value for each number.

15. +6, −3, +15, +6, −2

16. −4, +8, −9, + $\dfrac{1}{2}$, +3

17. −19, − $\dfrac{2}{3}$, 0, +6, −100

18. +13, +0.3, −6 $\dfrac{1}{2}$, +10,000, −4.8

Add.

19. (−9) + (+6) + (+5) + (−3)
20. (+14) + (−13) + (+12) + (−10)
21. (+65) + (−18) + (−13) + (−19)
22. (+28) + (−42) + (−36) + (+17)
23. (+14) + (−216) + (−82) + (+316)
24. (−425) + (+162) + (+138) + (+217)

Subtract.

25. (+325) − (−163)
26. (−428) − (+100)
27. (−62) − (−213)
28. (−426) − (−218)
29. (+261) − (+137)
30. (+126) − (+319)

Write the products.

31. (−6) × (−3) × (+4)
32. (−9) × (+6) × (+3)
33. (−3)(−6)(+5)(−4)
34. (−10)(−5)(+3)(−3)
35. (+5)(−4)(−60)(+2)
36. (+35)(−2)(+9)(+5)
37. (−6)(−4 − 3 + 2 − 3)
38. (−5 + 6 + 2)(−8)

Write the quotients.

39. $(+48) \div (-8)$ **40.** $(-35) \div (+7)$ **41.** $(-63) \div (-9)$

42. $(+81) \div (+9)$ **43.** $(-6) \div (0)$ **44.** $(+9) \div (-2)$

45. $\dfrac{(-4 + 2 - 3)}{+5}$ **46.** $\dfrac{(-10 + 2 - 8)}{-4}$

Decide which of the properties—commutative, associative, distributive—is suggested by each statement.

47. $(63 + 27) + 4 = 63 + (27 + 4)$

48. $9 \times (3 + 5) = (9 \times 3) + (9 \times 5)$

49. $116 \times 47 = 47 \times 116$ **50.** $\dfrac{5}{9} + \dfrac{2}{9} = \dfrac{2}{9} + \dfrac{5}{9}$

51. $(5 \times 32) + (5 \times 61) = 5 \times (32 + 61)$

52. $0.2 \times (1.1 \times 0.3) = (0.2 \times 1.1) \times 0.3$

53. $\dfrac{3}{4} + \dfrac{7}{8} = \dfrac{7}{8} + \dfrac{3}{4}$ **54.** $20 + 15 = 5 \times (4 + 3)$

55. Is $(16 \div 4) \div 2 = 16 \div (4 \div 2)$?

56. Is $9 \div (3 + 1) = (9 \div 3) + (9 \div 1)$?

Simplify.

57. $6 \div 3 + 3$ **58.** $8 - 4 \times 7$

59. $8 \div 4 - 6 \times 3$ **60.** $6 \times 2 \div 3 + 2$

61. $8[6(3 - 1) + 2] - 5$ **62.** $9 - [3(5 + 2) + 6]$

63. $4\{3 - [8 + (6 - 7)] + 5\}$ **64.** $3\{[8 - 2(6 - 3 + 7) + 4] - 7\}$

65. $\{3 - 7\}\{2 - 4(3 - 2)\}\{6 - (5 + 1)\}$

66. $\{9 - 11\}\{7 - [6 + (2 - 4) + 5]\}\{-3[2 - (6 - 6)]\}$

Add.

67. $(-39x) + (+46x)$ **68.** $(+42d) + (-17d)$

69. $(+42x) + (-18x) + (+7x)$ **70.** $(-63a) + (-72a) + (+41a)$

71. $(-103y) + (+27y) + (-63y)$ **72.** $(-206m) + (-172m) + (-304m)$

73. $(+31a) + (-42b) + (+16a)$ **74.** $(-15x) + (-12x) + (+15y)$

Subtract.

75. $(+623a) - (-418a)$ **76.** $(-637x) - (+242x)$

77. $(+962m) - (+1037m)$ **78.** $(+62p) - (+37q)$

79. $(-43a) - (+16b)$ **80.** $(+342a) - (-163a)$

Simplify.

81. $3x + 2x - 3y + y + 7$ **82.** $5a - 3a + b - 3b - 2$

83. $6p - 3d + 2p - 5d + p$ **84.** $96x - 362x - 421x + 37x$

Compute the products.

85. $a^7 \cdot a^3 \cdot a^5$ **86.** $a^3b^2 \cdot ab^3 \cdot ab \cdot a^2b \cdot a$

87. $(-6) \times (14xy)$ **88.** $(-8x)(-8x)$

89. $\left(\dfrac{1}{3} ab\right)(63a^2b^2)$ **90.** $(-6x)(-2y)(3x)(-y)$

Compute the quotients.

91. $x^9 \div x^3$ **92.** $x^{12} \div x^{15}$ **93.** $(xy)^3 \div (xy)$

94. $(-3)^4 \div (-3)^6$ **95.** $-18x \div 6$ **96.** $24x^2 \div 2x^2$

97. $(-76a^2b^3) \div (19ab^2)$ **98.** $320x^2y^4z^5 \div 16xyz^3$

99. $5a^2b \div 4ab$ **100.** $(-12x^2y) \div (4x)$

Add.

101.
$$\begin{array}{r} 32xy - 6xy^2 \\ 14xy + 9xy^2 \\ -3xy - 7xy^2 \\ (+)\ \underline{\ \ xy + \ \ xy^2} \end{array}$$

102.
$$\begin{array}{r} x + 2y - \ z \\ -x \qquad + \ z \\ 3y - 4z \\ (+)\ \underline{2x - 2y \qquad} \end{array}$$

103. $(4x - 2y); (3x - 7y + 2z); (5z - 3y); (2x + 2y)$

104. $(-3x - 2y + 5z); (-6x - 2z); (-5y + 3z); (-4x - 6z)$

Subtract the second polynomial from the first.

105. $(4x - 7y); (9y - 4x)$ **106.** $(3a + b^2); (b^2 - 5a)$

107. $(3a + 7b); (5a - 2b)$ **108.** $(30a - 2b); (30c + 2d)$

109. $(3a - 41b + 26c); (71c - 32d)$ **110.** $(4a^2 + 6a - 5); (-2a^2 + 5a + 3)$

111. Subtract $5x - 8y$ from -1.   **112.** Subtract $9a - 2b$ from 0.

113. What must be added to $(7m - 3k)$ to make the sum $(5a + 2k)$?

114. What must be added to $(-3d + 2f)$ to make the sum 0?

Multiply.

115. $(-3x)(-6x - 3y + 2)$ **116.** $(3m)(a - 3m + 2 + 4a)$

117. $(-9a^3)(3a^2 - 4a + 2m)$ **118.** $(-16a^2b)(-a + 4b - 3c)$

119. $(3a - 4x)(2a - 7x)$ **120.** $(3x - 2y)(3x - 3y + 7)$

121. $(x - 5y)(6x - 3y + 2z)$ **122.** $(-a + 4d)(-6a^3 + 3a^2 + 2a)$

123. $(a^2 + b - c)(a + b + c)$ **124.** $(3x^2 + 4x - 2)(12x^2 - 3x + 1)$

125. $(3a^2 - 2ab - 6b^2)(a^3 - 2a^2b)$

126. $(a^3 - 2ab + b^2)(a^3 + 2ab - b^2)$

Divide; express quotients in simplest terms.

127. $(m - 12m^2n^3) \div (mn)$ **128.** $(24a^2b^3 - 16ab^4) \div (4ab^2)$

129. $(-14x^2y + 21xy^2 - 63x^2y^3) \div (-7xy)$

130. $(36x^2y^3 + 12x^2y - 15x^2y^2) \div (-3xy)$

131. $(102x^4y^4z^2 - 51x^3y^2z^3 + 85x^2y^2z^2) \div (-17x^2y^2z^2)$

132. $(10a^2 - 19a + 6) \div (2a - 3)$ **133.** $(21x^2 - xy - 10y^2) \div (3x + 2y)$

134. $(4a^3 - 13a^2 + 11a - 2) \div (4a - 1)$

135. $(8x^2 + 10x - 65) \div (2x + 7)$ **136.** $(6m^2 + 14m - 15) \div (3m - 2)$

137. $(6a^3 - 4a^2 - 3a + 2) \div (2a^2 - 1)$

138. $(8x^3 + 28x^2 - 6x - 21) \div (4x^2 - 3)$

139. $(6x^4 + 3x^3 - 4x^2 + 3) \div (3x^3 - 2x + 1)$

140. $(8a^7 - 7a^4 + 2a^3 + 6a - 3) \div (a^4 - 2a + 1)$

SPECIAL PRODUCTS AND FACTORING

2

1. INTRODUCTION

In practice, it is often necessary to "rename" an algebraic expression, to express it equivalently but in different form. Some of the procedures developed in the previous chapter accomplish this equivalent change. For instance, the product of two algebraic factors can be *expanded* by Program **1.15** into an equivalent polynomial; and a polynomial can be *factored* by Program **1.17** into the product of two factors (i.e., the divisor and quotient).

It is useful to be able to expand and factor certain types of algebraic expressions on sight. In this chapter we make a study of "special products." They provide shortcuts for expanding and insight into the reverse process of factoring.

Note: Whether a polynomial is factorable depends on the domain of the coefficients. *In this chapter* a polynomial with integral coefficients is considered factorable if it can be expressed as the product of two or more factors that have integral coefficients and exponents that are whole numbers.

2. COMMON MONOMIAL

The distributive property lies at the heart of expanding and factoring.

$$\text{Expanding} \longrightarrow$$

$$a(b + c) = ab + ac$$

$$\longleftarrow \text{Factoring}$$

Its simplest expression is found in a polynomial in which each of the terms contains a common monomial.

PROGRAM 2.1

To factor a polynomial containing a common monomial:

Step 1: Identify the greatest number that will divide exactly every numerical coefficient of the polynomial.

Step 2: Identify the variables common to all terms of the polynomial and express each to the lowest power that it occurs in any one term.

Step 3: Divide the polynomial by the product of terms identified in Steps 1 and 2. The divisor and quotient are a pair of factors of the given polynomial.

EXAMPLES

1. Factor $8x^3 + 12x^2y + 4x^2$.

SOLUTION *Step 1:* The numerical coefficients are 8, 12, and 4. The greatest number that divides each of them is 4.

Step 2: The variables in the polynomial are x and y. Only x is common to all terms. The lowest power to which x appears is x^2.

Step 3: The product of the terms identified in Steps 1 and 2 is $4x^2$, which is used to divide the polynomial (recall Program **1.16**):

$$\frac{8x^3 + 12x^2y + 4x^2}{4x^2} = \frac{8x^3}{4x^2} + \frac{12x^2y}{4x^2} + \frac{4x^2}{4x^2}$$

$$= 2x + 3y + 1$$

Factors of the polynomial:

$$(4x^2)(2x + 3y + 1)$$

2. Factor $6a^2b^2c - 3ab^3 + 9ab^4$.

SOLUTION *Step 1:* The greatest common factor of 6, 3, and 9 is 3.

Step 2: Common variables are a and b. The lowest power to which a appears is a^1 or simply a. The lowest power to which b appears is b^2.

Step 3: Combine the terms into a product: $3ab^2$. Divide:

$$\frac{6a^2b^2c}{3ab^2} - \frac{3ab^3}{3ab^2} + \frac{9ab^4}{3ab^2} = 2ac - b + 3b^2$$

Factorization of $6a^2b^2c - 3ab^3 + 9ab^4$:

$$(3ab^2)(2ac - b + 3b^2)$$

3. Factor $5x^2y - 10xy^2 + 15xy$.

SOLUTION Often you can factor on sight.
Recognize 5 as the greatest common factor among the coefficients and write

$$5 \ (\qquad)$$

Recognize xy as the greatest common factor among the variables (or do it a letter at a time) and write

$$5xy(\qquad)$$

Mentally divide the terms of the polynomial, term by term, by $5xy$ and write the quotient within the parentheses

$$5xy(x - 2y + 3)$$

The result above is the polynomial, $5x^2y - 10xy^2 + 15xy$, in factor form.

4. Factor $a^3x^2 + x^2$.

SOLUTION The common factor is x^2:

$$a^3x^2 + x^2 = x^2(a^3 + 1)$$

EXERCISE 2-1

Write products for each of the following directly.

1. $(-3)(a - 6)$ **2.** $a(x - y)$ **3.** $4c(a^2 - c^2)$

4. $5x(a - b + c)$ **5.** $3xy(x - y + z)$ **6.** $-(4x - 3y + 2z)$

7. $-p^2(p^3 + p^2 - 3p)$ **8.** $-ab(ab - ab^2 + a^2b)$

9. $-4xy(2x - 3y - 5xy)$ **10.** $3m(n^2 + 2mn - m^2)$

11. $-5(4a^3 - 3a^2 + 2a + 1)$ **12.** $-4ab(2a^3b - 3a^2b^2 - ab^3)$

Factor.

13. $5x - 10y$

14. $6x - 2x^2$

15. $3a^2 - 12a$

16. $27a^2p^2 + 3ap$

17. $34pq - 51p$

18. $6a - 36a^2$

19. $am^2 - m^2$

20. $ax + x$

21. $p^2 + 3p^2q$

22. $2x - 2xy$

23. $14a^2xy - 2ax^2y$

24. $x^3 - 6x^2$

25. $2a^2 + a^2 - 3a$

26. $g^5 + 3g^4 - 2g^2$

27. $2mn - 4m + 6n$

28. $2x^2 + 4xy + 12y^2$

29. $2x^3y + 4x^2y^2 + 16xy^3$

30. $a^6 - 2a^4 + 3a^2$

31. $a^3b^2c^4 + a^2b^3c^3 + a^4b^2c^3$

32. $4a^2 - 6a^2b + 7ab^2$

33. $9x^2y - 6xy^2 + 3xy$

34. $2a^5 - 4a^4 + 8a^3 + 10a$

35. $h^2x^2 - 8px^2 + 4x^2y + 6x^2$

36. $24x^2y - 36xy^2 + 12xy$

3. SQUARE OF A BINOMIAL

A binomial is a two-term expression, such as $3a + b$. To **square** a binomial means to multiply it by itself. Thus

$$(3a + b)^2 = (3a + b)(3a + b)$$

which expands to $9a^2 + 6ab + b^2$.

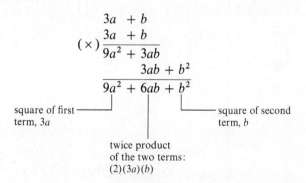

If we analyze the product of $(3a + b)^2$, we can see that it is the sum of

the square of the first term of the binomial,
plus twice the product of the two terms of the binomial,
plus the square of the second term of the binomial.

With a little practice, the computations can be done mentally and the product written directly.

PROGRAM 2.2

To square a binomial:

Step 1: Square the first term.

Step 2: Double the product of the first and second terms (consider the sign between the two terms as belonging to the second term).

Step 3: Square the second term.

Step 4: Express the desired product as the sum of the results of Steps 1, 2, and 3.

EXAMPLES

1. Square $4a + 3c$ [or "expand $(4a + 3c)^2$"].

SOLUTION *Step 1:* Square the first term:

$$(4a)^2 = 16a^2$$

Step 2: Double the product of the two terms:

$$2[(4a)(3c)] = 2[12ac] = 24ac$$

Step 3: Square the second term:

$$(3c)^2 = 9c^2$$

Step 4: Express the results of Steps 1, 2, and 3 as a sum:

$$(4a + 3c)^2 = 16a^2 + 24ac + 9c^2$$

2. Expand $(2a - b)^2$.

SOLUTION *Step 1:* Square the first term, $2a$:

$$(2a)^2 = 4a^2$$

Step 2: Double the product of the two terms, $2a$ and $-b$:

$$2[(2a)(-b)] = 2[-2ab] = -4ab$$

Step 3: Square the second term, $-b$:

$$(-b)^2 = b^2$$

Step 4: Add: $4a^2 - 4ab + b^2$.

3. Expand $(ab - c)^2$.

SOLUTION *Step 1:* $(ab)^2 = a^2b^2$

Step 2: $2[(ab)(-c)] = 2[-abc] = -2abc$

Step 3: $(-c)^2 = c^2$

Step 4: $a^2b^2 - 2abc + c^2$

4. Square 53.

Express 53 as $(50 + 3)$:

$$53^2 = (50 + 3)^2 = (50 + 3)(50 + 3)$$
$$= (50)^2 + 2(50)(3) + (3)^2$$
$$= 2500 + 300 + 9 = 2809$$

The square of a binomial is called a **square trinomial**, a three-term polynomial in which two of the terms are squared terms and the other term (disregarding the sign) is twice the product of the square roots* of the other two terms.

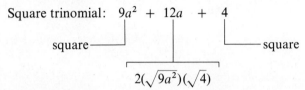

When the binomial to be squared is the *sum* of two terms, the sign of the middle term is always positive.

When the binomial to be squared is the *difference* of two terms, the sign of the middle term is always negative.

PROGRAM 2.3

To factor a square trinomial:

Step 1:   Find the square roots for each of the two terms that are squares.

Step 2:   Write two identical binomial factors whose terms are the two roots of Step 1, separated by the sign of the remaining trinomial term.

EXAMPLES

1.   Factor $4x^2 + 12x + 9$.

$4x^2 + 12x + 9$ is a square trinomial.

SOLUTION   *Step 1:* Find the square roots of each of the squared terms, $4x^2$ and 9:

$$\sqrt{4x^2} = 2x$$
$$\sqrt{9} = 3$$

Step 2: The remaining trinomial term $(+12x)$ is positive. Therefore, $(2x + 3)(2x + 3)$ is the factored form of $4x^2 + 12x + 9$.

---

* The square root of a number or polynomial is one of two identical factors whose product is the number or polynomial. Additional discussion of square roots will be found in Section 4 of Chapter 6.

2. Factor $25x^2 - 40x + 16$.

$25x^2 - 40x + 16$ is a square trinomial.

SOLUTION *Step 1:* Find the square roots of the squared terms:

$$\sqrt{25x^2} = 5x$$

$$\sqrt{16} = 4$$

Step 2: The remaining trinomial term $(-40x)$ is negative. Thus $(5x - 4)(5x - 4)$ is the factored form of $25x^2 - 40x + 16$.

3. Factor $16a^2 - 24ab + 9b^2$.

$16a^2 - 24ab + 9b^2$ is a square trinomial.

SOLUTION *Step 1:* $\sqrt{16a^2} = 4a$

$\sqrt{9b^2} = 3b$

Step 2: $(4a - 3b)(4a - 3b)$ is the factored form of $16a^2 - 24ab + 9b^2$.

4. Factor $4x^2 + 24xy + 36y^2$.

SOLUTION $4x^2 + 24xy + 36y^2$ is a square trinomial and is also a type covered by Program **2.1** (contains a common factor 4). Remove the factor 4 first:

$$4(x^2 + 6xy + 9y^2)$$

Since $x^2 + 6xy + 9y^2$ is also a square trinomial,

Step 1: $\sqrt{x^2} = x$

$\sqrt{9y^2} = 3y$

Step 2: $x^2 + 6xy + 9y^2 = (x + 3y)(x + 3y)$. Therefore, the factored form of $4x^2 + 24xy + 36y^2$ is

$$(4)(x + 3y)(x + 3y)$$

EXERCISE 2-2

Expand each, using Program **2.2.**

1. $(x + y)^2$
2. $(a + b)^2$
3. $(m + n)^2$
4. $(x - y)^2$
5. $(a - b)^2$
6. $(m - n)^2$
7. $(x + 3)^2$
8. $(b - 4)^2$
9. $(3x + 1)^2$
10. $(2y - 1)^2$
11. $(3a - 2)^2$
12. $(4x - 5)^2$
13. $(3a - 2b)^2$
14. $(2x + 3y)^2$
15. $(8x + 7y)^2$
16. $(pm + qn)^2$
17. $(ax - by)^2$
18. $(2p - st)^2$
19. $(2xy - y^2)^2$
20. $(12 - 5a^2b)^2$
21. $(12)^2$
22. $(15)^2$
23. $(25)^2$
24. $(61)^2$

In the parentheses, supply the coefficient necessary to make the expression a square trinomial.

25. $x^2 - (\ )xy + y^2$

26. $9x^2 + (\ )x + 16$

27. $25a^2 + (\ )ab + 81b^2$

28. $25x^2 - (\ )x + 144$

29. $4p^2 + (\ )pt + 9t^2$

30. $16a^4 + (\ )a^2 + 9$

31. $a^2x^2 + (\ )x + c^2$

32. $4a^2b^2 - (\ )d + 25c^2d^2$

Factor the trinomials completely.

33. $x^2 + 2xy + y^2$ **34.** $a^2 - 2ab + b^2$ **35.** $x^2 - 4x + 4$

36. $2x^2 - 12x + 18$ **37.** $16a^2 + 8a - 1$ **38.** $9x^2y^2 + 6xy + 1$

39. $18x^2 + 24xy + 8y^2$ **40.** $4a^2 - 12a + 9$

41. $25x^4 + 80x^2y^2 + 64y^4$ **42.** $16a^2 + 14a + 9$

43. $2x^2 - 12xy + 18y^2$ **44.** $20 - 20x + 5x^2$

45. $405x^2 + 360xy + 80y^2$ **46.** $75a^2 - 60ab + 12b^2$

47. $a^2x^2 + 2abx + b^2$ **48.** $c^2x^2y^2 + 2cxyz + z^2$

4. PRODUCT OF THE SUM AND DIFFERENCE OF TWO TERMS

Let the binomials $(x + y)$ and $(x - y)$ represent, respectively, the sum and difference of two terms x and y. Their product computed by Program **1.15** is shown here.

$$(x - y)(x + y) = x^2 - xy + xy - y^2$$

$$= x^2 - y^2$$

Note that the product is the square of the first term of the binomials (x) minus the square of the second term of the binomials (y). The so-called middle term of the product of the sum and difference of two terms adds to zero and does not appear in the final expression.

PROGRAM 2.4

To compute the product of the sum and difference of two terms:

Step 1:   Square the two terms.

Step 2:   Write the desired product as the square of the first term minus the square of the second term.

EXAMPLES

1. Write the product of $(3x + 4)(3x - 4)$.

SOLUTION *Step 1:* One factor is the sum of two terms, $3x$ and 4, and the other factor is their difference. Square the two terms:

$$(3x)^2 = 9x^2$$
$$(4)^2 = 16$$

Step 2: The product is the square of the first term less the square of the second term:

$$9x^2 - 16$$

2. Expand $(4 + 3x)(4 - 3x)$.

SOLUTION *Step 1:* $(4)^2 = 16$
$(3x)^2 = 9x^2$
Step 2: $16 - 9x^2$

3. Multiply $(5a - 2b)(5a + 2b)$.

SOLUTION
$$(5a)^2 - (2b)^2 = 25a^2 - 4b^2$$

4. Multiply 98×102.

SOLUTION Express 98 as $(100 - 2)$ and 102 as $(100 + 2)$.

$$98 \times 102 = (100 - 2)(100 + 2)$$
$$= (100)^2 - (2)^2$$
$$= 10,000 - 4 = 9996$$

The foregoing examples illustrate the fact that the product of the sum and difference of two terms is the difference of the squares of the two terms. Inversely, then, an algebraic expression involving the difference of two squares will have for factors the sum and difference of the square roots of the two terms.

PROGRAM 2.5

To factor the difference of two squared terms:

Step 1: Find the square root for each of the terms.
Step 2: Write one factor as the sum of the two square roots of Step 1 and the other factor as the difference of the two square roots.

Note: The domain of the variables in the following is limited to positive numbers.

1. Factor $a^2 - 4b^2$.

SOLUTION | *Step 1:* The two squared terms are a^2 and $4b^2$; find their respective square roots:

$$\sqrt{a^2} = a$$

$$\sqrt{4b^2} = 2b$$

Step 2: One factor is the sum of the two square roots, the other factor is their difference:

$$(a + 2b)(a - 2b)$$

2. Factor $1 - 16x^2y^2$

SOLUTION | *Step 1:* $\quad\sqrt{1} = 1$

$$\sqrt{16x^2y^2} = 4xy$$

Step 2: $(1 + 4xy)(1 - 4xy)$

3. Factor $12x^2y - 27y^3$.

SOLUTION | $12x^2y - 27y^3$ is an example of the type covered by Program **2.1**. Remove the common monomial $3y$:

$$12x^2y - 27y^3 = 3y(4x^2 - 9y^2)$$

Since

$$(4x^2 - 9y^2) = (2x + 3y)(2x - 3y)$$

the complete factorization of $12x^2y - 27y^3$ is

$$(3y)(2x + 3y)(2x - 3y).$$

4. Factor $1 - 16x^4$.

SOLUTION | By Program **2.5**:

$$1 - 16x^4 = (1 - 4x^2)(1 + 4x^2)$$

The $(1 - 4x^2)$ factor may be further factored by Program **2.5**:

$$1 - 4x^2 = (1 - 2x)(1 + 2x)$$

The complete factorization of $1 - 16x^4$ is

$$(1 - 2x)(1 + 2x)(1 + 4x^2)$$

Expand directly.

1. $(m - n)(m + n)$ **2.** $(s + t)(s - t)$
3. $(x - 3)(x + 3)$ **4.** $(y + 7)(y - 7)$
5. $(2x - y)(2x + y)$ **6.** $(3x - 2y)(3x + 2y)$
7. $(1 - n)(1 + n)$ **8.** $(8x - 9y)(8x + 9y)$
9. $(ab - c)(ab + c)$ **10.** $(15)(25) = (20 - 5)(20 + 5)$
11. $(37)(43)$ **12.** $(1 - xy)(1 + xy)$

Factor completely.

13. $a^2 - b^2$ **14.** $m^2 - n^2$ **15.** $a^2 - 9b^2$
16. $x^2 - 16y^2$ **17.** $16a^2 - 1$ **18.** $9b^2 - 1$
19. $9x^2y^2 - 64z^2$ **20.** $16x^2 - 81y^2z^2$ **21.** $25p^2 - 4q^2$
22. $1 - 100a^2b^2$ **23.** $x^2 - \dfrac{1}{4}y^2$ **24.** $\dfrac{1}{25}a^2 - c^2$
25. $a^4 - b^4$ **26.** $m^4 - n^4$ **27.** $12m^2 - 27n^2$
28. $18a^2 - 50$ **29.** $36a^2b^2 - a^2$ **30.** $16x^2 - x^2y^2$
31. $1 - 81x^4$ **32.** $16a^4 - 1$ **33.** $a^5b^2 - a^3$
34. $x^2y^3 - y^5$ **35.** $a^4 - b^4$ **36.** $m^4 - 16n^4$

5. PRODUCT OF ANY TWO BINOMIALS

The two preceding types of special products (Programs **2.2** and **2.4**) have been singled out because their forms are readily recognized. In this section we develop a general approach for finding the product of *any* two binomials directly, one that includes instances covered by those procedures.

Consider the following product of two binomials computed by Program **1.15** using the vertical form:

$$
\begin{array}{r}
2x + y \\
(\times)\ \underline{x - 3y} \\
2x^2 + xy \\
\underline{-6xy - 3y^2} \\
2x^2 - 5xy - 3y^2
\end{array}
$$

Notice:

1. The first term of the product $(2x^2)$ is the product of the first terms of the two binomials.
2. The last term of the product $(-3y^2)$ is the product of the second terms of the two binomials.

3. The middle term of the product $(-5xy)$ is the result of multiplying the first term of each binomial with the second term of the other binomial and adding the products.

Statement (3) is often referred to as **cross-multiplying** or computing the cross products.

Illustrated at the right below is a horizontal pattern for writing a product of two binomials directly. The corresponding steps are lettered in both forms. Note that the middle term in the horizontal form is computed mentally.

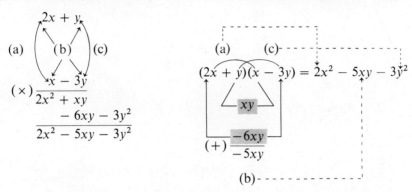

The pattern at the right is summarized in the following program.

PROGRAM 2.6

To multiply any two binomials directly:

Step 1: Compute the product of the first terms of the two binomials.

Step 2: Compute the sum of the two cross products.

Step 3: Compute the product of the second terms of the two binomials.

Step 4: Express the product of the two binomials as the sum of the results of Steps 1, 2, and 3.

EXAMPLES

1. Write the product directly for $(3x - 2y)(x - 4y)$.

SOLUTION *Step 1:* Product of the first terms, $3x$ and x:

$$(3x)(x) = 3x^2$$

Step 2: Sum of cross products:

$$\left.\begin{array}{l}(3x)(-4y) = -12xy \\ (-2y)(x) = -2xy\end{array}\right\} - 14xy$$

Step 3: Product of the second terms, $-2y$ and $-4y$:

$$(-2y)(-4y) = 8y^2$$

Step 4: Combining the results of the steps above:

$$(3x - 2y)(x - 4y) = 3x^2 - 14xy + 8y^2$$

2.   Write the product directly for $(4x - 5)(x + 6)$.

SOLUTION

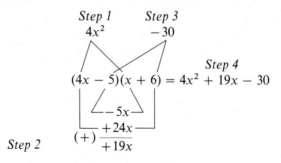

3.   Expand $(3x - 2y)(3x - 2y)$ directly.

SOLUTION

$$(3x - 2y)(3x - 2y) = 9x^2 - 12xy + 4y^2$$

Note: $(3x - 2y)(3x - 2y) = (3x - 2y)^2$ and, as can be expected, the expansion is a square trinomial.

| EXERCISE 2-4

Expand, using Program 2.6.

1. $(x + 3)(x + 2)$ **2.** $(5a + 1)(a + 7)$
3. $(3x + 4)(x + 3)$ **4.** $(2x - 5)(3x - 2)$
5. $(4y - 7)(2y - 5)$ **6.** $(3x + 1)(x - 5)$
7. $(2x - 7)(x + 4)$ **8.** $(5a - 4)(3a + 1)$
9. $(2m + 3)(2m + 5)$ **10.** $(8p - 3)(5p + 2)$
11. $(3x - 7)(3x - 7)$ **12.** $(3x - 7)(5x + 2)$
13. $(3 + 2a)(3 - a)$ **14.** $(4 - b)(4 + 3b)$
15. $(a - 2b)(a - 3b)$ **16.** $(m + b)(m - 2b)$
17. $(3x + 2y)(2x - 3y)$ **18.** $(4x - y)(x + 4y)$
19. $(3x + y)(3x + y)$ **20.** $(2x - y)(2x - y)$
21. $(3a - 2b)(3a - 2b)$ **22.** $(2a - c)(2a - c)$
23. $(4d - 3g)(2d + g)$ **24.** $(3x + y)(2x - 5y)$
25. $(ax + b)(cx + d)$ **26.** $(2px + 3)(x - p)$
27. $(2ax - 5b)(3ax + 2b)$ **28.** $(5xy - 3c)(2xy - c)$

58 Chapter 2 SPECIAL PRODUCTS AND FACTORING

6. FACTORING TRINOMIALS OF THE FORM $ax^2 + bxy + cy^2$

Although the product of two binomials is often a trinomial, all trinomials are not necessarily the product of a pair of binomials. Some are, of course. In this section we develop a procedure for finding the binomial factors with integral coefficients— when they exist—of trinomials of the form $ax^2 + bxy + cy^2$, where a, b, and c are integers.

There is a large element of trial and error in determining the factors. A few generalizations, however, coupled with conscious experience in computing binomial products will tend to minimize the number of necessary trials.

Recall the display of the previous section that demonstrated a horizontal procedure for expanding a pair of binomial factors into a trinomial. Here we repeat the display but with the arrows reversed. We are now interested in working *from* the trinomial *to* the factors.

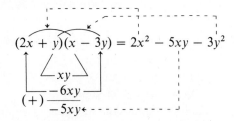

When we reverse the process—that is, go from the trinomial to its factors (right to left in the display above)—we notice that

1. The first term of the trinomial to be factored $(2x^2)$ must have as its factors the first terms of the desired binomials.
2. The last term of the trinomial to be factored $(-3y^2)$ must have as its factors the second terms of the desired binomials.
3. The middle term of the trinomial to be factored $(-5xy)$ results from the sum of the cross products of the first and second terms of the binomial factors.

A second generalization, as an aid to factoring trinomials, concerns the distribution of signs in the trinomial. We can use this set of products of pairs of binomials to confirm the following sign analysis:

$$(3x + 2y)(x + y) = 3x^2 + 5xy + 2y^2$$
$$(3x - 2y)(x - y) = 3x^2 - 5xy + 2y^2$$
$$(3x + 2y)(x - y) = 3x^2 - xy - 2y^2$$
$$(3x - 2y)(x + y) = 3x^2 + xy - 2y^2$$

1. When the sign of the last term of a factorable trinomial is plus, the signs of the second terms in both binomial factors are the same as that of the middle term of the trinomial.

2. When the sign of the last term of a factorable trinomial is minus, the signs of the second terms of the two binomial factors differ, and the prevailing sign of the sum of the resulting cross products is the same as that of the middle term of the trinomial.

We make use of the foregoing observations in the following procedure.

PROGRAM 2.7

To factor a trinomial of the form $ax^2 + bxy + cy^2$ that is the product of two binomials:

Step 1: Determine two factors of the first term of the trinomial (ax^2).

Step 2: Determine two factors of the last term of the trinomial (cy^2), apart from its sign.

Step 3: Use the factors of Steps 1 and 2 to form two binomials so that each binomial contains one factor from each step. Insert signs according to the sign analysis of the trinomial.

Step 4: Check the sum of the resulting cross products in Step 3. If it agrees with the middle term of the trinomial (bxy), the binomials of Step 3 are the desired factors. If it does not agree, repeat the steps with new factor combinations and with new sign arrangements if the binomials are to differ in sign.

EXAMPLES

1. Factor $x^2 - 5x + 6$.

SOLUTION *Step 1:* Factors of the first term of the trinomial (x^2) are x and x.*

Step 2: Factors of the last term of the trinomial (6) are 6 and 1; also 3 and 2.

Step 3: Coupling the x and x with the 6 and 1, respectively, we obtain $(x \quad 6)(x \quad 1)$. From analysis of the signs of the trinomial, the signs of the second terms of the binomials will be negative. Thus $(x - 6)(x - 1)$.

Step 4: The sum of the cross products for $(x - 6)(x - 1)$ is $(-6x) + (-x) = -7x$. This does not agree with the middle term of the trinomial, $-5x$; therefore, $(x - 6)$ and $(x - 1)$ are not the factors of $x^2 - 5x + 6$.

* $-x$ and $-x$ are also factors of x^2, but only the positive factors need be considered when the first term of the trinomial is positive.

60 Chapter 2 SPECIAL PRODUCTS AND FACTORING

Step 3: Couple the x and x factors of the first term with the factors 3 and 2 of the last term for $(x - 3)(x - 2)$.

Step 4: The sum of the cross products for $(x - 3)(x - 2)$ is $(-3x) + (-2x) = -5x$. Consequently, $(x - 3)(x - 2)$ is the factored form of $x^2 - 5x + 6$.

2. Factor $4x^2 + 12xy + 5y^2$.

SOLUTION | *Step 1:* Factor the first term of the trinomial, $4x^2$:

$$4x \text{ and } x, \qquad 2x \text{ and } 2x$$

Step 2: Factor the last term of the trinomial, $5y^2$:

$$5y \text{ and } y$$

Step 3: Before coupling the factors of Steps 1 and 2 in binomial form, note that all the signs of the trinomial are $+$. Therefore, the sign distribution in the binomial factors will be $(\cdots + \cdots)(\cdots + \cdots)$. First trial:

$$(4x + y)(x + 5y).$$

Step 4: Sum of the cross products $(4x)(5y)$ and $(y)(x)$ is $20xy + xy$, or $21xy$, which is not the middle term, $12xy$. Repeat Step 3 with a different arrangement of factors.

Step 3: $(x + y)(4x + 5y)$

Step 4: Check the cross products:

$$5xy + 4xy \neq 12xy$$

Repeat Step 3 with a different set of factors for $4x^2$.

Step 3: $(2x + y)(2x + 5y)$

Step 4: Check the cross products:

$$10xy + 2xy = 12xy$$

The factors of the given trinomial,

$$4x^2 + 12xy + 5y^2,$$

are

$$(2x + y)(2x + 5y).$$

3. Factor $2x^2 + xy - 3y^2$.

SOLUTION | *Step 1:* Factor $2x^2 : 2x \cdot x$.

Step 2: Factor $3y^2 : 3y \cdot y$.

Step 3: Couple these factors into binomial form. Because the last term of the trinomial is negative, the signs of the second terms of the binomials will differ; because the sign of the middle term of the trinomial (xy) is positive, the $+$ sign should go with the greater cross product. As a first trial: $(2x - y)(x + 3y)$.

Step 4: Check the sum of the cross products: $6xy - xy \neq xy$.

Step 3: Try a new combination: $(2x + 3y)(x - y)$.

Step 4: Check the sum of the cross products: $3xy - 2xy = xy$. Therefore, $(2x + 3y)(x - y)$ is the factored form of $2x^2 + xy - 3y^2$.

4.   Factor $6x^2 - x - 12$.

SOLUTION   The factors of $6x^2$ are $6x$ and x; $3x$ and $2x$. The factors of 12 are 12 and 1; 6 and 2; 4 and 3. From an analysis of the signs in the trinomial, establish the combination of signs in the binomials:

$$(\cdots + \cdots)(\cdots - \cdots)$$

Test the various factor combinations of the third term, 12, with the $6x$ and x factors of $6x^2$, keeping in mind that the sum of the cross products must be negative (because the middle term in the trinomial is negative).

Factors (?)	*Middle-term check*
$(6x + 1)(x - 12)$	$x - 72x \neq -x$
$(6x + 2)(x - 6)$	$2x - 36x \neq -x$
$(6x + 3)(x - 4)$	$3x - 24x \neq -x$
$(6x + 4)(x - 3)$	$4x - 18x \neq -x$
$(6x + 6)(x - 2)$	$6x - 12x \neq -x$

(The other combination leads to a positive middle term.)

Before abandoning the $6x$ and x combination, interchange them:

$$(x + 1)(6x - 12) \qquad 6x - 12x \neq -x$$

(The other combinations lead to a positive middle term.)

From this it is clear that $6x$ and x are *not* the proper pair of factors of $6x^2$. Try the pair $2x$ and $3x$:

$$(2x + 1)(3x - 12) \qquad 3x - 24 \neq -x$$
$$(2x + 2)(3x - 6) \qquad 6x - 12x \neq -x$$

(The other combinations lead to a positive middle term.)

Finally, interchange the factors $2x$ and $3x$:

$$(3x + 1)(2x - 12) \qquad 2x - 36x \neq -x$$
$$(3x + 2)(2x - 6) \qquad 4x - 18x \neq -x$$
$$(3x + 4)(2x - 3) \qquad 8x - 9x = -x$$

The factored form of $6x^2 - x - 12$ is $(3x + 4)(2x - 3)$.

Example 4 above was labored somewhat to demonstrate the general technique for factoring a trinomial. Increased experience with factoring tends to narrow the guesswork considerably. But often when we have difficulty striking the right combinations we are inclined to suspect that the trinomial is not factorable. The following statement provides a quick test as to whether a trinomial is indeed factorable (i.e., with integral coefficients).

Trinomials of the form $ax^2 + bxy + cy^2$ (in which the coefficients a, b, and c denote integers) *can be factored if the expression $b^2 - 4ac$ is the square of an integer.**

EXAMPLES

1. Can $3x^2 + 2x - 4$ be factored (i.e., with integral coefficients)?

SOLUTION If $3x^2 + 2x - 4 = ax^2 + bxy + cy^2$, then $a = 3, b = 2, c = -4$, and $y = 1$.

$$b^2 - 4ac = (2)^2 - 4(3)(-4)$$

$$= 4 - (-48) = +52$$

Since $+52$ is not the square of an integer, $3x^2 + 2x - 4$ cannot be factored into the product of two binomials with integral coefficients.

2. Can $3x^2 + 10x - 8$ be factored?

SOLUTION Here $a = 3, b = 10$, and $c = -8$.

$$b^2 - 4ac = (10)^2 - 4(3)(-8)$$

$$= 100 + 96 = 196 = (14)^2$$

Since 196 is the square of an integer, 14, we know that $3x^2 + 10x - 8$ can be factored. By Program **2.7**, the factors are found to be $(3x - 2)(x + 4)$.

3. Can $2x^2 + 11xy + 12y^2$ be factored?

SOLUTION $$b^2 - 4ac = 121 - 4(2)(12)$$

$$= 121 - 96 = 25 = (5)^2$$

Therefore, $2x^2 + 11xy + 12y^2$ can be factored. By Program **2.7**, $(2x + 3y)(x + 4y)$.

---

* Table I, page 429, lists squares of positive integers through 100. Why this test works will become evident in Chapter 7, when the quadratic formula is developed.

Factor completely. For those trinomials that you cannot factor, apply the $b^2 - 4ac$ test and write its numerical value instead of the factors.

1. $x^2 + 5x + 6$ 2. $y^2 - 7y + 12$ 3. $x^2 + x - 6$

4. $x^2 + 7x - 8$ 5. $x^2 - 2x - 15$ 6. $x^2 - 3x - 18$

7. $x^2 - 3x - 2$ 8. $x^2 - 7x + 6$ 9. $m^2 - 15m + 36$

10. $b^2 - 2b + 24$ 11. $a^2 + 9ab + 14b^2$ 12. $a^2 - 5ab + 6b^2$

13. $4a^2 - 8a + 3$ 14. $9a^2 + 11a + 2$ 15. $6y^2 + y - 5$

16. $2x^2 + 7x - 15$ 17. $2x^2 + 2x + 3$ 18. $18x^2 + 33x + 14$

19. $3x^2 + 29x + 40$ 20. $6a^2 + 11a - 10$ 21. $4x^2 - 16xy + 15y^2$

22. $3a^2 + 14a - 5$ 23. $2x^2 - 3x - 9$ 24. $100 - 20y + y^2$

25. $4 - 12b + 9b^2$ 26. $x^2 + 6x + 36$ 27. $3x^2 - x - 10$

28. $12a^2 - 35ab + 18b^2$ 29. $m^3 + 10m^2n + 24mn^2$

30. $9x^2 - 28xy + 3y^2$ 31. $18a^4 + 21a^2 - 4a$

32. $a^4 - 6a^2 + 8$ 33. $30x^2 + 4xy - 2y^2$ 34. $15r^2 + 20r + 8$

35. $12a^3 + 7a^2b - 10ab^2$ 36. $4apx^2 - 17apx + 18ap$

37. $18x^4 + 7x^2 - 1$ 38. $12x^4 - 11x^2 + 2$ 39. $a - 7a^2 - 18$

40. $5x^2 + x^4 - 12$ 41. $10x^2 - 3y^2 - xy$ 42. $-18 + 42x^2 - 24x$

7. FACTORING BY GROUPING

Some polynomials, without a common factor among the terms, can be factored by grouping the terms so as to generate a common factor. For example, the polynomial

$$ax + ay + bx + by$$

has no common monomial factor. However, if we group its terms in a certain way, say

$$ax + ay + bx + by$$

and factor the grouped parts using Program **2.1**, we have

$$[a(x + y)] + [b(x + y)]$$

In each bracketed part, $[a(x + y)]$ and $[b(x + y)]$, the binomial $(x + y)$ is a common factor, and as such it may be factored out:

$$(x + y)[a + b]$$

Thus

$$ax + ay + bx + by = (x + y)(a + b)$$

which can be checked by expanding the two factors on the right.

Note that the terms of such polynomials may be grouped differently, but the result does not change significantly:

$$ax + ay + bx + by = ax + bx + ay + by$$
$$= x(a + b) + y(a + b)$$
$$= (a + b)(x + y)$$

1.   Factor $xy + 5y + bx + 5b$.

SOLUTION   If we group and factor y out of the first *two* terms and b out of the last two terms, we can express the original polynomial as

$$\underbrace{xy + 5y}_{} + \underbrace{bx + 5b}_{}$$
$$y(x + 5) + b(x + 5)$$

Then if we factor out the common $(x + 5)$, we will have transformed the given polynomial into an equivalent product:

$$y(x + 5) + b(x + 5) = (x + 5)(y + b)$$

2.   Factor $3m - 3n + am - an$

SOLUTION   Group and factor 3 from the first two terms and a from the last two terms:

$$3m - 3n + am - an = 3(m - n) + a(m - n)$$
$$= (m - n)(3 + a)$$

3.   Factor $ab + b + ac + c$.

SOLUTION   Factor b from the first two terms, and c from the last two terms:

$$ab + b + ac + c = b(a + 1) + c(a + 1)$$
$$= (a + 1)(b + c)$$

An alternative approach:

$$ab + b + ac + c = ab + ac + b + c$$
$$= a(b + c) + 1(b + c)$$
$$= (b + c)(a + 1)$$

Recall that the negative of the polynomial $2a - b + c$ was demonstrated on page 30 to be $-2a + b - c$. That is,

$$2a - b + c = (-1)(-2a + b - c)$$

In a sense, then, if -1 were to be factored out of the polynomial $2a - b + c$, the remaining factor would have to be the negative of the original polynomial: $-2a + b - c$.

At times such a step is necessary when factoring by grouping, as in the case of

$$ab + ac - b - c = a(b + c) - b - c$$
$$= a(b + c) + (-1)(b + c)$$
$$= (b + c)(a - 1)$$

EXAMPLES

1. Factor $x^2 - x - xy + y$.

SOLUTION Factor x from the first two terms and y from the last two terms:

$$x(x - 1) + y(-x + 1)$$

Express $(-x + 1)$ as $(-1)(x - 1)$:

$$x(x - 1) + y(-1)(x - 1)$$
$$= x(x - 1) - y(x - 1)$$
$$= (x - 1)(x - y)$$

2. Factor $3m + 3n - am - an$.

SOLUTION Factor 3 from the first two terms and $-a$ from the last two terms:

$$3m + 3n - am - an = 3(m + n) - a(m + n)$$
$$= (m + n)(3 - a)$$

3. Factor $cx + y - x - cy$.

SOLUTION Rearrange and factor out c:

$$cx + y - x - cy = cx - cy + y - x$$
$$= c(x - y) + y - x$$

Express $y - x$ as $-1(x - y)$; then

$$c(x - y) + (y - x) = c(x - y) + (-1)(x - y)$$
$$= (x - y)(c - 1)$$

4. Factor $c^2d - 3c^2 - 9d + 27$.

SOLUTION Factor c^2 from the first two terms and -9 from the last two terms:

$$c^2(d - 3) - 9(d - 3) = (d - 3)(c^2 - 9)$$

But $(c^2 - 9)$ can be factored further, by Program **2.5**, as $(c - 3)(c + 3)$. Hence

$$c^2d - 3c^2 - 9d + 27 = (c - 3)(c + 3)(d - 3)$$

There are many other polynomials with terms other than monomials that are factorable by grouping. Generally, the more experience you have with factoring the different basic types, the more likely you are to recognize a prototype in these more complicated expressions.

EXAMPLES

1. Factor $(x - b)^2 + 9(x - b) + 14$.

SOLUTION Insight is gained if we let $a = (x - b)$; then

$$(x - b)^2 + 9(x - b) + 14 = a^2 + 9a + 14$$

The latter readily factors to $(a + 7)(a + 2)$. However, since $a = (x - b)$, then

$$(a + 7) = [(x - b) + 7]$$
$$(a + 2) = [(x - b) + 2]$$

and

$$(x - b)^2 + 9(x - b) + 14 = (x - b + 7)(x - b + 2)$$

2. Factor $(y + 2)^2 - (x + 3)^2$.

SOLUTION If we let $a = (y + 2)$ and $b = (x + 3)$, then

$$(y + 2)^2 - (x + 3)^2 = a^2 - b^2 = (a - b)(a + b)$$

Substituting back for a and b, we obtain

$$(y + 2)^2 - (x + 3)^2 = [(y + 2) - (x + 3)][(y + 2) + (x + 3)]$$
$$= [y + 2 - x - 3][y + 2 + x + 3]$$
$$= [y - x - 1][y + x + 5]$$

With experience, need for the substitution "crutch" will diminish, and factoring can proceed directly.

3. Factor $x^2 - 4xy + 4y^2 - 9$.

SOLUTION In this instance, group the first three terms and factor them as a trinomial, using Program **2.7**; then use Program **2.5**.

$$x^2 - 4xy + 4y^2 - 9 = (x^2 - 4xy + 4y^2) - 9$$
$$= (x - 2y)^2 - 9$$
$$= [(x - 2y) - 3][(x - 2y) + 3]$$
$$= (x - 2y - 3)(x - 2y + 3)$$

Factor by grouping.

1. $ax + bx + 3a + 3b$
2. $cx + cy - 4x - 4y$
3. $dx - c + x - dc$
4. $cn - 4c - bn + 4b$
5. $gx - 3gh + ghx - 3g$
6. $ab + 6a + 2b + 12$
7. $xy - 5x - 15 + 3y$
8. $3s - 2t - 6 + st$
9. $a^2x + a^2d - x - d$
10. $xy - hx + x^2 - hy$
11. $2at + bt - bs - 2as$
12. $2ay - 6ax - 3bx + by$
13. $5ac - ad - 2bd + 10bc$
14. $bc + 2c - 4b - 8$
15. $kx^2 + ky^2 - mx^2 - my^2$
16. $a^2x^2 - 3x^2 - 4a^2 + 12$

Factor.

17. $(x + 1)^2 + 2(x + 1) + 1$
18. $(a - b)^2 + 4(a - b) + 4$
19. $(x + 5)^2 - (x + 2)^2$
20. $4a^2(b + c) - b^2(b + c)$
21. $2(x + y)^2 - (x + y) - 3$
22. $4(m + n)^2 + 20(m + n) + 25$
23. $2(a^2 + 2ab + b^2) - 3(a + b) - 9$
24. $(x - y)(x + y)^2 + 6(x + y)(x - y) + 9(x - y)$
25. $(x^2 + 4x + 4) - (x^2 - 6x + 9)$
26. $(25x^2 + 10x + 1) - 2(x^2 - 5x - 1 - x^2) + 1$

Factor completely.

27. $8a^2 - 2ab - 3b^2$
28. $y^2 - 2xy^2 + 2x - 1$
29. $a + 3b - a^2 + 9b^2$
30. $a^4 - a^2 - 12$
31. $2a^3 + 5a^2 - 4a - 10$
32. $20x^2 + 63xy - 45y^2$
33. $pr - ps - qr + qs$
34. $ab - a - b + 1$
35. $(a - b)^2 - (a - b) - 12$
36. $2(x + y)^2 + 12(x + y) + 10$
37. $729x^2 - 961y^2$
38. $a^4 - 4a^2 + 4$
39. $(9a^2 + 30a + 25) - (a - 4)^2$
40. $2x^4 - 10x^2 - 12$
41. $16p^4 - 81q^4$
42. $(m + n)^2 - m^2$
43. $a^2 + 6ab + 9b^2 - 4$
44. $x^6 - x^4 - 16x^2 + 16$
45. $2x^4 + 6x^3 - 8x^2 - 24x$
46. $(x^2 + 4x - 6)^2 - 36$
47. $4x^2 - 12xy + 9y^2 - 16$
48. $a^2 + 2ab + b^2 + 4a + 4b + 3$
49. $2a^2 - 4ab + 2b^2 + 5a - 5b - 12$
50. $9a^2 - 6ab + b^2 - 25x^2 - y^2 + 10xy$

8. SUMS AND DIFFERENCES OF TWO CUBES★

The difference of two squares, we noted in Section 4, can be factored but not the sum of two squares, at least not with integral coefficients.† However, both the sum and the difference of two cubed terms can be factored.

By multiplying, it can be verified that

$$(x + y)(x^2 - xy + y^2) = x^3 + y^3$$

and

$$(x - y)(x^2 + xy + y^2) = x^3 - y^3$$

There is a fairly consistent and easily remembered pattern among the terms of the factors of these two binomials. By using

$$x^3 + y^3 = (x + y)(x^2 - xy + y^2)$$

and

$$x^3 - y^3 = (x - y)(x^2 + xy + y^2)$$

as formulas, it is possible to factor any sum or difference of two cubes.

EXAMPLES

1. Factor $a^3 + b^3$.

SOLUTION Using $x^3 + y^3 = (x + y)(x^2 - xy + y^2)$ as a formula, and replacing x with a, and y with b, we get

$$a^3 + b^3 = (a + b)(a^2 - ab + b^2)$$

2. Factor $27p^3 - 8$.

SOLUTION Using $x^3 - y^3 = (x - y)(x^2 + xy + y^2)$ as a formula, and substituting $3p$ for x, and 2 for y [since $(3p)^3 = 27p^3$ and $(2)^3 = 8$], we get

$$27p^3 - 8 = (3p)^3 - (2)^3 = [3p - 2][(3p)^2 + (3p)(2) + (2)^2]$$
$$= (3p - 2)(9p^2 + 6p + 4)$$

Note: We could have achieved the same result by considering $27p^3 - 8$ as $27p^3 + (-8)$ and

$$x^3 + y^3 = (x + y)(x^2 - xy + y^2)$$

as the formula. Thus $x = 3p$ and $y = -2$, and

$$27p^3 - 8 = (3p)^3 + (-2)^3$$
$$= [3p + (-2)][(3p)^2 - (3p)(-2) + (-2)^2]$$
$$= (3p - 2)(9p^2 + 6p + 4)$$

★ The content of this section, in more basic courses of study, may be postponed without significant effect upon the remaining parts of the book.

† Apply the $b^2 - 4ac$ test to $9x^2 + y^2$, for example. Here $a = 9, b = 0, c = 1$, and $b^2 - 4ac = -36$, not the square of some integer.

A somewhat more direct method for generating the factors of the sum or difference of two cubes is presented in the following program.

PROGRAM 2.8

To factor the sum or difference of two cubed terms:

Step 1: Find the cube root of each term; consider the sign between the two terms as belonging to the second term.

Step 2: Write one factor as the sum of the two cube roots of Step 1.

Step 3: Write the other factor as a trinomial in which the terms are the square of the first cube root of Step 1, minus the product of the two cube roots of Step 1, plus the square of the second cube root of Step 1.

EXAMPLES

1. Factor $m^3 + n^3$.

SOLUTION   *Step 1:* Find the cube root of each term:

$$\sqrt[3]{m^3} = m \quad [\text{since } (m)(m)(m) = m^3]$$

$$\sqrt[3]{n^3} = n \quad [\text{since } (n)(n)(n) = n^3]$$

Step 2: One factor is the sum of the two cube roots:

$$m + n$$

Step 3: The other factor:

$$(m)^2 - mn + (n)^2$$

or

$$m^2 - mn + n^2$$

The factorization of $m^3 + n^3$ is

$$(m + n)(m^2 - mn + n^2)$$

2. Factor $a^3 - 27b^3$.

SOLUTION   *Step 1:* Find the cube root of each term, a^3 and $-27b^3$.

$$\sqrt[3]{a^3} = a \quad [\text{since } (a)(a)(a) = a^3]$$

$$\sqrt[3]{-27b^3} = -3b \quad [\text{since } (-3b)(-3b)(-3b) = -27b^3]$$

Step 2: One factor: $a + (-3b) = (a - 3b)$

Step 3: Other factor: $(a)^2 - (a)(-3b) + (-3b)^2$
Simplified: $a^2 + 3ab + 9b^2$
Thus

$$a^3 - 27b^3 = (a - 3b)(a^2 + 3ab + 9b^2)$$

3. Factor $8m^3 + 125n^3$.

SOLUTION *Step 1:* $\sqrt[3]{8m^3} = 2m$

$\sqrt[3]{125n^3} = 5n$

Steps 2 and 3:

$$(2m + 5n)[(2m)^2 - (2m)(5n) + (5n)^2]$$

or

$$(2m + 5n)(4m^2 - 10mn + 25n^2)$$

4. Factor $27p^3 - 8$.

SOLUTION
$$27p^3 - 8 = (3p)^3 + (-2)^3$$
$$= [3p + (-2)][(3p)^2 - (3p)(-2) + (-2)^2]$$
$$= (3p - 2)(9p^2 + 6p + 4)]$$

5. Factor $(a - 3)^3 + 64$.

SOLUTION
$$(a - 3)^3 + 64 = [(a - 3) + 4][(a - 3)^2 - (a - 3)(4) + (4)^2]$$
$$= [a - 3 + 4][a^2 - 6a + 9 - 4a + 12 + 16]$$
$$= (a + 1)(a^2 - 10a + 37)$$

EXERCISE 2-7

Factor.

1. $x^3 - y^3$	**2.** $p^3 + q^3$	**3.** $x^3 + 8$
4. $m^3 + 125$	**5.** $27 - y^3$	**6.** $8s^3 - 1$
7. $8r^3 + 27t^3$	**8.** $x^3y^3 + 1$	**9.** $64a^3b^3 - c^3$
10. $125a^3 + 64b^3$	**11.** $x^6 - y^3$	**12.** $64x^3 + y^6$
13. $27x^9 - y^6$	**14.** $216b^{12} + 27c^6$	**15.** $64t^{21} - 8t^3$
16. $(a + 1)^3 - y^3$	**17.** $(a - 1)^3 + 27$	**18.** $(a - 1)^3 - (a - 3)^3$
19. $(x + 3)^3 + (x - 3)^3$	**20.** $(2x + y)^3 - (3x - y)^3$	

9. SUMMING UP

Here are some guidelines for greater efficiency when factoring a polynomial.

1. Look for common monomial(s). Use Program **2.1**.
2. Consider the number of terms.
   a. Two terms
      Difference of two squares? Use Program **2.5**.
      Sum or difference of two cubes? Use Program **2.8**.

b. Three terms

Is it of the type $ax^2 + bx + c$? Use the $b^2 - 4ac$ test of Section 6 to determine factorability.

If factorable, is it a square trinomial? Use Program **2.3**.

If factorable and not a square trinomial, use Program **2.7**.

c. Four or more terms

Consider factoring by grouping, Section 7.

3. Factor completely.

Inspect each factor to determine whether it can be factored further.

4. Check by multiplying the factors.

EXERCISE 2-8

Factor completely.

1. $(t + 2)^2 - (t - 3)^2$ **2.** $5m^4 + 15m^3 - 20m^2 - 60m$

3. $6(2m + n)^2 - (2m + n) - 7$ **4.** $a^3 - 2a^2 - 4a + 8$

5. $4(9x^2 + 6xy + y^2) - 8(3x + y) - 5$

6. $3x^2 - 6xy + 3y^2 + 13x - 13y - 30$

7. $(a^2 + 6a + 9) - (a^2 - 10a + 25)$

8. $m^3 - n^3$ **9.** $16m^2 - 2mn - 5n^2$

10. $m^3 + 64$ **11.** $3c^4 + 5c^3 - 15c - 25$

12. $125 - x^3$ **13.** $4s + 3t - 16s^2 + 9t^2$

14. $64a^3 + 27b^3$ **15.** $16m^4 - n^4$ **16.** $125s^3t^3 - u^3$

17. $x^4y^4 - 81z^4$ **18.** $a^9 - b^3$ **19.** $16a^4b^2 - 1$

20. $m^{12} - 125n^6$ **21.** $(9m^2 + 30m + 25) - (m - 5)^2$

22. $27x^{15} - 64y^9$ **23.** $256x^4 - 81y^4$ **24.** $(m - 4)^3 + 8$

25. $9x^2 + 12xy + 4y^2 - 16$ **26.** $(s + 5)^3 - (s - 4)^3$

REVIEW

PART A

Answer True or False.

1. *Factoring* and *expanding* are reverse processes of one another.

2. The product of the sum and difference of any two numbers is the difference of their squares.

3. The product of every pair of binomial factors is a polynomial of three terms or more.

4. The square of every binomial is a square trinomial.

5. The binomial factors of $48x^2 - 200x - 625$ will all have the same sign.

6. Trial and error has no part in factoring.

7. In trinomials of the form $ax^2 + bxy + cy^2$, the middle term is a consequence of the cross-products of its binomial factors.

8. If the $b^2 - 4ac$ test is applied to a trinomial and the result is an integer, then it is certain that the trinomial can be factored into a pair of binomials with integral coefficients.

9. -1 may be considered to be a factor of any polynomial.

10. $x^3 + y^3$ cannot be factored with integral coefficients.

PART B

Write the products for each of the following directly.

1. $-4(m + 7)$ 2. $-5d(c^4 - d^5)$

3. $-4km(k - 3m + 2n)$ 4. $-d^2(d^2 - 4d + 9)$

5. $7ab(-3a - 4b + 5ab)$ 6. $-10(3x^3 - 5x^2 + x - 15)$

Factor.

7. $14a - 21b$ 8. $5m^2 - 30m$ 9. $54xy - 90y$

10. $30d^2pq + 6dpq^2$ 11. $7m^3 + 5m^2 - 9m$ 12. $3ab + ba - 12b$

13. $8a^3b + 24a^2b^2 + 16ab^3$ 14. $k^6m^5n^4 + k^4m^4n^4 - k^4m^5n^3$

15. $6a^2m^3 + 8bm^3 - 3m^3n^2 - m^3$

Expand, using Program **2.2.**

16. $(a + b)^2$ 17. $(x + 5y)^2$ 18. $(s - 2t)^2$

19. $(8 + b)^2$ 20. $(5m + 1)^2$ 21. $(7x - 5)^2$

22. $(3s - 7t)^2$ 23. $(4a + 9b)^2$ 24. $(bm - 2cn)^2$

25. $(3ac + 2b)^2$ 26. $(17)^2$ 27. $(31)^2$

In the parentheses, supply the coefficient necessary to make the expression a square trinomial.

28. $a^2 - (\ )ab + b^2$ 29. $9x^2 + (\ )xy + 100y^2$

30. $25a^2 - (\ )ab + b^2$ 31. $a^4z^2 - (\ )z + 9$

Factor those which are square trinomials.

32. $a^2 - 2ab + b^2$ 33. $x^2 + 10x + 25$

34. $49m^2 + 42m - 7$ 35. $9m^2 - 6mn + n^2$

36. $16s^4 + 40s^2t^2 + 25t^4$ 37. $4x^2 - 3xy + 9y^2$

38. $36m^2 + 168m + 196$ 39. $a^4m^2 + 4a^2my^2 + 4y^4$

Expand.

40. $(c + d)(c - d)$ 41. $(a - 5)(a + 5)$

42. $(3m + n)(3m - n)$ 43. $(1 - y)(1 + y)$

44. $(mn + t)(mn - t)$ 45. $(18)(22)$

Factor completely.

46. $m^2 - n^2$ 47. $36t^2 - 1$ 48. $9a^2 - 64b^2$

49. $121a^2 - 81b^2$ 50. $8a^2 - 32b^2$ 51. $m^5 - m^7n^2$

Expand, using Program **2.6**.

52. $(m + 2)(m + 7)$

53. $(3t + 7)(t + 4)$

54. $(5a - 2)(3a - 8)$

55. $(3a - 8)(4a + 11)$

56. $(7t + 4)(7t + 1)$

57. $(8a - 3)(8a - 3)$

58. $(5m - 7)(5m + 7)$

59. $(4mn - 3)(5mn + 3)$

60. $(3x - 5y)(2x + y)$

61. $(2x - y)(3x + 2y)$

62. $(3xy - a)(xy - 2a)$

63. $(2ab - 1)(1 + 5ab)$

Factor completely. For these trinomials which you cannot factor, apply the $b^2 - 4ac$ *test and write its numerical value instead of the factors.*

64. $a^2 + 10a + 9$ **65.** $t^2 - 9t + 20$ **66.** $a^2 + 10a - 11$

67. $x^2 - 4x + 5$ **68.** $5s^2 + 64s + 48$ **69.** $15m^2 - 26m + 8$

70. $3t^2 - 14t - 49$ **71.** $9 - 21a + 6a^2$ **72.** $b^2 - 13b + 14$

73. $x^3 - 10x^2 + 21x$ **74.** $4m^4 + 2m^2 - 5m$

75. $40t^2 + 28st - 48s^2$ **76.** $8x^2 - 18x + 9$

77. $4x^2 - 33x + 8$ **78.** $21m^2 - 2n^2 + mn$

79. $b^2 + 4b - 10$

Factor by grouping.

80. $mx + nx + 5m + 5n$ **81.** $ax - x - y + ay$

82. $adx + 3dx - 5ad - 15d$ **83.** $mn + 7n - 5m - 35$

84. $a^2x + a^2y - 4x - 4y$ **85.** $2ab - dy - 2by + ad$

86. $3am - an - 2bn + 6bm$ **87.** $3as^2 + 3at^2 - bs^2 - bt^2$

Factor.

88. $(a + 3)^2 + 4(a + 3) + 4$ **89.** $(m + 3)^2 - (m - 2)^2$

90. $6(x + y)^2 + (x + y) - 2$ **91.** $(x^2 + 2xy + y^2) - 3(x + y) + 2$

92. $(4a^2 + 2a + 1) - (a^2 + 6a + 9)$

93. $15x^2 - 61xy + 4y^2$ **94.** $2x^4 + 5x^3 - 4x - 10$

95. $2k - 3a - 4k^2 + 9a^2$ **96.** $81a^4 - 1$

97. $a^6 - 16a^2$ **98.** $x^2 - 4x^2y^4$

99. $(4m^2 - 4m + 1) - (m + 2)^2$ **100.** $81x^4y^2 - 256y^2$

101. $4x^2 - 20xy + 25y^2 - 9$ **102.** $a^4 - 2a^3 + a^2 - 2a$

103. $x^3 + 2x^2 - 3x - 6$ **104.** $3a^2 + 4ab - 2b^2$

105. $y^3 + x^3$ **106.** $1 - 64x^3$ **107.** $54x^3 - 2y^3$

108. $8xy^6 - 27xy^3$ **109.** $x^3 + y^9$ **110.** $(p - q)^3 + 27$

111. $(a - b)^3 - (a + b)^3$

3 FRACTIONS

1. EQUIVALENT FRACTIONS

A fraction expresses a ratio of one term, **numerator**, to another term, **denominator**:

$$\frac{\text{numerator}}{\text{denominator}}$$

and is called a *ratio*nal expression.

A fraction may also be thought of as a quotient, the result of an implied division of its numerator by its denominator.

$$\text{numerator} \div \text{denominator} = \frac{\text{numerator}}{\text{denominator}}$$

$$3 \quad \div \quad 4 \quad = \quad \frac{3}{4}$$

$$(x + y) \quad \div \quad (2x^2 + 3y^2) = \frac{x + y}{2x^2 + 3y^2}$$

Each division has a unique quotient value. But there are an infinite number of ways to express that quotient. For instance,

$$2 \div 3 = \frac{2}{3}, \frac{4}{6}, \frac{8}{12}, \text{etc.}$$

$$a \div x = \frac{a}{x}, \frac{3a}{3x}, \frac{a^2}{ax}, \frac{a(x+y)}{x(x+y)}, \text{etc.}$$

A simple way to test whether two fractions are equivalent is to apply the following generalization.

Two fractions, $\dfrac{a}{b}$ and $\dfrac{p}{q}$, are equivalent if and only if $(a)(q) = (b)(p)$, provided that $b, q \neq 0$.

EXAMPLES

1. **(a)** $\dfrac{3}{5} \overset{?}{=} \dfrac{9}{15}$ **(b)** $\dfrac{2}{3} \overset{?}{=} \dfrac{3}{5}$

SOLUTION Apply the generalization.

(a) $\dfrac{3}{5} \times \dfrac{9}{15} \Longrightarrow 45$

$\dfrac{3}{5} \times \dfrac{9}{15} \Longrightarrow 45$

$45 = 45$

So $\dfrac{3}{5} = \dfrac{9}{15}$

(b) $\dfrac{2}{3} \times \dfrac{3}{5} \Longrightarrow 10$

$\dfrac{2}{3} \times \dfrac{3}{5} \Longrightarrow 9$

$10 \neq 9$

So $\dfrac{2}{3} \neq \dfrac{3}{5}$

2. $\dfrac{x+y}{3} \overset{?}{=} \dfrac{2x+2y}{6}$

SOLUTION $6(x+y) \overset{?}{=} 3(2x+2y)$

$6x + 6y = 6x + 6y$

So the given fractions are equivalent.

---

* Continually making statements to rule out denominators or divisors of zero becomes tedious. Hereafter we assume that no denominator or divisor under discussion is zero and omit the restricting statement.

The following statements are related to the previous generalization about equivalent fractions. They are valid for both arithmetic and algebraic fractions.

- The value of a fraction is unchanged when the numerator and denominator are multiplied or divided by the same number, not zero.
- Reducing to lower terms is a form of division of numerator and denominator by a common divisor or factor.
- A fraction is in lowest terms when numerator and denominator are relatively prime to one another.

Note: Two integers are *relatively prime* to one another when they have no common integral factor except 1. Thus 25 and 28 are not prime numbers, but they are relatively prime to one another because each contains among its prime factors no factor in common with the other:

$$25 = (5)(5)$$

$$28 = (2)(2)(7)$$

On the other hand, 24 and 45 are *not* relatively prime, for they do have a factor in common.

$$24 = (2)(2)(2)\,(3)$$

$$45 = (3)\,(3)(5)$$

Polynomials are relatively prime to one another when there is no common factor having integers for coefficients among the factors of each polynomial. Thus $x^2 - 5x + 6$ and $x^2 + 4x + 4$ are factorable, yet are relatively prime to one another because there is no common factor among the factors of each.

$$x^2 - 5x + 6 = (x - 2)(x - 3)$$

$$x^2 + 4x + 4 = (x + 2)(x + 2)$$

On the other hand, $x^2 - 5x + 6$ and $x^2 - 4$ are not relatively prime because in $(x - 2)$ each has a factor in common:

$$x^2 - 5x + 6 = (x - 2)(x - 3)$$
$$x^2 - 4 = (x - 2)(x + 2)$$

PROGRAM 3.1

To express the equivalent of an algebraic fraction in lowest terms:

Step 1: Factor the numerator completely; factor the denominator completely.

Step 2: Divide out all common factors that exist in both the numerator and the denominator.

Caution: You may eliminate *factors only* by division, *never* parts of a factor or individual terms of polynomials.

1. Express $\dfrac{x^2 + 2x - 8}{x^2 + x - 12}$ in lowest terms.

SOLUTION

Step 1: Factor the numerator and denominator completely.

$$x^2 + 2x - 8 = (x - 2)(x + 4)$$

$$x^2 + x - 12 = (x + 4)(x - 3)$$

Step 2: $\dfrac{x^2 + 2x - 8}{x^2 + x - 12} = \dfrac{(x - 2)(x + 4)}{(x + 4)(x - 3)} = \dfrac{x - 2}{x - 3}$

2. Express $\dfrac{3x^2 - 9ax + 6a^2}{6x^2 - 6a^2}$ in lowest terms.

SOLUTION

Step 1: Factor numerator and denominator:

$$3x^2 - 9ax + 6a^2 = (3)(x - a)(x - 2a)$$

$$6x^2 - 6a^2 = (2)(3)(x - a)(x + a)$$

Step 2: Simplify:

$$\dfrac{3x^2 - 9ax + 6a^2}{6x^2 - 6a^2} = \dfrac{(3)(x - a)(x - 2a)}{(2)(3)(x - a)(x + a)} = \dfrac{x - 2a}{2(x + a)}$$

3. Express $\dfrac{cp - 2a - ap + 2c}{p^2 + 4p + 4}$ in lowest terms.

SOLUTION

Step 1: $cp - 2a - ap + 2c = (c - a)(p + 2)$

$$p^2 + 4p + 4 = (p + 2)(p + 2)$$

Step 2: $\dfrac{cp - 2a - ap + 2c}{p^2 + 4p + 4} = \dfrac{(c - a)(p + 2)}{(p + 2)(p + 2)} = \dfrac{c - a}{p + 2}$

EXERCISE 3-1

Tell whether the rational expressions are equivalent.

1. $\dfrac{3a}{4b} \overset{?}{=} \dfrac{18a}{24b}$ **2.** $\dfrac{2p}{5q} \overset{?}{=} \dfrac{35p}{14q}$ **3.** $\dfrac{32m}{6n} \overset{?}{=} \dfrac{48m}{9n}$

4. $\dfrac{40p}{15ab} \overset{?}{=} \dfrac{16p}{6ab}$ **5.** $\dfrac{x - y}{a} \overset{?}{=} \dfrac{(x + y)^2}{a(x + y)}$ **6.** $\dfrac{b}{x + y} \overset{?}{=} \dfrac{b^2}{(x + y)^2}$

7. $\dfrac{x + y}{x - y} \overset{?}{=} \dfrac{x - y}{x + y}$ **8.** $\dfrac{x^2 - y^2}{4} \overset{?}{=} \dfrac{4}{x^2 - y^2}$ **9.** $\dfrac{aq + ap}{p + q} \overset{?}{=} \dfrac{ab}{b}$

10. $\dfrac{m}{mn} \overset{?}{=} \dfrac{a + b}{an + bn}$

Express in lowest terms.

11. $\dfrac{18a^2}{30a^2 - 12a}$

12. $\dfrac{3x - 9}{2x - 6}$

13. $\dfrac{ab - 4b}{5a - 20}$

14. $\dfrac{ab - ac}{b^2 - c^2}$

15. $\dfrac{2 + 3b}{x + 3b}$

16. $\dfrac{b - 5x}{1 - 5x}$

17. $\dfrac{xy}{x^2y^2 - 3xy}$

18. $\dfrac{x^4 - 16}{x^3 + 4x}$

19. $\dfrac{x^2 - 5x + 6}{x^2 - 4x + 4}$

20. $\dfrac{x^2 + 6x + 9}{x^2 - 9}$

21. $\dfrac{6x^2 - 11x - 10}{3x^2 - 19x - 14}$

22. $\dfrac{2x^2 - x - 10}{x^2 - 2x - 8}$

23. $\dfrac{ax - ay - 2x + 2y}{a^2 - 7a + 10}$

24. $\dfrac{x^3 - 5x^2 + 6x}{2x^3 - 8x}$

25. $\dfrac{4x^3y - 12x^2y + 9xy}{2axy - 3ay - 2bxy + 3by}$

26. $\dfrac{4a^2 - b^2}{(2a - b)(2a^2 + ab)}$

27. $\dfrac{(x + y)^2 - 3(x + y) - 10}{x^2 + xy + 2x}$

28. $\dfrac{(a - b)^2 - 9}{bc - ac - 3c}$

29. $\dfrac{a^2x + b^2y - b^2x - a^2y}{ax^2 - ay^2 - by^2 + bx^2}$

30. $\dfrac{x^2 - 2xy + y^2 - 16}{(x^2 - 8x + 16) - y^2}$

2. NEGATIVE OF A RATIONAL EXPRESSION

A fraction or rational expression, we noted in the previous section, may be interpreted as the quotient of numerator divided by denominator, that is, $n \div d = \dfrac{n}{d}$.

It follows from Program **1.6**, on division of real numbers, that the sign of the fraction (quotient) would have to be positive when both numerator and denominator are of like sign, and negative when numerator and denominator differ in sign. For instance:

$$\frac{+3}{+5} = (+3) \div (+5) = +\frac{3}{5}$$

$$\frac{+3}{-5} = (+3) \div (-5) = -\frac{3}{5}$$

$$\frac{-3}{+5} = (-3) \div (+5) = -\frac{3}{5}$$

$$\frac{-3}{-5} = (-3) \div (-5) = +\frac{3}{5}$$

In general, each fraction may be considered to have three signs associated with it:

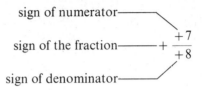

sign of numerator

sign of the fraction $+\dfrac{+7}{+8}$

sign of denominator

If any *pair* of the three signs is reversed, the value of the fraction is *unchanged*

1. Changing the signs in pairs keeps the equality:

$$+\frac{+2}{+3} = \begin{cases} +\dfrac{-2}{-3} \\[2ex] -\dfrac{+2}{-3} \\[2ex] -\dfrac{-2}{+3} \end{cases}$$

2. Reversing one or all three signs reverses the value of the fraction; that is, it becomes the negative of its former value.

$$+\frac{+3}{+5} \Rightarrow +\frac{-3}{+5} = -\frac{+3}{+5} \quad \text{(reversing one sign)}$$

$$+\frac{+3}{+5} \Rightarrow -\frac{-3}{-5} = -\frac{+3}{+5} \quad \text{(reversing three signs)}$$

Polynomials cannot be classified as positive or negative because the variables in them ordinarily represent numbers that may be positive or negative, depending on the replacement set. For instance, $(c - a)$ may represent either a positive or a negative number, depending on the relative values of the numbers that replace c and a.

When $c > a$, the expression $(c - a)$ represents a positive number. When $c < a$, the expression $(c - a)$ represents a negative number. On the other hand, the *negative of* $(c - a)$ does exist. It is $-(c - a)$, whatever the real number replacements for c and a may be.

Recognizing the negatives of some polynomials, such as $(p - q)$, is a fairly simple matter when the expression is preceded by a minus sign: $-(p - q)$. When expressed in expanded form, however, the fact is not so evident.

$$-(p - q) = (-1)(p - q) = (-1)(p) + (-1)(-q)$$

$$= -p + q$$

$$= q - p$$

Thus the negative of $p - q$ is $q - p$; conversely, the negative of $q - p$ is $p - q$.

EXAMPLES

1. Write the negative of $(x - y)$.

SOLUTION The negative of $(x - y)$ is $-(x - y) = -x + y = (y - x)$.

2. Is $(a - b)$ the negative of $(b - a)$?

SOLUTION The negative of $(a - b)$ is $-(a - b)$, which in expanded form is $-a + b$. It, in turn, can be expressed as $b - a$. So $(b - a) = -(a - b)$, and $(a - b) = -(b - a)$.

3. Write the negative of $x^2 - x - 6$.

SOLUTION The negative of $x^2 - x - 6$ is

$$-(x^2 - x - 6) = -x^2 + x + 6 = 6 + x - x^2$$

4. Express $\dfrac{1 - 3x}{6x - 2}$ in lowest terms.

SOLUTION The given fraction in factored form is

$$\frac{1 - 3x}{2(3x - 1)}$$

Note that $1 - 3x$ is the negative of $3x - 1$.

One alternative is to express $1 - 3x$ as $(-1)(3x - 1)$, and apply Program **3.1**:

$$\frac{-1(\overset{1}{\cancel{3x - 1}})}{2(\underset{1}{\cancel{3x - 1}})} = \frac{(-1)(1)}{(2)(1)} = \frac{-1}{2} = -\frac{1}{2}$$

A second alternative is to express $3x - 1$ as $(-1)(1 - 3x)$, and apply Program **3.1**:

$$\frac{(\overset{1}{\cancel{1 - 3x}})}{(2)(-1)(\underset{1}{\cancel{1 - 3x}})} = \frac{1}{2(-1)(1)} = -\frac{1}{2}$$

A third alternative is simply to divide $1 - 3x$ by $3x - 1$ for a quotient of -1 (the quotient of any number or expression divided by its negative); then simplify:

$$\frac{(\overset{-1}{\cancel{1 - 3x}})}{2(\underset{1}{\cancel{3x - 1}})} = \frac{-1}{2} = -\frac{1}{2}$$

5. Express $\dfrac{x^2 + x - 12}{15 - 2x - x^2}$ in lowest terms.

SOLUTION | According to Program **3.1**,

Step 1: $x^2 + x - 12 = (x - 3)(x + 4)$

$15 - 2x - x^2 = (5 + x)(3 - x)$

Step 2: $\dfrac{\overset{-1}{\cancel{(x - 3)}}(x + 4)}{(5 + x)\cancel{(3 - x)}} = \dfrac{(-1)(x + 4)}{5 + x} = \dfrac{-(x + 4)}{(5 + x)}$

$$\text{or } -\dfrac{x + 4}{5 + x} \text{ or } \dfrac{x + 4}{-(5 + x)}$$

EXERCISE 3-2

Insert the proper sign in the empty parentheses to make the fractions equivalent.

1. $-\dfrac{-2}{+5} = (\quad)\dfrac{+2}{+5}$ 　　　　 **2.** $+\dfrac{+2}{+7} = (\quad)\dfrac{-2}{-7}$

3. $+\dfrac{+1}{-4} = (\quad)\dfrac{-1}{-4}$ 　　　　 **4.** $-\dfrac{+5}{+8} = (\quad)\dfrac{-5}{-8}$

5. $-\dfrac{+2}{+9} = (\quad)\dfrac{-2}{+9}$ 　　　　 **6.** $+\dfrac{+3}{-5} = (\quad)\dfrac{+3}{+5}$

7. $+\dfrac{-1}{-6} = (\quad)\dfrac{+1}{+6}$ 　　　　 **8.** $-\dfrac{+2}{-3} = (\quad)\dfrac{+2}{+3}$

Make the statements true by inserting the proper sign (+ or −) in the parentheses.

9. $\dfrac{x}{x - y} = (\quad)\dfrac{x}{y - x}$ 　　　　 **10.** $-\dfrac{a}{3x - 4} = (\quad)\dfrac{-a}{3x - 4}$

11. $\dfrac{m - n}{a - b} = (\quad)\dfrac{n - m}{b - a}$ 　　　　 **12.** $\dfrac{1}{(x - y)(s - t)} = (\quad)\dfrac{1}{(y - x)(s - t)}$

Express in lowest terms.

13. $\dfrac{4 - 3x}{3x - 4}$ 　　　 **14.** $\dfrac{ab - c}{c - ab}$ 　　　 **15.** $\dfrac{a - ax}{x - 1}$

16. $\dfrac{4 - 2t}{3t - 6}$ 　　　 **17.** $-\dfrac{a - 1}{p - ap}$ 　　　 **18.** $\dfrac{x - 2}{2a - ax}$

19. $\dfrac{6x - 15}{-(20y - 8xy)}$

20. $\dfrac{5x - 15ax}{6a - 2}$

21. $\dfrac{x^3 - 4x}{4 - x^2}$

22. $\dfrac{4x^2 - 9}{-(9x^2 - 4x^4)}$

23. $\dfrac{3x - 5}{25 - 9x^2}$

24. $\dfrac{2a - 3}{9 - 4a^2}$

25. $\dfrac{3x^2 + 2x - 8}{4 + x - 3x^2}$

26. $-\dfrac{6x^2 - 23x + 21}{12 - 5x - 2x^2}$

27. $\dfrac{(x - 2)(x - 3)(x - 4)}{(2 - x)(3 - x)(4 - x)}$

28. $\dfrac{(a - b)(c + d)(a - c)}{(c - a)(d + c)(b - a)}$

29. $\dfrac{6x^2 - x^3 - x^4}{xy - 2y - x + 2}$

30. $\dfrac{12x^2y - 4x^3 - 9xy^2}{4x^3 + 4x^2y - 15xy^2}$

31. $\dfrac{x^3 - y^3}{y^2 - x^2}$

32. $\dfrac{a^3 - b^3}{b^3 + a^3}$

3. PRODUCTS

As with arithmetic fractions, the product of two or more algebraic fractions is computed by multiplying numerators together and denominators together. Thus the factors of the numerators will be the factors of the product's numerator. And the factors of the denominators will be the factors of the product's denominator.

We can generally save time and effort by dividing out like factors that occur in *a* numerator and *a* denominator of the fractions being multiplied. When this process is done thoroughly, the resulting product is automatically in lowest terms.

PROGRAM 3.2

To multiply algebraic fractions:

Step 1:   Factor completely the numerators and denominators of the given fractions.

Step 2:   Divide out any factor common to any numerator and any denominator.

Step 3:   Write as the desired product a fraction whose numerator is the product of the remaining numerator factors and whose denominator is the product of the remaining denominator factors.

EXAMPLES

1. Compute the product for $\dfrac{x^2 + x - 6}{2x^2 - 7x - 15} \cdot \dfrac{2x^2 - 3x - 9}{x^2 - 5x + 6}$.

Step 1: Factor numerators and denominators:

$$\left. \begin{aligned} x^2 + x - 6 &= (x + 3)(x - 2) \\ 2x^2 - 3x - 9 &= (2x + 3)(x - 3) \end{aligned} \right\} \text{ numerators}$$

$$\left. \begin{aligned} 2x^2 - 7x - 15 &= (2x + 3)(x - 5) \\ x^2 - 5x + 6 &= (x - 2)(x - 3) \end{aligned} \right\} \text{ denominators}$$

Step 2: Divide out factors common to numerators and denominators:

$$\frac{(x + 3)\cancel{(x - 2)}}{\cancel{(2x + 3)}(x - 5)} \cdot \frac{\cancel{(2x + 3)}\cancel{(x - 3)}}{\cancel{(x - 2)}\cancel{(x - 3)}}$$

Step 3: $\dfrac{x + 3}{x - 5}$ is the desired product in lowest terms.

2. Compute the product for $\dfrac{2x^2 - 18}{x^2 + 6x - 7} \cdot \dfrac{x^2 - 1}{8x^2 + 4x - 24}$.

SOLUTION

$$\frac{2x^2 - 18}{x^2 + 6x - 7} \cdot \frac{x^2 - 1}{8x^2 + 4x - 24} = \frac{\overset{}{2}(x + 3)(x - 3)}{(x + 7)\cancel{(x - 1)}} \cdot \frac{\cancel{(x - 1)}(x + 1)}{\underset{2}{4}(x + 2)(2x - 3)}$$

$$= \frac{(x + 3)(x - 3)(x + 1)}{2(x + 7)(x + 2)(2x - 3)}$$

Note: The product may be expanded or left in factor form; the latter is the usual practice.

3. Compute the product $(x^2 - 1)\left(\dfrac{3x - 1}{x^2 + 2x + 1}\right)$.

SOLUTION The polynomial, $x^2 - 1$, may be thought to have a denominator of 1. Apply Program **3.2**:

$$\frac{(x - 1)\cancel{(x + 1)}}{1} \cdot \frac{3x - 1}{\cancel{(x + 1)}(x + 1)} = \frac{(x - 1)(3x - 1)}{x + 1}$$

$$= \frac{3x^2 - 4x + 1}{x + 1}$$

4. Multiply $\dfrac{ac - bc + ad - bd}{c^2 - d^2} \cdot \dfrac{d^2 + 2cd + c^2}{b^2 - a^2}$.

SOLUTION

$$\frac{ac - bc + ad - bd}{c^2 - d^2} \cdot \frac{d^2 + 2cd + c^2}{b^2 - a^2} = \frac{\cancel{(a - b)}(c + d)}{(c - d)\cancel{(c + d)}} \cdot \frac{\overset{-1}{(d + c)}(d + c)}{\cancel{(b - a)}(b + a)}$$

$$= -\frac{(c + d)^2}{(c - d)(b + a)}$$

Multiply; express products in lowest terms.

1. $\dfrac{24}{3x - 6} \cdot \dfrac{x^2 - 2}{4}$

2. $\dfrac{a^2 - b^2}{a^2} \cdot \dfrac{ab}{a + b}$

3. $\dfrac{x(x - y)^3}{y(x^2 - y^2)} \cdot \dfrac{y(x + y)}{x^2(x - y)^2}$

4. $\dfrac{m(m + 2n)^2}{m - 3n} \cdot \dfrac{3n - m}{m^2 + 2mn}$

5. $\dfrac{9xy}{x + 4} \cdot \dfrac{2x^2 + 5x - 12}{3x^2y^2 - 6xy}$

6. $\dfrac{4x^2 - y^2}{y - 2x} \cdot \dfrac{3y}{xy + 2x^2}$

7. $\dfrac{6a - 18}{9a^2 + 6a - 24} \cdot \dfrac{12a - 16}{8a - 12}$

8. $\dfrac{a + b}{2a + b} \cdot \dfrac{4a^2 - b^2}{a^2 - b^2}$

9. $\dfrac{3ay^2 - 9ay}{10y^2 + 5y} \cdot \dfrac{2y^3 + y^2}{a^2y - 3a^2}$

10. $\dfrac{x^2 - 8x + 15}{x^2 + 4x - 21} \cdot \dfrac{x^2 - 6x - 16}{x^2 + 9x + 14}$

11. $\left(\dfrac{3x - 2}{x - 7}\right)(7 - x)$

12. $(4a^2 - 49)\left(\dfrac{6}{2a + 7}\right)$

13. $(m^2 - n^2)\left(\dfrac{m + n}{m - n}\right)$

14. $(2x - y)\left(\dfrac{2x + y}{4x^2 - y^2}\right)$

15. $\dfrac{x^2 - 2x - 3}{x^2 - 1} \cdot \dfrac{x^2 + 6x + 9}{x^2 - 9}$

16. $\dfrac{xy - ay - bx + ab}{2x - a} \cdot \dfrac{a - 2x}{a - x}$

17. $\dfrac{ab + 3b + 2a + 6}{3b + 6} \cdot \dfrac{a^2 - 5a + 6}{a^2 - 9}$

18. $\dfrac{xy - 2y - 3x + 6}{2y - 4x + 8 - xy} \cdot \dfrac{4x - 2y - 8 + xy}{xy - 2y - 3x + 6}$

19. $\left(\dfrac{x^2 - 2x + 1}{7x^2 - 7x}\right)\left(\dfrac{42x^2}{x^2 - 4x + 3}\right)$

20. $\left(\dfrac{9a + 9b}{3a^2 + 6ab + 3b^2}\right)\left(\dfrac{2a - 2b}{4a^2 - 16b^2}\right)$

21. $\left(\dfrac{2x - y}{x + y}\right)\left(\dfrac{x - y}{y - 2x}\right)\left(\dfrac{x + y}{y - x}\right)$

22. $\left(\dfrac{x - 2xy + y^2 - 16}{2x - y}\right)\left(\dfrac{2y - 4x}{2x - 8 - 2y}\right)$

23. $\dfrac{a^2 - b^2}{a^3 - 3a^2b + 3ab^2 - b^3} \cdot \dfrac{b^2 - 2ab + a^2}{3a^2 + 2ab - b^2}$

24. $\dfrac{x - 3y}{x^2y - 9y^3} \cdot \dfrac{x^2 + 2xy - 3y^2}{x - y}$

25. $\dfrac{a^3 - b^3}{a^2 + ab - 2b^2} \cdot \dfrac{a^2 - 4b^2}{a^2 + ab + b^2}$

26. $\dfrac{2x^2 - xy - 3y^2}{x^3 + y^3} \cdot \dfrac{3x^2 - xy - 2y^2}{6x^2 - 5xy - 6y^2}$

27. $\dfrac{6ac - 3bc - 2ad + bd}{b - a + 3} \cdot \dfrac{(a - 3)^2 - b^2}{8a^3 - b^3}$

28. $\dfrac{4c^2 - d^2}{3c^2d^2 - 6c^3d} \cdot \dfrac{(d - 2c)^3}{d^3 - 8c^3} \cdot \dfrac{3c^2d^3 + 6c^3d^2 + 12c^4d}{4c^2 + 4cd + d^2}$

4. QUOTIENTS

In arithmetic the usual procedure for dividing by a fraction is to multiply the dividend by the reciprocal of the divisor. Thus

$$(\text{dividend}) \div (\text{divisor}) = (\text{dividend}) \times (\text{reciprocal of divisor})$$

$$6 \div \frac{2}{3} = 6 \times \frac{3}{2} = \frac{18}{2} = 9$$

$$\frac{3}{4} \div \frac{2}{5} = \frac{3}{4} \times \frac{5}{2} = \frac{15}{8} = 1\frac{7}{8}$$

The **reciprocal** of a fraction is a related fraction in which the original numerator and denominator have been interchanged (i.e., the original fraction "inverted").

Fraction	Reciprocal
$\dfrac{3}{4}$	$\dfrac{4}{3}$
$\dfrac{p}{q}$	$\dfrac{q}{p}$
$\dfrac{x - 2y}{x^2 + y^2}$	$\dfrac{x^2 + y^2}{x - 2y}$

The product of a fraction and its reciprocal is always 1.

$$\frac{3}{4} \cdot \frac{4}{3} = 1$$

$$\frac{p}{q} \cdot \frac{q}{p} = 1$$

$$\frac{x - 2y}{x^2 + y^2} \cdot \frac{x^2 + y^2}{x - 2y} = 1$$

Note: The reciprocal of a fraction is also called the **multiplicative inverse** of the fraction. Every real number except zero has a unique multiplicative inverse—a number which, when multiplied with the given number, has 1 for a product.

PROGRAM 3.3

To divide algebraic fractions:

Step 1: Determine the reciprocal of the divisor.

Step 2: Multiply the dividend and the reciprocal of Step 1 for the desired quotient.

EXAMPLES

1. Divide $\dfrac{6a^5}{25b^4} \div \dfrac{a}{5b^2}$.

SOLUTION *Step 1:* The reciprocal of the divisor $\dfrac{a}{5b^2}$ is the divisor inverted: $\dfrac{5b^2}{a}$ (i.e., with numerator and denominator interchanged).

Step 2: Multiply the dividend, $\dfrac{6a^5}{25b^4}$, and the reciprocal of the divisor:

$$\frac{6a^5}{25b^4} \cdot \frac{5b^2}{a} = \frac{6 \cdot \overset{a^4}{\cancel{a^5}} \cdot \cancel{5} \cdot \cancel{b^2}}{\underset{5}{\cancel{25}} \cdot \underset{b^2}{\cancel{b^4}} \cdot \cancel{a}} = \frac{6a^4}{5b^2}$$

2. Divide $\dfrac{3x^2}{5y^2}$ by $2xy$.

SOLUTION *Step 1:* The reciprocal of the divisor, $2xy$, is $\dfrac{1}{2xy}$.

Step 2: Multiply the dividend and the reciprocal of the divisor:

$$\frac{3x^2}{5y^2} \cdot \frac{1}{2xy} = \frac{3x^{\overset{1}{\cancel{2}}}}{10xy^3} = \frac{3x}{10y^3}$$

3. Compute $\dfrac{2a + 6}{a^2 - 6a + 9} \div \dfrac{a + 3}{a^2 - 5a + 6}$.

SOLUTION *Step 1:* Invert $\dfrac{a + 3}{a^2 - 5a + 6}$ and factor:

$$\frac{a^2 - 5a + 6}{a + 3} = \frac{(a - 3)(a - 2)}{(a + 3)}$$

Step 2: Carry out the multiplication:

$$\frac{2\cancel{(a + 3)}}{\cancel{(a - 3)}(a - 3)} \cdot \frac{\cancel{(a - 3)}(a - 2)}{\cancel{(a + 3)}} = \frac{2(a - 2)}{a - 3}$$

4. Compute $\dfrac{c^2 - 2cd + d^2}{b^2 + 4} \div \dfrac{db - cb + 2d - 2c}{b^4 - 16}$.

SOLUTION *Step 1:* Invert $\dfrac{db - cb + 2d - 2c}{b^4 - 16}$ and factor:

$$\frac{b^4 - 16}{db - cb + 2d - 2c} = \frac{(b^2 + 4)(b - 2)(b + 2)}{(d - c)(b + 2)}$$

Step 2: $\dfrac{(c-d)(\cancel{c-d})}{\cancel{b^2+4}} \cdot \dfrac{\overset{-1}{(\cancel{b^2+4})}(b-2)(\cancel{b+2})}{(\cancel{d-c})(\cancel{b+2})}$

$$= (-1)(c-d)(b-2) = [(-1)(c-d)][(b-2)]$$

$$= (d-c)(b-2)$$

5.   Determine the quotient: $\dfrac{2a^2 - 3a - 2}{3a - 1} \div (2 - a)$.

SOLUTION   Factor the dividend; invert the divisor; multiply:

$$\dfrac{\overset{-1}{(\cancel{a-2})}(2a+1)}{3a-1} \cdot \dfrac{1}{(\cancel{2-a})} = -\dfrac{2a+1}{3a-1}$$

EXERCISE 3-4

Divide; express quotients in lowest terms.

1. $\dfrac{4x^3}{3a^2} \div 6$ **2.** $\dfrac{2x}{4y} \div 2$ **3.** $\dfrac{2ab^2}{4c} \div 3ac$

4. $\dfrac{6mn}{5n^3} \div 3n$ **5.** $9x^2y^3 \div \dfrac{3x}{2y^2}$ **6.** $24a^2b^3 \div \dfrac{6ac}{b}$

7. $\dfrac{x-3}{x+2} \div (x^2 - 6x + 9)$ **8.** $\dfrac{a-2b}{a+2b} \div (a^2 - 4b^2)$

9. $(p^2 - 2pq - 3q^2) \div \dfrac{p+q}{p-q}$ **10.** $(a+b)^3 \div \dfrac{3}{a+b}$

11. $\dfrac{x^2 + x - 6}{x - a} \div \dfrac{x^2 - 4}{x - a}$ **12.** $\dfrac{x^2 - 6x + 8}{x - 3} \div \dfrac{x - 4}{9 - x^2}$

13. $\dfrac{x - 3y}{x^2 - 6x + 8} \div \dfrac{3x - 9y}{x^2 - 16}$ **14.** $\dfrac{3c - bc}{5x - ax} \div \dfrac{3a - ab}{5m - am}$

15. $\dfrac{3x + 6}{x^2 + 4x} \div \dfrac{3x^2 - 9}{x^2 + 2x}$ **16.** $\dfrac{x^2 - 9}{3x - 3y} \div \dfrac{x^2 + 9 - 6x}{y^2 - x^2}$

17. $\dfrac{2m + 4}{m^2 - 7m - 18} \div \dfrac{4m - 16}{m^2 - 81}$ **18.** $\dfrac{x^2 - 16}{x^2 - 9} \div \dfrac{x - 4x^2}{9x + 27}$

19. $\dfrac{2x^2 - 7x + 6}{2x^2 - 13x + 15} \div \dfrac{4x^2 - 12x + 5}{2x^2 - 11x + 5}$

20. $\dfrac{6x^2 + 5xy - 6y^2}{9x^2 - 21xy + 10y^2} \div \dfrac{6x^2 - 25xy + 14y^2}{6x^2 - 31xy + 35y^2}$

21. $(a^2 - b^2) \div \dfrac{a^2 - 4ab + 3b^2}{a + b}$

22. $\dfrac{as - bs + bt - at}{3m^3 - 3mn^2} \div \dfrac{as - at}{n^2 - 2mn + m^2}$

23. $\dfrac{6a^2 + 11ab - 10b^2}{a^2 + b^2} \div (3a^2 + ab - 2b^2)$

24. $\dfrac{3x^2 - 14x + 8}{2x^2 - 3x - 20} \div \dfrac{6 - 25x + 24x^2}{15 - 34x - 16x^2}$

25. $\dfrac{x^2 - 5x + 6}{x - 2} \div (x - 2)$

26. $\dfrac{ab - 3b + 2a - 6}{ac - 5a - 3c + 15} \div \dfrac{2b - 6d - 12 + bd}{-5d + 2c - 10 + cd}$

27. $\dfrac{a^2 - 2ab + b^2 - 16}{a^2 - 5ba + 6b^2} \div \dfrac{a - b + 4}{6a - 18b}$

28. $\dfrac{a^2 - 18 + 3a}{30 - a^2 + a} \div \dfrac{15 - 2a - a^2}{-(a^4 - 36a^2)}$

29. $\dfrac{x^2 - y^2}{x^3 + 3x^2y + 3xy^2 + y^3} \div \dfrac{3y - 3x}{x^2 + 2xy + y^2}$

30. $\dfrac{27x^3 - 54x^2y + 36xy^2 - 8y^3}{9x^2 - 12xy + 4y^2} \div \dfrac{9x^2 - 3xy - 2y^2}{27x^3 + 27x^2y + 9xy^2 + y^3}$

31. $\dfrac{a^2 - b^2}{a^2 + 2ab + b^2} \div \dfrac{3a^3b^2 + 3a^2b^3 + 3ab^4}{a^3 - b^3}$

32. $\dfrac{2x^2 - 7xy + 6y^2}{2x^2 - xy - 3y^2} \div \dfrac{x - 2y}{x^3 + y^3}$

33. $\dfrac{xz - 2y + 2x - yz}{2x^2 + xy - 3y^2} \div \dfrac{8 + z^3}{3xz^2 - 6xz + 12x}$

34. $\dfrac{4a^2 - b^2}{6a - 3b} \div \dfrac{6a^2 + ab - b^2}{27a^3 - b^3}$

5. LEAST COMMON DENOMINATOR

In arithmetic we learned how to add and subtract fractions that have different denominators. The first step is to replace one or more fractions with equivalent fractions that have a denominator in common with all the other fractions. Any common denominator will do, although the **least common denominator** (LCD) of all the fractions under consideration generally results in a simpler computation.

This is also true when it comes to computing sums and differences with algebraic fractions. Remember that the LCD for a set of fractions is the smallest multiple that is *exactly* divisible by *all* the denominators.

PROGRAM 3.4

To determine the least common denominator (LCD) of several algebraic fractions:

Step 1: Factor each denominator completely.

Step 2: Form a product of factors of the denominators such that each different factor of Step 1 is represented at the highest power it appears in any of the denominators. This is the least common denominator (usually left in factor form).

EXAMPLES

1. Find the LCD.

$$\frac{x + 2}{x^2 + 2x - 3}, \frac{7}{2x^2 - 8x + 6}, \frac{x^2 - 2}{9x^2 - 54x + 81}, \frac{2 - x}{3x^2 - 15x + 18}$$

SOLUTION *Step 1:* Factor each denominator:

$$x^2 + 2x - 3 = (x + 3)(x - 1)$$
$$2x^2 - 8x + 6 = (2)(x - 3)(x - 1)$$
$$9x^2 - 54x + 81 = (3)(3)(x - 3)(x - 3)$$
$$= (3)^2(x - 3)^2$$
$$3x^2 - 15x + 18 = (3)(x - 2)(x - 3)$$

Step 2: Construct the least common denominator so that each factor of each denominator is present, and to the highest power it appears in any one denominator:

$$(x + 3)(x - 1)(2)(3)^2(x - 3)^2(x - 2)$$
$$LCD = (2)(3)^2(x + 3)(x - 1)(x - 3)^2(x - 2)$$

or

$$18(x + 3)(x - 1)(x - 3)^2(x - 2).$$

Note: Sometimes a chart like the one below is helpful in constructing an LCD. The guiding rule is to place like factors, and those that are negatives of one another, in the same column wherever possible. The "bottom line" is the LCD.

$x^2 + 2x - 3$	$x + 3$	$x - 1$					
$2x^2 - 8x + 6$		$x - 1$	2	$x - 3$			
$9x^2 - 54x + 81$				$x - 3$	3	3	$x - 3$	
$3x^2 - 15x + 18$				$x - 3$	3			$x - 2$
LCD	$x + 3$	$x - 1$	2	$x - 3$	3	3	$x - 3$	$x - 2$

2. Find the LCD for $\dfrac{x + 5}{x^2 - 4}, \dfrac{3}{x^2 + x - 6}, \dfrac{8 - x}{6 - x - x^2}$

SOLUTION *Step 1:* Factor each denominator:

$$x^2 - 4 = (x - 2)(x + 2)$$

$$x^2 + x - 6 = (x - 2)(x + 3)$$

$$6 - x - x^2 = (2 - x)(3 + x)$$

Step 2: Construct the least common denominator:

$$(x - 2)(x + 2)(x + 3) = \text{LCD}$$

Note: $(2 - x)$ is the negative of $(x - 2)$ and divides that factor exactly (-1) times; therefore, there is no need to introduce the factor $(2 - x)$ among the factors of the LCD.

Once the least common denominator for a set of fractions has been determined, the next step in computing their sum or difference is to express each of the fractions as equivalent fractions having this same common denominator. Contained in the LCD is the necessary "appropriate" multiplier—for both the numerator and denominator of each fraction—to produce an equivalent fraction that has the desired least common denominator.

PROGRAM 3.5

To express an algebraic fraction, with a specified denominator, equivalent to a given fraction:

Step 1: Divide the specified denominator by the denominator of the given fraction.

Step 2: Multiply the numerator and denominator of the given fraction by the quotient of Step 1.

EXAMPLES

1. Express a fraction equivalent to $\dfrac{3x + 2}{x - 7}$ that has a denominator of $(x - 7)(x + 2)$.

SOLUTION *Step 1:* Divide $(x - 7)(x + 2)$ by $(x - 7)$:

$$\frac{(x - 7)(x + 2)}{(x - 7)} = (x + 2)$$

Step 2: Multiply numerator and denominator of the given fraction $\dfrac{3x + 2}{x - 7}$ by $(x + 2)$:

$$\frac{(3x + 2)(x + 2)}{(x - 7)(x + 2)}$$

This fraction is equivalent to $\dfrac{3x + 2}{x - 7}$.

2. Express each of the three fractions:

(1) $\dfrac{x + 2}{x^2 + x - 6}$, (2) $\dfrac{3x}{20 - 6x - 2x^2}$, (3) $\dfrac{x - 7}{x^2 + 8x + 15}$

as equivalent fractions, each having as its denominator the LCD of the three fractions.

(a) Factor the three denominators.

(1) $x^2 + x - 6$	$x - 2$	$x + 3$		
(2) $20 - 6x - 2x^2$	$2 - x$		2	$5 + x$
(3) $x^2 + 8x + 15$		$x + 3$		$x + 5$
LCD	$x - 2$	$x + 3$	2	$x + 5$

(b) Identify the LCD of the denominators:

$$(2)(x - 2)(x + 3)(x + 5)$$

(c) Divide the LCD by each denominator:

(1) $\dfrac{(2)(x - 2)(x + 3)(x + 5)}{(x - 2)(x + 3)} = 2(x + 5)$

(2) $\dfrac{(2)(x - 2)(x + 3)(x + 5)}{(2)(2 - x)(5 + x)} = (-1)(x + 3)$ or $-(x + 3)$

(3) $\dfrac{(2)(x - 2)(x + 3)(x + 5)}{(x + 3)(x + 5)} = 2(x - 2)$

(d) Multiply the numerator and denominator of each fraction by its respective quotient in (c):

(1) $\dfrac{(x + 2)}{(x - 2)(x + 3)} \dfrac{(2)(x + 5)}{(2)(x + 5)} = \dfrac{(2)(x + 2)(x + 5)}{(2)(x - 2)(x + 3)(x + 5)}$

(2) $\dfrac{(3x)}{(2)(2-x)(5+x)}\dfrac{(-1)(x+3)}{(-1)(x+3)}$

$$= \dfrac{(-1)(3x)(x+3)}{(-1)(2)(2-x)(x+3)(5+x)}$$

$$= \dfrac{-3x(x+3)}{(2)(x-2)(x+3)(x+5)}$$

$$\underset{\longleftarrow}{}(-1)(2-x)$$

(3) $\dfrac{(x-7)}{(x+3)(x+5)}\dfrac{(2)(x-2)}{(2)(x-2)} = \dfrac{(2)(x-7)(x-2)}{(2)(x-2)(x+3)(x+5)}$

EXERCISE 3-5

Write the least common denominator, in factor form, for each set of fractions.

1. $\dfrac{x}{3x-1}, \dfrac{x+1}{9x-3}$

2. $\dfrac{3}{a^2-2a}, \dfrac{m}{a-2}$

3. $\dfrac{t}{x-y}, \dfrac{k}{y-x}$

4. $\dfrac{7}{p-3}, \dfrac{9}{3-p}$

5. $\dfrac{m}{x-3}, \dfrac{n}{2x^2-7x+3}, \dfrac{p}{2x-1}$

6. $\dfrac{a}{2x-1}, \dfrac{b}{3x+4}, \dfrac{c}{6x^2+5x-4}$

7. $\dfrac{1}{x-3}, \dfrac{2}{x^2+x-6}, \dfrac{3}{x^2-9}$

8. $\dfrac{x}{ab-3b}, \dfrac{y}{2a^2-7a+3}, \dfrac{z}{2ab-b}$

9. $\dfrac{a}{2m^2+m-3}, \dfrac{b}{m^2-2m+1}, \dfrac{c}{1-m}$

10. $\dfrac{p}{2b-a}, \dfrac{q}{a^2-4ab+4b^2}, \dfrac{r}{2a+b}$

11. $\dfrac{w}{a^2+b^2}, \dfrac{x}{a^2-b^2}, \dfrac{y}{a^3-a^2b+ab^2-b^3}, \dfrac{z}{b-a}$

12. $\dfrac{a}{x^2-5x+6}, \dfrac{b}{x^2-4x+4}, \dfrac{c}{2x^2-5x+2}, \dfrac{d}{2x^2-7x+3}$

Write equivalent fractions for each fraction in the set, using the LCD of the fractions.

13. $\dfrac{x}{x-2}, \dfrac{x+1}{x^2+x-6}$

14. $\dfrac{a}{a-4}, \dfrac{a-1}{a^2-16}$

15. $\dfrac{x-1}{2x^2-x}, \dfrac{x-2}{4x^2+4x-3}$

16. $\dfrac{2a-1}{9a^2-1}, \dfrac{3a-2}{3a^2-a}$

17. $\dfrac{x-4}{2x^2-3x-2}, \dfrac{3x-1}{x^2-5x+6}$

18. $\dfrac{m-4}{4m^2+4m-3}, \dfrac{2m-1}{2m^2-7m-15}$

19. $\dfrac{a-b}{2a^2+2ab-3a-3b}, \dfrac{b-1}{2ab-3b+4a-6}$

20. $\dfrac{x+4}{xy-3y+6-2x}, \dfrac{y-1}{2y-2x-y^2+xy}$

21. $\dfrac{x+1}{6x^2-5x-6}, \dfrac{x+2}{4x^2-12x+9}, \dfrac{x+3}{3x^2-x-2}$

22. $\dfrac{2x-1}{9x^2-1}, \dfrac{2x-3}{6x^2-13x-5}, \dfrac{2x-5}{9x^2+6x+1}$

23. $\dfrac{2x-1}{6x^2-11x+3}, \dfrac{3x+4}{9x^2+9x-4}, \dfrac{3x-7}{12x^2-19x+5}$

24. $\dfrac{3a-4}{2a^2-9a+9}, \dfrac{a^2+4a}{2a^2+5a-12}, \dfrac{13a+5}{a^2-9}$

6. SUMS AND DIFFERENCES

We are now ready to organize the discussion of the previous section into a single program for adding and subtracting algebraic fractions.

PROGRAM 3.6

To add (subtract) algebraic fractions:

Step 1: Find the LCD of the terms if the denominators are different.

Step 2: Express each fraction as an equivalent fraction whose denominator is the LCD of Step 1.

Step 3: Write the sum (difference) of the numerators of the fractions of Step 2 as the numerator of the sum (difference), and write the LCD as its denominator.

Step 4: (Optional but usual.) Simplify the numerator of Step 3 and express the resulting fraction in lowest terms.

EXAMPLES

1. Add $\dfrac{x-23}{x^2-x-20}+\dfrac{x-3}{x-5}$.

SOLUTION *Step 1:* Determine the LCD of the denominators.

$$x^2-x-20=(x-5)(x+4)$$
$$x-5=(x-5)$$
$$\text{LCD}=(x-5)(x+4)$$

Step 2: Express each of the fraction addends as equivalent fractions having the LCD of Step 1.

$$\frac{x - 23}{(x - 5)(x + 4)} + \frac{(x - 3)(x + 4)}{(x - 5)(x + 4)}$$

Step 3: Add numerators, and express the sum over the LCD:

$$\frac{(x - 23) + (x - 3)(x + 4)}{(x - 5)(x + 4)}$$

Step 4: Simplify:

$$\frac{x - 23 + x^2 + x - 12}{(x - 5)(x + 4)} = \frac{x^2 + 2x - 35}{(x - 5)(x + 4)}$$

$$= \frac{(x + 7)(x - 5)}{(x - 5)(x + 4)} = \frac{(x + 7)}{(x + 4)}$$

Thus

$$\frac{x - 23}{x^2 - x - 20} + \frac{x - 3}{x - 5} = \frac{x + 7}{x + 4}$$

2. Simplify $\dfrac{x - 2}{x + 2} + 3x + \dfrac{1}{x^2 - 4}$.

SOLUTION | *Step 1:* Determine the LCD of the denominators ($3x$ has a denominator of 1):

$$x + 2 = (x + 2)$$

$$1 = 1$$

$$x^2 - 4 = (x + 2)(x - 2)$$

$$\text{LCD} = (1)(x + 2)(x - 2) \quad \text{or simply} \quad (x + 2)(x - 2)$$

Step 2: Rewrite each addend as an equivalent fraction having the LCD as denominator:

$$\frac{(x - 2)(x - 2)}{(x + 2)(x - 2)} + \frac{3x(x + 2)(x - 2)}{(x + 2)(x - 2)} + \frac{1}{(x + 2)(x - 2)}$$

$$= \frac{(x^2 - 4x + 4) + 3x(x^2 - 4) + 1}{(x + 2)(x - 2)}$$

Steps 3 and 4: Add, and simplify:

$$= \frac{x^2 - 4x + 4 + 3x^3 - 12x + 1}{(x + 2)(x - 2)}$$

$$= \frac{3x^3 + x^2 - 16x + 5}{(x + 2)(x - 2)}$$

3. Simplify $\dfrac{7x - 8}{x^2 - 9} - (3x + 2) + \dfrac{3x^2}{x - 3}$.

SOLUTION *Step 1:* Compute the LCD of the denominators:

$$x^2 - 9 = (x - 3)(x + 3)$$

$$x - 3 = (x - 3)$$

$$\text{LCD} = (x - 3)(x + 3) \quad \text{or} \quad (x^2 - 9)$$

Step 2: Rewrite terms as equivalent fractions having the LCD of the denominators:

$$\frac{7x - 8}{(x - 3)(x + 3)} - \frac{(3x + 2)(x^2 - 9)}{(x - 3)(x + 3)} + \frac{3x^2(x + 3)}{(x - 3)(x + 3)}$$

Steps 3 and 4: Add and simplify:

$$\frac{(7x - 8) - [(3x + 2)(x^2 - 9)] + 3x^2(x + 3)}{(x - 3)(x + 3)}$$

$$= \frac{(7x - 8) - (3x^3 + 2x^2 - 27x - 18) + (3x^3 + 9x^2)}{(x - 3)(x + 3)}$$

$$= \frac{7x - 8 - 3x^3 - 2x^2 + 27x + 18 + 3x^3 + 9x^2}{(x - 3)(x + 3)}$$

$$= \frac{7x^2 + 34x + 10}{(x - 3)(x + 3)}$$

Note: $7x^2 + 34x + 10$ is not factorable with integral coefficients because $b^2 - 4ac = 1156 - 280 = 876 \neq$ square of an integer; therefore, the result is in lowest terms. See page 63.

4. Add $\dfrac{3x}{2x - 1} + \dfrac{2x}{1 - 2x}$.

SOLUTION *Step 1:* LCD $= 2x - 1$

Note: $(2x - 1) \div (1 - 2x) = -1$; or $(2x - 1) = -(1 - 2x)$.

Step 2: $\dfrac{3x}{2x - 1} + \dfrac{-2x}{2x - 1}$

Step 3: $\dfrac{3x - 2x}{2x - 1} = \dfrac{x}{2x - 1}$

96 Chapter 3 FRACTIONS

Simplify.

1. $x + \dfrac{3}{x}$

2. $\dfrac{7}{y} + y^2$

3. $\dfrac{2}{a} + \dfrac{3}{b}$

4. $\dfrac{3}{m} - \dfrac{2}{n^2}$

5. $\dfrac{x + 3y}{18} + \dfrac{x - 2y}{24}$

6. $\dfrac{s - t}{14s} - \dfrac{s - t}{21t}$

7. $\dfrac{3a - 6b}{12a^2b} + \dfrac{3a - 5b}{18ab^2}$

8. $\dfrac{1}{f_1} + \dfrac{1}{f_2}$

9. $\dfrac{5}{r} - \dfrac{6}{t} + \dfrac{8}{r}$

10. $a - 3 + \dfrac{5}{a}$

11. $\dfrac{3}{x} - y + 2x$

12. $\dfrac{4}{a} + \dfrac{3}{b} - \dfrac{2}{c}$

13. $2x - \dfrac{6x^2}{3x - 2}$

14. $1 - \dfrac{2x}{3x + y}$

15. $x - 2 + \dfrac{4x + 3}{x - 7}$

16. $x + 3 - \dfrac{2x - 5}{7 - x}$

17. $2x + 5y + \dfrac{25y^2}{2x - 5y}$

18. $\dfrac{3b}{2b - 1} - b + 2$

19. $\dfrac{3x - 5}{x + 3} + \dfrac{2x - 1}{x - 4}$

20. $\dfrac{1}{p + q} - \dfrac{1}{p - q}$

21. $\dfrac{3}{a - b} - \dfrac{4}{b - a}$

22. $\dfrac{x - 2}{4x} - \dfrac{3x + 5}{6x}$

23. $\dfrac{3x}{x^2 - 4} - \dfrac{2}{x + 2}$

24. $\dfrac{8y}{y^2 - 9} + \dfrac{4}{3 - y}$

25. $\dfrac{3a - 4}{a^2 - a - 20} - \dfrac{3}{a - 5}$

26. $\dfrac{x + 2}{x^2 - 9} + \dfrac{3x - 1}{x^2 + x - 12}$

27. $\dfrac{3c - d}{2c - d} + \dfrac{3c^2}{d^2 - 4c^2}$

28. $\dfrac{x - 4}{x - 2} + \dfrac{3 + 5x}{2 - x}$

29. $\dfrac{1}{x^2 - 5x + 6} - \dfrac{4}{4 - x^2}$

30. $\dfrac{2x}{x^2 + xy} - \dfrac{3}{xy + y^2}$

31. $2 - \dfrac{x}{x - 2} + \dfrac{3(2x - 9)}{x^2 - 5x + 6}$

32. $2 - \dfrac{3 - 2x}{2x - 3} + 2x$

33. $\dfrac{x + 2}{2x^2 - 3x - 9} - \dfrac{x - 2}{3x^2 - 11x + 6}$

34. $\dfrac{2}{x^2 - 9} - \dfrac{3}{x^2 - 1} + \dfrac{1}{x^2 + 2x - 3}$

35. $\dfrac{8}{x^2 - x - 2} + \dfrac{3 + x}{x^2 - 4} + \dfrac{4 - 2x}{x^2 + 3x + 2}$

36. $\dfrac{3}{6x^2 - 11x + 3} + \dfrac{3x + 1}{12x - 4} - \dfrac{2x - 3}{9x^2 - 1}$

7. COMPLEX FRACTIONS★

Most of the fractions that we have seen so far can be classified as **simple fractions**—those with numerators and denominators that do not contain fractions. In contrast, **complex fractions** are fractions in which either the numerator or denominator or both involve fractions.

Complex fractions can be simplified (i.e., expressed by a simpler equivalent) if we apply some of the procedures learned up to this point. The usual approach is to express the complex fraction as a division of numerator by denominator.

EXAMPLES

1. Simplify $\dfrac{\dfrac{3}{xy^2}}{\dfrac{2}{x^2y}}$.

SOLUTION Consider the numerator fraction, $\dfrac{3}{xy^2}$, as the dividend and the denominator fraction, $\dfrac{2}{x^2y}$, as the divisor, then carry out the division:

$$\frac{\dfrac{3}{xy^2}}{\dfrac{2}{x^2y}} = \frac{3}{xy^2} \div \frac{2}{x^2y} = \frac{3}{xy^2} \cdot \frac{x^2y}{2} = \frac{3x}{2y}$$

2. Simplify $\dfrac{\dfrac{1}{s} + \dfrac{1}{t}}{\dfrac{s}{t} - \dfrac{t}{s}}$.

SOLUTION Express numerator and denominator as fractions, then express as dividend and divisor:

$$\frac{\dfrac{1}{s} + \dfrac{1}{t}}{\dfrac{s}{t} - \dfrac{t}{s}} = \frac{\dfrac{t+s}{st}}{\dfrac{s^2-t^2}{st}} = \frac{t+s}{st} \div \frac{s^2-t^2}{st}$$

$$= \frac{t+s}{st} \cdot \frac{st}{s^2-t^2} = \frac{1}{s-t}$$

★ The content of this section, in more basic courses of study, may be postponed without significant effect upon the remaining parts of the book.

3. Simplify $\dfrac{1 - \dfrac{3y}{x+y}}{1 - \dfrac{y}{x-y}}$.

SOLUTION First, simplify the numerator and denominator separately:

$$1 - \frac{3y}{x+y} = \frac{x+y}{x+y} - \frac{3y}{x+y} = \frac{x-2y}{x+y}$$

$$1 - \frac{y}{x-y} = \frac{x-y}{x-y} - \frac{y}{x-y} = \frac{x-2y}{x-y}$$

Replace the original numerator and denominator by their equivalents:

$$\frac{1 - \dfrac{3y}{x+y}}{1 - \dfrac{y}{x-y}} = \frac{\dfrac{x-2y}{x+y}}{\dfrac{x-2y}{x-y}} = \frac{\cancel{x-2y}}{x+y} \cdot \frac{x-y}{\cancel{x-2y}} = \frac{x-y}{x+y}$$

Note: The numerator and denominator of $\dfrac{\dfrac{x-2y}{x+y}}{\dfrac{x-2y}{x-y}}$ can be divided by $(x-2y)$ directly:

$$\frac{\dfrac{\cancel{x-2y}}{x+y}}{\dfrac{\cancel{x-2y}}{x-y}} = \frac{1}{x+y} \cdot \frac{x-y}{1} = \frac{x-y}{x+y}$$

4. Simplify $1 - \dfrac{2}{3 - \dfrac{1}{2 + \dfrac{3}{x-1}}}$.

SOLUTION (a) $\quad 1 - \dfrac{2}{3 - \dfrac{1}{2 + \dfrac{3}{x-1}}} \qquad \left[= \dfrac{2x - 2 + 3}{x-1} = \dfrac{2x+1}{x-1} \right]$

(b) $\quad 1 - \dfrac{2}{3 - \dfrac{1}{\dfrac{2x+1}{x-1}}} \qquad \left[= \dfrac{x-1}{2x+1} \right]$

(c) $1 - \dfrac{2}{3 - \dfrac{x-1}{2x+1}}$ $\left[= \dfrac{6x+3-x+1}{2x+1} = \dfrac{5x+4}{2x+1} \right]$

(d) $1 - \dfrac{2}{\dfrac{5x+4}{2x+1}}$ $\left[= \dfrac{2(2x+1)}{5x+4} = \dfrac{4x+2}{5x+4} \right]$

(e) $1 - \dfrac{4x+2}{5x+4} = \dfrac{5x+4-4x-2}{5x+4} = \dfrac{x+2}{5x+4}$

EXERCISE 3-7

Simplify.

1. $\dfrac{\dfrac{4}{ab^2}}{\dfrac{2}{a^2b}}$

2. $\dfrac{\dfrac{x^2y^2}{12}}{\dfrac{xy}{15}}$

3. $\dfrac{1}{\dfrac{1}{x}+\dfrac{1}{y}}$

4. $\dfrac{3}{\dfrac{1}{a}-\dfrac{1}{b}}$

5. $\dfrac{a}{\dfrac{2}{a}-\dfrac{a}{2}}$

6. $\dfrac{m}{\dfrac{m}{2}+\dfrac{n}{3}}$

7. $\dfrac{\dfrac{1}{x}}{\dfrac{x}{2}-1}$

8. $\dfrac{\dfrac{2}{c}}{3-\dfrac{1}{c}}$

9. $\dfrac{\dfrac{1}{a}+\dfrac{1}{b}}{\dfrac{2}{a}}$

10. $\dfrac{\dfrac{2}{c}-\dfrac{c}{2}}{\dfrac{1}{c}}$

11. $\dfrac{\dfrac{2}{x}-1}{3+\dfrac{2}{x}}$

12. $\dfrac{a-\dfrac{b}{c}}{b+\dfrac{a}{c}}$

13. $\dfrac{\dfrac{x-5}{x^2-25}}{2x+3}$

14. $\dfrac{\dfrac{a-2}{a^2-4}}{2a+1}$

15. $\dfrac{\dfrac{x^2-y^2}{(x+y)^2}}{\dfrac{3x+3y}{x-y}}$

16. $\dfrac{\dfrac{8x-4}{9x^2-1}}{\dfrac{2x-1}{1-3x}}$

17. $\dfrac{\dfrac{1}{a}-\dfrac{1}{b}}{\dfrac{a}{b}+\dfrac{b}{a}}$

18. $\dfrac{\dfrac{2}{x}+\dfrac{3}{y}}{\dfrac{2y}{3x}-\dfrac{3x}{2y}}$

19. $\dfrac{\dfrac{a}{b^2} - \dfrac{b^2}{a}}{\dfrac{b^2}{3a^2} - \dfrac{1}{3a}}$

20. $\dfrac{\dfrac{2}{x} + \dfrac{3}{y} + 2}{6 + \dfrac{6}{x} + \dfrac{9}{y}}$

21. $\dfrac{1 - \dfrac{2x}{3 + x}}{4 - \dfrac{3x}{x + 3}}$

22. $\dfrac{3a + \dfrac{2a}{a - 6}}{a - \dfrac{4a}{a + 3}}$

23. $\dfrac{1 - \dfrac{1}{a}}{1 + \dfrac{1}{1 - \dfrac{1}{a}}}$

24. $\dfrac{2 + \dfrac{1}{x}}{1 - \dfrac{2}{2 - \dfrac{1}{x}}}$

25. $\dfrac{a + \dfrac{1}{a - 2}}{a + \dfrac{3}{a - \dfrac{4}{a}}}$

26. $\dfrac{1 - \dfrac{10a - 5}{3a - 4}}{2a - \dfrac{3a + 1}{1 + \dfrac{a - 3}{2a - 1}}}$

REVIEW

PART A

Answer True or False.

1. The prime factors of 9 are 3 and 3; the prime factors of 10 are 2 and 5; so 9 and 10 are relatively prime to one another.

2. Two fractions, $\dfrac{x}{y}$ and $\dfrac{m}{n}$, are equivalent if $(x)(m) = (y)(n)$.

3. The fraction $\dfrac{x^2}{xy}$ becomes $\dfrac{x}{y}$ when reduced to lowest terms.

4. It is clear that $-\dfrac{a}{x}$ represents a negative number and $\dfrac{a}{x}$ represents a positive number.

5. $a^2 + 2a - 5$ is the negative of $5 - 2a - a^2$.

6. The reciprocal of $\dfrac{a - b}{7}$ is $\dfrac{7}{a - b}$.

7. The multiplicative inverse of $\dfrac{3}{4}$ is $\dfrac{4}{3}$.

8. The least common denominator for the fractions

$$\frac{1}{x - a}, \quad \frac{1}{x + a}, \quad \frac{1}{a - x},$$

is $(x - a)(x + a)(a - x)$.

9. $\dfrac{\frac{1}{2}}{xy}$ is a complex rather than a simple fraction.

10. $\dfrac{\frac{1}{2}}{xy}$ implies $\left(\dfrac{1}{2}\right)(xy)$.

PART B

Insert the proper sign $(+$ *or* $-)$ *in the parentheses to make the statements true.*

1. $-\dfrac{2}{7} = (\ \ )\dfrac{-2}{+7}$

2. $\dfrac{m}{a-b} = (\ \ )\dfrac{-m}{b-a}$

3. $\dfrac{a-5}{(a-4)(a-2)} = (\ \ )\dfrac{5-a}{(2-a)(4-a)}$

Express in lowest terms.

4. $\dfrac{x^2 - 2x}{x^2 - 4}$

5. $\dfrac{3ab + 6a^2}{a + 2b}$

6. $\dfrac{2x^2 - 17x + 21}{x^2 - 12x + 35}$

7. $\dfrac{6x^2 - 23x + 20}{6x^2 - 5x - 4}$

8. $\dfrac{(3-a)(a^2+9)}{a^4 - 81}$

9. $\dfrac{xy - ay - ab + bx}{ab - bx + ay - xy}$

10. $\dfrac{ab + 6 - 3b - 2a}{4b + 2a - 8 - ab}$

11. $\dfrac{x^2 - 4x + 4}{4a - 4ax + ax^2}$

12. $\dfrac{x^2 - y^2 + x - y}{x^2 + 2xy + y^2 - 1}$

13. $\dfrac{x^3 + 2x^2 - 9x - 18}{6 + x - x^2}$

14. $\dfrac{3(m^2 + 2mn + n^2) + 7m + 7n - 6}{m + n + 3}$

15. $\dfrac{(x^2 - 8x + 16) - 4}{1 - (x^2 - 2x + 1)}$

Simplify.

16. $\left(\dfrac{a - 2b}{b - 3}\right) \cdot (3 - b)$

17. $(d^2 - c^2) \cdot \left(\dfrac{c + d}{c - d}\right)$

18. $\dfrac{2x + y}{4x^2 - y^2} \cdot \dfrac{x^3 - 2x^2 y}{x^2 - 2xy}$

19. $\dfrac{3 + 2x}{2 + 35x^2 + 19x} \cdot \dfrac{49x^2 - 1}{4x^2 + 12x + 9}$

20. $\dfrac{2a - 3}{a^2 - 1} \cdot \dfrac{2a^2 + a - 3}{9 - 4a^2}$

21. $\dfrac{a^2 + 5a + 6}{a^2 - 1} \cdot \dfrac{a^2 - 2a - 3}{a^2 - 9}$

22. $\dfrac{m^2 - 5m + 6}{m^2 - 4m + 3} \cdot \dfrac{m^2 - 5m + 4}{m^2 - 6m + 8}$

23. $\dfrac{1 - 2x + x^2}{1 - x^2} \cdot \dfrac{x}{x - 1} \cdot \dfrac{1 + x}{1 + x^2}$

24. $\dfrac{ab - b + 2a - 2}{cd - 4d + 3c - 12} \cdot \dfrac{d^2 - d - 12}{b^2 + 4b + 4}$

25. $\dfrac{am - na + bm - bn}{ax + bx - ay - by} \cdot \dfrac{mx - nx - my + ny}{m^2 - n^2}$

26. $\dfrac{4m^2 - 4mn + n^2}{3m^2 + 13mn - 10n^2} \cdot \dfrac{9m^2 - 12mn + 4n^2}{3m^2 + 4mn - 4n^2} \cdot \dfrac{m^2 - 4n^2}{2m^2 - 5mn + 2n^2}$

27. $\dfrac{a^2 - b^2}{a^3 - 3a^2b + 3ab^2 - b^3} \cdot \dfrac{3a^2 - 6ab + 3b^2}{a^2 + 3ab + 2b^2}$

28. $\dfrac{35a^2b^3}{3ab^2} \div 7ab^4c$

29. $(a^3 - b^3) \div \dfrac{a^2 + ab + b^2}{a + b}$

30. $\dfrac{8a^3b^2}{7xy^2} \div \dfrac{a^4b^2 - 3a^2b^3}{2x^3y - x^2y^2}$

31. $\dfrac{4a^2 - 28a + 49}{12a^2 - 17a + 6} \div \dfrac{4a^2 - 49}{12a^2 - a - 6}$

32. $\dfrac{2x^2 - 13x + 15}{3x^2 - 17x + 10} \div \dfrac{4x^2 - 9}{3px - 2p}$

33. $\dfrac{a^2 + 4a + 3}{a^2 + a - 6} \div \dfrac{a^2 + 3a + 2}{pa^2 - 2pa}$

34. $\dfrac{9a^2 + 6ab + b^2 - 4}{2a^2 - 5ab + 3b^2} \div \dfrac{10 + 5b + 15a}{a^2 - 7ab + 6b^2}$

35. $\dfrac{ab + a + b + 1}{mn + 2n + 2m + 4} \div \dfrac{a^2 - 1}{n^2 + 4n + 4}$

36. $\dfrac{6x^2 - 19x + 10}{8x^2 - 14x - 15} \div \dfrac{15x^2 - 16x + 4}{15x^2 - 11x + 2}$

37. $\left(2 - \dfrac{2 + 8m - 3m^2}{9 - m^2}\right) \div \left(6 - \dfrac{14 + 7m}{3 + m}\right)$

38. $\dfrac{x^3 - 1}{x^4 + x^2 + 1} \cdot \dfrac{x^3 + 1}{2x^2 + x - 1} \div \dfrac{2x^2 - x - 1}{2x^2 + x - 1}$

39. $\dfrac{a^2 - 1}{a^2 - a(a + 1)} \cdot \dfrac{a^2 - 3a}{a^2 - 3a + 2} \div \dfrac{a^2 - 3a - 4}{a - 2}$

40. $\dfrac{3}{a} + \dfrac{2b}{3} - \dfrac{4}{c}$

41. $\dfrac{a}{x} - \dfrac{1}{y} + \dfrac{b}{3x} - \dfrac{1}{2}$

42. $\dfrac{2}{a - 2} - \dfrac{3}{5a - 10}$

43. $\dfrac{3}{x + y} - \dfrac{5}{xy - y^2} + \dfrac{2}{x^2 - y^2}$

44. $\dfrac{3}{x - 2} - \dfrac{x}{x^2 - 4} - \dfrac{2}{x + 2}$

45. $\dfrac{2a - 1}{a + 3} - (a - 2) + \dfrac{1}{6 + 2a}$

46. $\dfrac{2m + 6}{m^2 - m - 12} - \dfrac{3(m + 5)}{m^2 + 3m - 10}$

47. $\dfrac{y}{2y - x} + \dfrac{1}{x - 2y} + \dfrac{x}{2y^2 - xy}$

48. $\dfrac{m^2 + 1}{m^3 - m^2 + m - 1} - \dfrac{m^2 - 1}{m^3 + m^2 - m - 1}$

49. $\dfrac{3a^2}{a^4 - 4} + \dfrac{5a^2 - 3}{2a^4 + a^2 - 6}$

50. $\dfrac{1}{ab + a + b + 1} + \dfrac{1}{2 + ab + 2a + b} + \dfrac{1}{2 + 2b + a + ab}$

51. $\dfrac{a - 1}{a + 1} - 1 + \left(\dfrac{a - 1}{a^2 - 1} - \dfrac{5a - 1}{a - 1} + 2 \right)$

52. $\dfrac{2 - 2x}{x^3 - 3x^2 + 3x - 1} + \dfrac{2}{1 - x} + \dfrac{x + 1}{x^2 - 2x + 1}$

53. $\dfrac{\dfrac{2}{x} - \dfrac{1}{y}}{\dfrac{3}{xy}}$

54. $\dfrac{3a - \dfrac{a}{b}}{\dfrac{b}{a} - 2b}$

55. $\dfrac{\dfrac{4a^2 - 4ab + b^2}{4a^2 - b^2}}{\dfrac{b^2 - 2ab}{b^2 + 2ab}}$

56. $\dfrac{\dfrac{1}{a + 1} - \dfrac{1}{a - 1}}{1 - \dfrac{a - 1}{a + 1}}$

57. $\dfrac{a - \dfrac{a}{a - 2}}{a - \dfrac{a}{a + \dfrac{1}{a + 2}}}$

58. $\dfrac{1 + \dfrac{1}{1 + x(x - 1)}}{3 - \dfrac{3x}{2x + \dfrac{2}{x - 1}}}$

59. $\dfrac{-3 + \dfrac{2}{7 - a}}{5a - \dfrac{4a - 1}{1 - \dfrac{2a + 5}{3a - 2}}}$

FIRST-DEGREE EQUATIONS AND INEQUALITIES

1. TERMINOLOGY

Various kinds of number sentences are used in algebra. The most familiar is the **equation**—a number sentence that has an equal sign ($=$) for its verb. The expressions on either side of the $=$ sign are called **members**. Most of the equations we meet in algebra are **conditional equations**, those that make a true statement for at least one, but not all number replacements for the variable. An **identity**, on the other hand, is an equation that is true for *all* number replacements for the variable.

EXAMPLES

1.

$$\overbrace{x + 3}^{\text{member}} = \overbrace{5}^{\text{member}} \text{ is an equation.}$$

variable ⟶ ↑   ↑ ⟵ equal sign

2. $x + 3 = 5$ is a conditional equation. Replace x by 1, 2, 3, successively:

$$x = 1: \quad \boxed{1} + 3 = 5 \quad \text{(false)}$$
$$x = 2: \quad \boxed{2} + 3 = 5 \quad \text{(true)}$$
$$x = 3: \quad \boxed{3} + 3 = 5 \quad \text{(false)}$$

$x + 3 = 5$ is a conditional equation because we obtain a true statement when x is replaced by 2, but not by 1 and 3 (or, in this instance, by any other number except 2).

3. $2 - x = 4 - x - 2$ is an identity. Any number that replaces the variable, x, results in a true statement. For instance, 6 and -3:

$$2 - \boxed{x} = 4 - \boxed{x} - 2$$
$$x = 6: \quad 2 - \boxed{6} = 4 - \boxed{6} - 2$$
$$-4 = -4$$
$$x = -3: \quad 2 - (\boxed{-3}) = 4 - (\boxed{-3}) - 2$$
$$5 = 5$$

When the variable of an equation has 1 for an exponent, the equation is said to be of **first-degree**. For example:

$$x + 2 = 7 \quad \text{and} \quad 3a + a = 2a - 1$$

are first-degree equations in x and a, respectively. But

$$x^2 + 2 = 7 \quad \text{and} \quad 4m^3 - 5 = m$$

are not first-degree because the variable in each carries an exponent greater than 1.

Replacements for the variable that make an equation true are called **solutions**. Equations that have the same solution(s) are called **equivalent equations**. Each of the four equations on the left below is true for the same variable replacement, 3. Therefore, each of the equations is an equivalent of the others.

Equation	Test: $x = 3$
$x + 3 = 6$	$\boxed{3} + 3 = 6$
$x - 3 = 0$	$\boxed{3} - 3 = 0$
$2x - 5 = x - 2$	$2(\boxed{3}) - 5 = \boxed{3} - 2$
$x = 3$	$\boxed{3} = 3$

Solutions are said to *satisfy* an equation. For an identity, all numbers in the replacement set result in a true statement when substituted for the variable, and therefore satisfy the equation. For a conditional equation, only some numbers in the replacement set satisfy the equation.

2. SOLVING FIRST-DEGREE EQUATIONS

The usual technique for solving a first-degree equation is given in Program **4.1**. It involves working through a sequence of ever simpler equivalent equations until we arrive at an equation in which the solution is obvious (such as $x = -3$, as in Example 1 below).

PROGRAM 4.1

To solve a first-degree equation:

Step 1: Multiply both members of the equation by the least common denominator (LCD) of the fractions if the equation involves fractions; remove all parentheses and simplify.

Step 2: Add to, subtract from, divide both members equally, as necessary, to produce an equivalent equation in which one member contains only the variable with a coefficient of $+1$. The other member is a possible solution of the equation.

Step 3: Substitute the possible solution of Step 2 for the variable in the given equation; it is a solution if it satisfies the given equation.

EXAMPLES

1. Solve for x: $2x - 14 = 4x - 8$.

SOLUTION

Step 1: No fractions or parentheses in this equation.

Step 2: Subtract $4x$ from both members and simplify.

$$2x - 14 - 4x = 4x - 8 - 4x$$

$$-2x - 14 = -8$$

Add 14 to both members and simplify.

$$-2x - 14 + 14 = -8 + 14$$

$$-2x = 6$$

Divide both members by -2 and simplify.

$$\frac{-2x}{-2} = \frac{6}{-2}$$

$$x = -3 \quad \text{(possible solution)}$$

Step 3: Substitute the possible solution, -3, for the variable in the given equation.

$$2x - 14 = 4x - 8$$

$$x = -3: \quad 2(-3) - 14 \overset{?}{=} 4(-3) - 8$$

$$-6 - 14 \overset{?}{=} -12 - 8$$

$$-20 = -20$$

The "possible solution" of Step 2 satisfies the given equation; therefore, it *is* the solution of the equation.

2. Solve for x: $5x - \dfrac{2}{15} = \dfrac{2}{3} + 3x$.

Step 1: LCD = 15: $\quad 15\left(5x - \dfrac{2}{15}\right) = 15\left(\dfrac{2}{3} + 3x\right)$

$$75x - 2 = 10 + 45x$$

Step 2: Add 2 to each member; subtract 45x from each member; simplify.

$$75x - 2 + 2 - 45x = 10 + 45x + 2 - 45x$$

$$30x = 12$$

Divide each member by 30.

$$\frac{30x}{30} = \frac{12}{30}$$

$$x = \frac{12}{30} = \frac{2}{5} \quad \text{(possible solution)}$$

Step 3: Check by substituting $\dfrac{2}{5}$ for x wherever it appears in the given equation.

$$5x - \frac{2}{15} = \frac{2}{3} + 3x$$

$$x = \frac{2}{5}: \quad 5\left(\frac{2}{5}\right) - \frac{2}{15} \overset{?}{=} \frac{2}{3} + 3\left(\frac{2}{5}\right)$$

$$2 - \frac{2}{15} \overset{?}{=} \frac{2}{3} + \frac{6}{5}$$

$$\frac{30}{15} - \frac{2}{15} \overset{?}{=} \frac{10}{15} + \frac{18}{15}$$

$$\frac{28}{15} = \frac{28}{15}$$

Thus $\dfrac{2}{5}$ is the solution of $5x - \dfrac{2}{15} = \dfrac{2}{3} + 3x$.

Note: In some instances it may be simpler to start with Step 2, and leave the clearing of fractions to later. In this example:

$$5x - \frac{2}{15} = \frac{2}{3} + 3x$$

Step 2. Add $\frac{2}{15}$ to each member; subtract $3x$ from each member; simplify:

$$5x - \frac{2}{15} + \frac{2}{15} - 3x = \frac{2}{3} + 3x + \frac{2}{15} - 3x$$

$$2x = \frac{2}{3} + \frac{2}{15}$$

$$2x = \frac{12}{15}$$

$$x = \left(\frac{1}{2}\right)\left(\frac{12}{15}\right)$$

$$= \frac{2}{5}$$

3. Solve for y: $\dfrac{5(y + 1)}{6} = \dfrac{y}{2} - \dfrac{2y + 5}{4}$.

SOLUTION | *Step 1:* LCD = 12:

$$12\left[\frac{5(y + 1)}{6}\right] = 12\left[\frac{y}{2} - \frac{2y + 5}{4}\right]$$

$$10(y + 1) = 6y - 3(2y + 5)$$

$$10y + 10 = 6y - 6y - 15$$

$$10y + 10 = -15$$

Step 2: Subtract 10 from both members:

$$10y + 10 - 10 = -15 - 10$$

$$10y = -25$$

Divide each member by 10:

$$\frac{10y}{10} = \frac{-25}{10}$$

$$y = \frac{-25}{10} = -\frac{5}{2}$$

Step 3: Check by replacing y in the given equation with $-\dfrac{5}{2}$:

$$\frac{5(y+1)}{6} = \frac{y}{2} - \frac{2y+5}{4}$$

$$y = -\frac{5}{2}: \quad \frac{5}{6}\left(-\frac{5}{2}+1\right) \overset{?}{=} \frac{1}{2}\left(-\frac{5}{2}\right) - \frac{1}{4}\left[2\left(-\frac{5}{2}\right)+5\right]$$

$$\frac{5}{6}\left(-\frac{3}{2}\right) \overset{?}{=} \frac{1}{2}\left(-\frac{5}{2}\right) - \frac{1}{4}(-5+5)$$

$$-\frac{5}{4} \overset{?}{=} -\frac{5}{4} - \frac{1}{4}(0)$$

$$-\frac{5}{4} = -\frac{5}{4}$$

Thus $-\dfrac{5}{2}$ is the solution.

4. Solve $\dfrac{3x}{4} - \dfrac{1}{2} = \dfrac{1}{4}(3x-2)$.

SOLUTION The equation is an identity. As with all identities, when Program **4.1** is applied, the final equivalent equation is $0 = 0$:

$$\frac{3x}{4} - \frac{1}{2} = \frac{1}{4}(3x-2)$$

$$\frac{3x}{4} - \frac{1}{2} = \frac{3}{4}x - \frac{2}{4}$$

$$\frac{3x}{4} - \frac{1}{2} + \frac{1}{2} = \frac{3}{4}x - \frac{2}{4} + \frac{1}{2}$$

$$\frac{3x}{4} = \frac{3}{4}x$$

$$\frac{3x}{4} - \frac{3x}{4} = \frac{3}{4}x - \frac{3x}{4}$$

$$0 = 0$$

EXERCISE 4-1

For each of the following equations, a single operation on both members will result in an equivalent equation, but one in which the solution is immediately evident. Find the solution for each.

1. $x - 3 = 7$ 2. $x + 4 = 6$ 3. $x + 5 = 9$
4. $x - 2 = 5$ 5. $4 + x = 8$ 6. $7 + x = 10$
7. $x + 6 = 4$ 8. $x + 3 = 1$ 9. $4 = x - 3$
10. $2 = x + 1$ 11. $2x = 12$ 12. $5x = 15$
13. $4x = -16$ 14. $3x = -9$ 15. $-2x = 8$

16. $-4x = 20$ 17. $5x = \dfrac{3}{5}$ 18. $\dfrac{1}{2}x = 16$

19. $\dfrac{3}{4}x = 24$ 20. $\dfrac{2}{3}x = -3$

Solve for the variable.

21. $2x + 3 = 4$ 22. $5 - 2x = 3$ 23. $2x - 4 = x$
24. $3x + 8 = x$ 25. $x = 5x - 8$ 26. $x = 14 - 6x$
27. $a - 4 + 2a = 0$ 28. $3b - 5 + b = 0$ 29. $2t + 1 = t - 1$
30. $p + 6 = -4p + 1$ 31. $3x + 8 = 12 + x$
32. $3x - 12 = 2x - 10$ 33. $3x + 7 = 6x$
34. $4x - 11 = 2x - 7$ 35. $3x - 8 = 12x - 7$
36. $4x - 3 = 3(6 - x)$ 37. $7x = -2(x + 9)$
38. $(2a - 9) - (a - 3) = 0$ 39. $4 - (2 - b) = 2 + 2b$
40. $(1 - t) - 6 = 3(7 - 2t)$ 41. $3x + 4(3x - 5) = 12 - x$
42. $3(6a - 5) = 7(3a + 10)$ 43. $3(2x - 1) = 4(x - 3) - 5$

44. $6(x - 5) = 15 + 5(7 - 2x)$ 45. $\dfrac{1}{2}x - x = 12$

46. $a - 5 = \dfrac{1}{4}a$ 47. $\dfrac{p}{3} - \dfrac{7}{6} + \dfrac{2p}{5} = \dfrac{3p}{4}$

48. $\dfrac{1}{3}(x + 2) - \dfrac{1}{2}(x + 8) = -2$ 49. $\dfrac{5}{7}y - y - \dfrac{5}{3} = \dfrac{1}{21}(3y + 1)$

50. $1 - \dfrac{1}{15}(a + 7) = -\dfrac{1}{12}(3a - 2)$

51. $(a - 1) - (a + 4) = (2a - 5)$ 52. $\dfrac{x}{2} + \dfrac{3x + 1}{5} = \dfrac{x + 3}{10}$

53. $\dfrac{2x + 3}{4} - \dfrac{3x + 2}{12} = 1$ 54. $\dfrac{3x - 1}{4} + \dfrac{2x + 5}{6} = 1$

55. $\dfrac{2a - 1}{3} - \dfrac{5a - 10}{4} = 1$ 56. $\dfrac{x + 4}{3} - \dfrac{x + 3}{5} = 1$

57. $\dfrac{m - 3}{2} - \dfrac{2m + 4}{3} = m - \dfrac{m - 4}{3}$

58. $\dfrac{3k - 2}{5} + \dfrac{5 - 3k}{4} - 2 = \dfrac{k + 6}{10}$

59. Is 1 the only solution of the following equation?

$$\frac{x-3}{5} + 2x = \frac{11x-3}{5}$$

What are equality sentences of this type called?

60. Without solving the equation, decide whether $\frac{1}{2}$ is a solution of

$$(x-2) + \frac{x-3}{3} = \frac{x+7}{15}$$

3. LITERAL EQUATIONS

Equations containing several variables are sometimes called **literal equations**; for example,

$$2x - p = 4$$

$$mx + 3 = 7b$$

$$a = lw$$

Formulas of science, engineering, business, and so on are usually equations of this type. For instance,

$$I = Prt \qquad \text{(simple interest)}$$

$$A = \frac{1}{2}h(b_1 + b_2) \quad \text{(area of a trapezoid)}$$

$$C = \frac{5}{9}(F - 32) \quad \text{(Fahrenheit–Celsius conversion)}$$

To solve such equations for a certain variable means to treat the equation as though it is one in a single variable (the variable to be solved for) and the other variables are constants. The equation is then solved like any other equation in a single variable.

EXAMPLES

1. Solve $2x - p = 4$ for x.

SOLUTION Here x is treated as the variable, p and 4 as constants.

$$2x - p = 4$$

$$2x = 4 + p$$

$$x = \frac{4 + p}{2}$$

2. Solve $2x - p = 4$ for p.

SOLUTION Here p is treated as the variable, x and 4 as constants.

$$2x - p = 4$$
$$-p = 4 - 2x$$
$$p = 2x - 4$$

3. Solve $A = \frac{1}{2}h(b_1 + b_2)$ for b_1.

SOLUTION

$$A = \frac{1}{2}h(b_1 + b_2)$$
$$2A = hb_1 + hb_2$$
$$2A - hb_2 = hb_1$$
$$\frac{2A - hb_2}{h} = b_1$$

4. Solve for x: $2x + 4 = a^2 - ax$.

SOLUTION Collect terms involving the variable, x, in one member, the remaining terms in the other:

$$2x + ax = a^2 - 4$$

Factor the variable out of the left member:

$$x(2 + a) = a^2 - 4$$

Solve for x:

$$x = \frac{a^2 - 4}{2 + a}$$
$$x = \frac{(a - 2)(a + 2)}{2 + a}$$
$$x = a - 2$$

EXERCISE 4-2

Solve for x.

1. $5x = b$ **2.** $4a^2x = 16$ **3.** $x + 6 = d$

4. $x - 8 = p$ **5.** $3 + x = t$ **6.** $a + x = -a$

7. $x(a + 3) = a - 5$ **8.** $x(y - 1) = y + 3$

9. $x(m - n) = m - n$ **10.** $x(b - a) = a - b$

11. $2x - 6b = 8a$ **12.** $6x + 3m = 9m$ **13.** $3a - x = x - 4b$

14. $3x + c = 2x$ **15.** $\dfrac{2x + 3b}{7} = \dfrac{x + b}{3}$ **16.** $\dfrac{3x + 2y}{3} = \dfrac{x + 4y}{6}$

17. $\dfrac{1}{2}(x - 6a) = \dfrac{1}{3}(6a - x)$ **18.** $\dfrac{3}{4}\left(\dfrac{x}{m} + 4\right) = \dfrac{2}{3}\left(\dfrac{x}{m} - 1\right)$

19. $3(cx - 2ac) = c(x - 4a)$ **20.** $2(px + pd) - p(2d - x) = 0$

21. $\dfrac{6x - 5a}{2} = \dfrac{5x + 2a}{8} + x$ **22.** $5a(bx - 3bc) = 7b(ax - 5ac)$

23. $xp + 4 = p^2 - 2x$ **24.** $a^2 - 3x = 9 + ad$

25. $2ax + 7a = 2a^2 + x + 3$ **26.** $2m^2 + 3x = 2mx + 5m - 3$

27. $a^3 + bx = b^3 + ax$ **28.** $3x + 3a + ad = 2d - dx + 6$

29. Solve $y = mx + b$ (equation of a straight line) for b, for x, for m. When $m = \dfrac{1}{2}$, $b = 5$, and $y = 7$, what is x?

30. Solve $P_2 = P_1 T_2/T_1$ (Gay–Lussac's law of gases) for P_1, T_1, T_2. When $P_1 = 15$ lb, $T_1 = 300°$ Kelvin (K), and $T_2 = 280°$ K, what is P_2?

31. Solve $S = \dfrac{n}{2}(a + l)$ (sum of arithmetic progression) for n, a, l. When $n = 8$, $a = 2$, and $S = 48$, what is l?

32. Solve $\dfrac{x^2}{a^2} + \dfrac{y^2}{b^2} = 1$ (equation of an ellipse) for x^2 and for y^2. When $x = 2$, $a = 4$, and $b = 2$, what is y^2?

4. FRACTIONAL EQUATIONS

An equation in which the variable appears in the denominator of at least one term is called a **fractional equation**. When both members of the equation are multiplied by a common denominator of the fraction terms, a new equation results, called the **derived equation**. The derived equation will be without a fraction, but the equation may or may not be an equivalent equation (i.e., have the same solution).

When the derived equation is first degree, it can be solved as any other first-degree equation (by Program **4.1**). *It is very important that the solution obtained be substituted for the variable in the given equation to ensure that it satisfies the given equation.*

PROGRAM 4.2

To solve a fractional equation whose derived equation is first degree:

Step 1: Find the LCD of the terms of the equation that are in fraction form.

Step 2: Multiply both members of the equation by the LCD of Step 1 to obtain a derived equation without fractions.

Step 3: Solve the derived equation of Step 2.

Step 4: Substitute the solution obtained in Step 3 for the variable of the given equation; if it satisfies the given equation, it is the solution of that equation.

1. Solve for x: $\dfrac{2}{3x} = \dfrac{3}{4x} - \dfrac{1}{2}$.

SOLUTION *Step 1:* Find the LCD of $3x$, $4x$, and 2; it is $12x$.

Step 2: Multiply the terms of both members of the equation by the LCD, $12x$; then simplify:

$$\left(\overset{4}{\cancel{12x}} \cdot \frac{2}{\cancel{3x}} \right) = \left(\overset{3}{\cancel{12x}} \cdot \frac{3}{\cancel{4x}} \right) - \left(\overset{6x}{\cancel{12x}} \cdot \frac{1}{\cancel{2}} \right)$$

$$8 \quad = \quad 9 \quad - \quad 6x$$

Step 3: Solve the derived equation of Step 2:

$$8 = 9 - 6x$$
$$6x = 1$$
$$x = \frac{1}{6} \quad \text{(possible solution of the given equation)}$$

Step 4: Check the solution, $\dfrac{1}{6}$, by substituting it for the variable in the given equation:

$$\frac{2}{3x} = \frac{3}{4x} - \frac{1}{2}$$

$$x = \frac{1}{6}: \quad \frac{2}{3\left(\frac{1}{6}\right)} \overset{?}{=} \frac{3}{4\left(\frac{1}{6}\right)} - \frac{1}{2}$$

$$\frac{2}{\frac{1}{2}} \overset{?}{=} \frac{3}{\frac{2}{3}} - \frac{1}{2}$$

$$4 \overset{?}{=} \frac{9}{2} - \frac{1}{2}$$

$$4 = 4$$

The given equation is satisfied; its solution is $\dfrac{1}{6}$.

2. Solve $\dfrac{8}{3x + 1} + 2 = \dfrac{2x}{x - 1}$.

SOLUTION *Step 1:* Determine the least common denominator:

$$\text{LCD} = (3x + 1)(x - 1)$$

Step 2: Clear fractions by multiplying both members by the LCD:

$$\left[(3x+1)(x-1)\cdot\frac{8}{3x+1}\right] + [(3x+1)(x-1)\cdot 2]$$

$$= (3x+1)(x-1)\cdot\frac{2x}{(x-1)}$$

$$8(x-1) + 2(3x+1)(x-1) = 2x(3x+1)$$

Step 3: Solve the derived equation:

$$8x - 8 + 2(3x^2 - 2x - 1) = 6x^2 + 2x$$

$$8x - 8 + 6x^2 - 4x - 2 = 6x^2 + 2x$$

$$8x - 4x - 10 = 2x$$

$$2x = 10$$

$$x = 5 \quad \text{(possible solution of the given equation)}$$

Step 4: Check the solution of Step 3, using the original equation:

$$\frac{8}{3x+1} + 2 = \frac{2x}{x-1}$$

$$x = 5: \quad \frac{8}{3(5)+1} + 2 \overset{?}{=} \frac{2(5)}{(5)-1}$$

$$\frac{8}{16} + 2 \overset{?}{=} \frac{10}{4}$$

$$2\frac{1}{2} = 2\frac{1}{2} \quad \text{(solution checks)}$$

3. Solve $\dfrac{3x}{x-2} = 4 + \dfrac{6}{x-2}$.

Step 1: LCD $= (x-2)$

Step 2: $(x-2)\cdot\dfrac{3x}{(x-2)} = [(x-2)\cdot 4] + \left[(x-2)\cdot\dfrac{6}{x-2}\right]$

$$3x = 4(x-2) + 6$$

Step 3: $\qquad\qquad\qquad 3x = 4x - 8 + 6$

$$-x = -2$$

$$x = 2 \quad \text{(possible solution)}$$

116 Chapter 4 FIRST-DEGREE EQUATIONS AND INEQUALITIES

Step 4: Substitute 2 for the variable x in the given equation:

$$\frac{3x}{x-2} = 4 + \frac{6}{x-2}$$

$$x = 2: \quad \frac{3(2)}{(2)-2} \overset{?}{=} 4 + \frac{6}{(2)-2}$$

$$\frac{6}{0} \overset{?}{=} 4 + \frac{6}{0}$$

Since $\frac{6}{0}$ is undefined, 2 is not a solution of the given equation. In this case, the given equation has no solution.

EXERCISE 4-3

Solve (when x is present, consider it to be the variable).

1. $3 = \dfrac{1}{4x}$

2. $\dfrac{7}{3x+1} = \dfrac{3}{2x-1}$

3. $\dfrac{x-5}{x+4} = \dfrac{x-1}{x-4}$

4. $\dfrac{3}{x-2} + \dfrac{2}{x-1} = 2$

5. $\dfrac{2x}{x+1} + \dfrac{1}{3x-2} = 2$

6. $\dfrac{a+x}{a-x} = \dfrac{a+b}{a-b}$

7. $\dfrac{3x-2}{4x-1} = \dfrac{3x+1}{4x-7}$

8. $\dfrac{3x-2}{2x-3} = \dfrac{6x-6}{4x-7}$

9. $\dfrac{m}{m-2} - 5 = \dfrac{2}{m-2}$

10. $4 - \dfrac{x}{x+5} = \dfrac{5}{x+5}$

11. $\dfrac{p^2}{p^2-4} = \dfrac{p}{p-2}$

12. $\dfrac{2}{a+2} = \dfrac{1}{a^2-4}$

13. $\dfrac{2}{x-2} = \dfrac{x}{x^2-x-2}$

14. $\dfrac{2}{m-2} = \dfrac{m+5}{m^2-4m+4}$

15. $\dfrac{2p}{x+1} - \dfrac{p}{x} = \dfrac{p}{x^2+x}$

16. $\dfrac{x+13}{x^2-x} = \dfrac{6}{x-1} + \dfrac{3}{x}$

17. $\dfrac{a}{a+4} - \dfrac{4}{a-4} = \dfrac{a^2+16}{a^2-16}$

18. $\dfrac{2}{x+2} = \dfrac{x}{2-x} + \dfrac{x^2+4}{x^2-4}$

19. $\dfrac{6}{x+2} - \dfrac{5}{x} = \dfrac{5-4x}{x^2+2x}$

20. $\dfrac{3}{x} - \dfrac{2}{x+1} = \dfrac{x-2}{x^2-1}$

21. $\dfrac{3}{x+2} + \dfrac{2x}{4-x^2} = \dfrac{m}{x-2}$

22. $\dfrac{2}{4-x^2} = \dfrac{x}{4-x^2} + \dfrac{3}{x+2}$

23. $\dfrac{9}{x+1} - \dfrac{2x+3}{x-2} = \dfrac{7x-2x^2}{x^2-x-2}$

24. $\dfrac{x+a}{x-b} - \dfrac{x-a}{x+b} - \dfrac{a^2-b^2}{x^2-b^2} = 0$

25. $\dfrac{5x}{x^2 + x - 6} - \dfrac{3x}{x^2 + 2x - 8} = \dfrac{2}{x + 3}$

26. $\dfrac{1 - \dfrac{2}{x}}{3 - \dfrac{4}{x}} = \dfrac{2}{7}$

27. $\dfrac{3 - \dfrac{2}{x}}{6 - \dfrac{3}{x}} = \dfrac{x + 4}{2x + 9}$

28. $\dfrac{2x + 4}{x^2 + 2x - 8} - \dfrac{x - 4}{x^2 - 3x + 2} = \dfrac{x + 6}{x^2 + 3x - 4}$

5. SOLVING FIRST-DEGREE INEQUALITIES

An **inequality** is another type of mathematical sentence. It states that two number expressions (members) are unequal. For instance,

$$3 + x > 5$$

$$x + 2 < 4$$

As with equations, two inequalities are equivalent if they have the same solutions, although inequalities generally have many more solutions than do equations.

Let us assume that the variable replacements for the first inequality above, $3 + x > 5$, is restricted to the set of integers. Because every integer greater than 2 will satisfy the inequality, all integers greater than 2 are said to belong to the solution set.

Thus for

$$3 + x > 5$$

the solution set is

$$\{3, 4, 5, 6, 7, \ldots\}$$

since

$$3 + 3 > 5$$

$$3 + 4 > 5$$

$$3 + 5 > 5$$

etc.

For the inequality

$$x + 2 < 4$$

when the replacement set is the set of integers, the solution set is

$$\{\ldots, -3, -2, -1, 0, 1\}$$

1.　What is the solution set for $8 + x > 12$ if the replacement set is the set of integers?

SOLUTION　The integer 4 is not in the solution set because $8 + 4$ *equals* 12 and is not *greater than* 12. Integers less than 4 are not in the solution set either because when each is added to 8, the resulting sum is less than 12. Integers greater than 4, however, are in the solution set because when each is added to 8, the resulting sum is greater than 12. So the set of integers greater than 4, or $\{5, 6, 7, 8, 9, \ldots\}$, satisfies $8 + x > 12$.

2.　What is the solution set for $x + 3 < 6$ if the replacement set is the set of integers?

SOLUTION　All integers less than 3, or $\{\ldots, -4, -3, -2, -1, 0, 1, 2\}$, satisfy the inequality $x + 3 < 6$.

3.　What is the solution set for $5 \leq x - 1$ if the replacement set is the set of integers?

SOLUTION　The symbol $\leq$ means "is less than or equal to." The sentence $5 \leq x - 1$ is really a double or compound sentence that states "$5 < x - 1$ or $5 = x - 1$." The solution set for such compound sentences contains all those numbers that make *either* sentence true. Thus substituting 6 for x makes $5 = x - 1$ true; and the integers $7, 8, 9, 10, \ldots$ make $5 < x - 1$ true when each replaces the variable. So

$$\{6, 7, 8, 9, 10, \ldots\}$$

is the solution set for $5 \leq x - 1$.

4.　What is the solution set for $2x \geq -6$ if the replacement set is the set of integers?

SOLUTION　The symbol $\geq$ means "is greater than or equal to." The compound sentence $2x \geq -6$ is equivalent to the two sentences, $2x = -6$ and $2x > -6$. The solution set for $2x \geq -6$ is

$$\{-3, -2, -1, 0, 1, \ldots\}$$

Had the replacement set in the preceding examples been the set of real numbers instead of the set of integers, the various solution sets would have contained many more numbers. Because solution sets that range over the real numbers cannot be readily tabulated, a graph is frequently used, as illustrated below.

EXAMPLES

1. Graph the solution set for $8 + x > 12$ if the replacement set is the set of real numbers.

SOLUTION $8 + x > 12$:

The graph is based on the number line for real numbers. The open dot, O, at 4 on the number line means that 4 is *not* in the solution set. The bolder half-line to the right of 4 means that all real numbers greater than 4 are in the solution set.

2. Graph the solution set for $x + 3 < 6$ if the replacement set is the set of real numbers.

SOLUTION $x + 3 < 6$:

Interpretation: 3 (open dot) is not in the solution set, but all real numbers less than 3 are in the solution set.

3. Graph the solution set for $5 \leq x - 1$ if the replacement set is the set of real numbers.

SOLUTION $5 \leq x - 1$:

Interpretation: 6 (solid dot) is in the solution set, and so is every real number greater than 6 in the solution set.

4. Graph the solution set for $2x \geq -6$ if the replacement set is the set of real numbers.

SOLUTION $2x \geq -6$:

Interpretation: -3 (solid dot) is in the solution set, and so is every real number greater than -3 in the solution set.

Simple inequalities, such as those in the examples above, can often be solved by inspection. More complicated inequalities are not so easily solved. As with equations, however, it is possible to proceed through a series of equivalent but simpler inequalities until we arrive at an inequality for which the solution is obvious.

In general, the following operations always result in an inequality that is equivalent to the original inequality—that is, they both have identical solution sets:

- Addition or subtraction of the same number (which may be in the form of the variable) to both members.

- Multiplication or division of both members by the same positive number (variable excepted).
- Multiplication or division of both members by the same negative number (variable excepted) when the sense of the inequality is reversed.

Two inequalities are said to have the same **sense** when their symbols of inequality both "point" in the same direction. For instance, the inequalities $x < 4$ and $y < 8$ are said to have the same sense, while $a > 6$ and $b < 2$ are opposite in sense because their inequality symbols point in opposite directions.

So when two members of an inequality are multiplied or divided by a positive number, the sense of the inequality remains the same. But when two members of an inequality are multiplied or divided by a negative number, the sense of the inequality is reversed.

EXAMPLES

(Unless otherwise indicated, the replacement set for the variable is the set of real numbers.)

1. Solve $8 + x > 12$.

SOLUTION Subtract 8 from both members:

$$8 + x - 8 > 12 - 8$$

Simplify:

$$x > 4$$

The inequality $x > 4$ is equivalent to $8 + x > 12$. The solution set for $x > 4$ is immediately evident but not for $8 + x > 12$. (Compare: Example 1, page 120.)

2. Solve $5 \leq x - 1$.

SOLUTION Add 1 to both members:

$$5 + 1 \leq x - 1 + 1$$

Simplify:

$$6 \leq x$$

(Compare: Example 3, page 120.)

Note: $6 \leq x$ and $x \geq 6$ are equivalent.

3. Solve $2x \geq -6$.

SOLUTION Divide both members by 2:

$$\frac{2x}{2} \geq \frac{-6}{2}$$

$$x \geq -3$$

(Compare: Example 4, page 120.)

4. Solve $3 - 2x < 5$.

SOLUTION Subtract 3 from both members:

$$3 - 2x - 3 < 5 - 3$$

$$-2x < 2$$

Divide both members by -2 (reverse the sense or direction of the inequality):

$$\frac{-2x}{-2} > \frac{2}{-2}$$

$$x > -1$$

Graph of the solution set:

5. Solve $2x - 2 < 4x + 6$; variable replacements restricted to the set of integers.

SOLUTION

$$2x - 2 < 4x + 6$$

$$2x - 2 + 2 < 4x + 6 + 2$$

$$2x < 4x + 8$$

$$2x - 4x < 4x - 4x + 8$$

$$-2x < 8$$

$$\frac{-2x}{-2} > \frac{8}{-2} \quad \text{(note sign reversal)}$$

$$x > -4 \quad (x \text{ an integer})$$

Graph of the solution set:

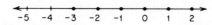

Note: -4 is not included in the solution set because x must be *greater than* -4.

6. Solve $4p + 2 \leq 3p + 3$.

$$4p + 2 \leq 3p + 3$$
$$4p + 2 - 2 \leq 3p + 3 - 2$$
$$4p \leq 3p + 1$$
$$4p - 3p \leq 3p + 1 - 3p$$
$$p \leq 1$$

Graph of the solution set:

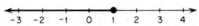

Note: $p = 1$ is included in the solution set because p is less than *or equal to* 1.

Solution sets can also be expressed by what is called **set-builder notation**. For example,

$\{x \,|\, x < 3,\, x \in I\}$ means the set of all integers less than 3.

$\{m \,|\, m > 2,\, m \in R\}$ means the set of all real numbers greater than 2.

The symbolism, taken apart and expressed literally, may be read:

$\{x$	$\mid$	$x < 3,$	$x \in I\}$
The set of numbers x ...	such that ...	x is less than 3 ...	and x is a member ($\in$) of the set of integers (I).

$\{m$	$\mid$	$m > 2,$	$m \in R\}$
The set of numbers m ...	such that ...	m is greater than 2 ...	and m is a member ($\in$) of the set of real numbers (R).

The solution sets for the last three examples can be expressed in set-builder notation as follows.

Example 4: $\{x \,|\, x > -1,\, x \in R\}$
Example 5: $\{x \,|\, x > -4,\, x \in I\}$
Example 6: $\{p \,|\, p \leq 1,\, p \in R\}$

As a practical matter, however, when the domain of the variable is the set of real numbers, solution sets for inequalities are usually given by a simple inequality; for example:

$$\{x \,|\, x > -1,\, x \in R\} \Rightarrow x > -1$$
$$\{p \,|\, p \leq 1,\, p \in R\} \Rightarrow p \leq 1$$

That convention will be generally followed in this book.

Graph the solution set for each inequality. The replacement set is the set of real numbers.

1. $x > 3$ 2. $x < 5$ 3. $x < -3$

4. $x > -4$ 5. $x \geq 6$ 6. $x \leq 5$

7. $x \geq -2$ 8. $x \leq -7$ 9. $x \leq 0$

10. $x \geq -1$

Solve; the replacement set is the set of real numbers.

11. $x + 1 < 6$ 12. $6 < x - 1$ 13. $p + 3 \geq 4$

14. $5 > t - 6$ 15. $x + 3 \leq 0$ 16. $3x < 12$

17. $144 \leq 12m$ 18. $\dfrac{1}{2}k \leq -5$ 19. $3x > -15$

20. $14 \leq \dfrac{1}{2}x$ 21. $2a - 3 \leq 5a$ 22. $3z - 4 > 2z - 9$

23. $k + 3 < 2k + 1$ 24. $6 - x > 2x + 5$ 25. $2p - 3p + 7 \geq 0$

26. $0 < 2x - 3 + 4x$ 27. $3m - 2 + m \leq 5m + 6$

28. $5p - 2 + 7p > 4p - 6 + 2p$ 29. $3x + 4(3x - 5) > 12 - x$

30. $3(2a - 1) < 4(a - 3) - 5$ 31. $\dfrac{x}{3} + \dfrac{2x}{5} \geq \dfrac{3x}{4} + \dfrac{7}{6}$

32. $\dfrac{a}{4} - \dfrac{2a}{3} \leq \dfrac{5a}{6} + 1$ 33. $\dfrac{2x + 3}{4} - \dfrac{4x - 1}{12} \leq 1\dfrac{1}{3}$

34. $\dfrac{2c + 1}{2} - \dfrac{3c - 2}{5} > 1$ 35. $\dfrac{3x + 1}{5} - \dfrac{1 - 3x}{2} - 6 \geq 0$

36. $x - 4 + \dfrac{2x - 3}{3} + 10 < 0$

6. ABSOLUTE-VALUE EQUATIONS*

Recall that the absolute value of a number may be thought of as its distance from zero on the number line, without regard to direction. The absolute value of both $+3$ and -3, for example, is 3:

$$|+3| = |-3| = 3$$

In the case of the equation

$$|x| = 3$$

---

* The content of this section, in more basic courses of study, may be postponed without significant effect upon the remaining parts of the book.

the solutions are obviously 3 and -3. That is, the equation is satisfied by the solutions of these two equations

$$x = 3 \qquad x = -3$$

Equations such as

$$|x| = 3 \quad \text{and} \quad |x - 3| = 5$$

are called **absolute-value equations**. Although the algebraic expression within the absolute-value symbol is first-degree, the equation is *not* first-degree. It has two solutions. In the case of $|x - 3| = 5$, the solutions are -2 and 8, as we can verify.

$$|x - 3| = 5$$
$$x = 8: \qquad |8 - 3| \overset{?}{=} 5$$
$$|5| = 5 \quad \text{(true)}$$
$$x = -2: \quad |-2 - 3| \overset{?}{=} 5$$
$$|-5| = 5 \quad \text{(true)}$$

In general, if

$$|ax + b| = k \quad (k \text{ a positive number})$$

then

$$(ax + b) = k \quad \text{or} \quad -(ax + b) = k$$

which may also be expressed

$$ax + b = k \quad \text{or} \quad ax + b = -k$$

PROGRAM 4.3

To solve an absolute-value equation of the form $|ax + b| = k$, k a positive number:

Step 1:   Solve the equation $ax + b = k$.

Step 2:   Solve the equation $ax + b = -k$.

Step 3:   Write the solutions obtained in Steps 1 and 2 as the solution for the given equation.

EXAMPLES

1.   Solve $|x + 1| = 3$.

SOLUTION   *Step 1:* Solve $x + 1 = 3$
$$x = 2$$

Step 2: Solve $x + 1 = -3$
$$x = -4$$

Step 3: Write the solutions of Steps 1 and 2 as the solution set for the given equation. Thus the solution set for $|x + 1| = 3$ is $\{2, -4\}$.

2.   Solve $|2x - 5| = 1$.

SOLUTION *Step 1:* $2x - 5 = 1$
$$2x = 6$$
$$x = 3$$

Step 2: $2x - 5 = -1$
$$2x = 4$$
$$x = 2$$

Step 3: The solution set is $\{2, 3\}$.

3.   Solve $|3p + 5| = 1$.

SOLUTION $3p + 5 = 1$ $3p + 5 = -1$

$3p = -4$ $3p = -6$

$p = -\dfrac{4}{3}$ $p = -2$

The solution set is $\left\{-\dfrac{4}{3}, -2\right\}$.

4.   Solve $3|5 - 6x| - 7 = 8$.

SOLUTION Transform the equation to the model for Program **4.3**, $|ax + b| = k$:

$$3|5 - 6x| - 7 = 8$$

Add 7 to both members:

$$3|5 - 6x| = 8 + 7$$

Simplify:

$$3|5 - 6x| = 15$$

Divide both members by 3:

$$|5 - 6x| = 5$$

Now, solve by Program **4.3**:

$$5 - 6x = 5 \qquad 5 - 6x = -5$$
$$-6x = 0 \qquad -6x = -10$$
$$x = 0 \qquad x = \frac{5}{3}$$

The solution set is $\left\{0, 1\dfrac{2}{3}\right\}$.

Solve for the variable.

1. $|x| = 8$ 2. $|m| = 2$ 3. $|x - 3| = 1$
4. $|x - 5| = 2$ 5. $|a - 4| = 3$ 6. $|b - 7| = 2$
7. $|k + 2| = 4$ 8. $|x + 6| = 2$ 9. $|x + 3| = 1$
10. $|p + 4| = 10$ 11. $|5 - 3x| = 8$ 12. $|6 - 2x| = 14$
13. $|3 - 2x| = 0$ 14. $|6x + 2| = 0$ 15. $|6 + 5x| = 1$

16. $|9 - 4x| = 9$ 17. $\left|\dfrac{1}{2}x - 3\right| = 3$ 18. $\left|6 - \dfrac{1}{2}x\right| = 5$

19. $\left|\dfrac{3}{2}m + 5\right| = \dfrac{1}{2}$ 20. $\left|6 - \dfrac{3}{4}k\right| = \dfrac{3}{4}$ 21. $3|2x - 5| = 9$

22. $4|3x + 7| = 20$ 23. $5|3x + 7| = -10$ 24. $-3\left|6x + \dfrac{1}{3}\right| = -27$

25. $4 - 3|3x - 5| = 0$ 26. $2 + 7|2 - 3x| = 0$

27. $4|2x - 3| = \dfrac{1}{2}$ 28. $\dfrac{3}{5} = 5|3x - 1|$

29. $3|5x + 2| + 7 = 19$ 30. $2|6 - 3x| - 5 = 17$

7. ABSOLUTE-VALUE INEQUALITIES*

We saw in the previous section that an absolute-value equation is, in effect, two equations. Similarly, an absolute-value inequality may be considered two inequalities. For instance,

$$|x| > 3$$

may be thought of as the pair of inequalities:

$$-x > 3 \qquad x > 3$$

or (because $-x > 3$ may be expressed equivalently as $x < -3$)

$$x < -3 \qquad x > 3$$

If we graph these two alternative inequalities on a single number line, the result is as shown in Figure 4.1.

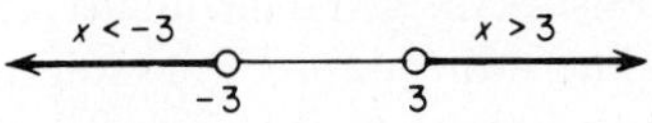

Figure 4.1

*The content of this section, in more basic courses of study, may be postponed without significant effect upon the remaining parts of the book.

The solution set for $|x| > 3$, therefore, is

$$\{x < -3 \text{ or } x > 3\}$$

Next, let us consider the same inequality with its sense reversed:

$$|x| < 3$$

It, too, may be thought of as a pair of inequalities:

$$-x < 3 \qquad x < 3$$

or

$$x > -3 \qquad x < 3$$

Combining the graphs of the two inequalities, against the same number line, we obtain the effect shown in Figure 4.2.

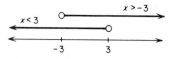

Figure 4.2   **Figure 4.3**

The overlap of Figure 4.2 may be plotted as shown in Figure 4.3, and represents that *both* inequalities are satisfied *only* by real numbers between -3 and 3. Those same numbers make up the solution set for $|x| < 3$ and can be expressed as

$$\{-3 < x < 3\}$$

In effect, "x is a number between -3 and 3."

PROGRAM 4.4

To solve an absolute value inequality of the form $|ax + b| < k$, or $|ax + b| > k$, k a positive number:

Step 1:   Replace the inequality symbol with an equals sign and solve the equation for the variable by using Program **4.3**.

Step 2:   Record the solutions on a number line.

Step 3:   Select any number from those represented by the middle segment of the number line of Step 2, and substitute it for the variable in the given inequality.

    (a) If the selected number satisfies the given inequality, the middle segment of the number line is the graph of the solution set.

    (b) If the selected number does not satisfy the given inequality, the two outer segments of the number line is the graph of the solution set.

1. Solve $|x + 2| < 8$.

SOLUTION | *Step 1:* Replace the inequality symbol $<$ with $=$ and solve by using Program **4.3**.

$$|x + 2| = 8$$

$$x + 2 = 8 \qquad x + 2 = -8$$

$$x = 6 \qquad x = -10$$

Step 2: Record the solutions of Step 1 on a number line.

Step 3: Select any number represented by the middle segment of the number line of Step 2, say 0. Does 0 satisfy the given inequality?

$$|x + 2| < 8$$

$$x = 0: \quad |0 + 2| \overset{?}{<} 8$$

$$2 < 8 \quad \text{(true)}$$

So the middle segment of the number line is the graph of the solution set of the given inequality:

Thus the solution set for $|x + 2| < 8$ is

$$\{-10 < x < 6\}$$

2. Solve $|x + 2| > 8$.

SOLUTION | *Steps 1 and 2:* Same as Example 1.

Step 3: Select any number represented by the middle segment of the number line of Step 2. This time, -2. Does -2 satisfy the given inequality?

$$|x + 2| > 8$$

$$x = -2: \quad |-2 + 2| \overset{?}{>} 8$$

$$0 > 8 \quad \text{(false)}$$

So the graph of the solution set of the given inequality is represented by the two outer segments of the number line of Step 2:

Thus the solution set for $|x + 2| > 8$ is

$$\{x < -10 \text{ or } x > 6\}$$

3. Solve $|2m - 5| \leq 3$.

SOLUTION *Step 1:* Replace $\leq$ with $=$:

$$|2m - 5| = 3$$

$$2m - 5 = 3 \qquad 2m - 5 = -3$$

$$2m = 8 \qquad\qquad 2m = 2$$

$$m = 4 \qquad\qquad m = 1$$

Step 2: Record the solutions of Step 1:

Step 3: Choose a number from the middle segment of the number line of Step 2 (say 3), and test:

$$|2m - 5| \leq 3$$

$$m = 3: \quad |2(3) - 5| \overset{?}{\leq} 3$$

$$|6 - 5| \leq 3$$

$$1 \leq 3 \quad \text{(true)}$$

Thus the graph of the solution set of the given inequality, $|2m - 5| \leq 3$, is

The sign of the inequality is $\leq$. The equality part means that the solutions of the equation of Step 1 will also be included in the solution set:

$$\{1 \leq x \leq 4\}$$

4. Solve $7 - 2|2x - 5| \leq -9$.

SOLUTION Transform the given inequality into the form $|ax + b| \leq$ or $\geq k$, k a positive number.

$$7 - 2|2x - 5| \leq -9$$

$$-2|2x - 5| \leq -9 - 7$$

$$-2|2x - 5| \leq -16$$

Divide both members by -2 (reverse the sense of the inequality):

$$|2x - 5| \geq 8$$

That is, $|2x - 5| \geq 8$ is an equivalent form of the given inequality, $7 - 2|2x - 5| \leq -9$. The solution set of the former will also be the solution set of the latter.

Next, solve the equivalent form by Program **4.3** after replacing $\geq$ by $=$:

$$|2x - 5| = 8$$

$$2x - 5 = 8 \qquad 2x - 5 = -8$$

$$2x = 13 \qquad 2x = -3$$

$$x = 6\frac{1}{2} \qquad x = -1\frac{1}{2}$$

Trial number: 0.

$$|2x - 5| \geq 8$$

$$x = 0: \quad |2(0) - 5| \overset{?}{\geq} 8$$

$$5 \ngeq 8$$

Thus

or

$$\left\{ x \leq -1\frac{1}{2} \text{ or } x \geq 6\frac{1}{2} \right\}$$

is the solution set of

$$7 - 2|2x - 5| \leq -9$$

EXERCISE 4-6

Solve and graph the solution sets. The variables represent real numbers.

1. $|x| > 4$ 2. $|x| > 3$ 3. $|m| \geq 3$
4. $|a| \geq 2$ 5. $|x - 3| > 7$ 6. $|y + 2| > 7$
7. $|a + 2| \geq 8$ 8. $|x - 4| \geq 9$ 9. $|p + 8| \geq 2$
10. $|m + 6| \geq 4$ 11. $|3p - 2| \geq 4$ 12. $|6 - 3x| > 3$
13. $|4 - 5x| > 0$ 14. $|3a| > 0$ 15. $|x| < 4$
16. $|x| < 3$ 17. $|m| \leq 2$ 18. $|a| \leq 7$
19. $|a - 3| < 4$ 20. $|x - 6| < 8$ 21. $|x - 2| < 1$
22. $|x - 4| < 3$ 23. $|m + 3| \leq 2$ 24. $|c + 6| \leq 5$
25. $|4 + 2x| < 3$ 26. $|6 + 3x| < 1$ 27. $|2x + 2| \leq 2$
28. $|3m - 4| \leq 2$ 29. $|6 - 2x| < 2$ 30. $|3 - x| < 3$
31. $|2 - 3x| \leq 1$ 32. $|5 - 2p| \leq 15$ 33. $2|3 - x| < 8$
34. $3|2x + 1| > 9$ 35. $-3|2x - 1| \geq -15$

36. $-2|3x - 5| \leq -14$ **37.** $8 + 6|5x - 2| > 10$
38. $4 - 2|3x + 1| > 0$ **39.** $2 - 3|2x + 1| \leq 1$
40. $6 - 5|-2x + 3| \leq 1$

| REVIEW

PART A

Answer True or False.

1. When all members of the domain of the variable satisfy an equation, the equation is called an identity.
2. The equation, $x + 1 = 3$, is a conditional equation.
3. The equations, $x = 4$ and $x + 1 = 4$, are equivalent equations because they have the same variable.
4. Dividing both members of an equation by -2 produces a new equation equivalent to the first.
5. Every equation has at least one solution.
6. Many formulas of science, engineering, business, and so on, are in effect equations.
7. The two inequalities, $x > 2$ and $y > -4$, have the same sense.
8. If $|px + c| = k$, k cannot be negative.
9. $-x > 6$ and $x < 6$ are equivalent.
10. The notation, $\{-2 < x < 5\}$, may be interpreted to mean that the value of x lies between -2 and 5.

PART B

Solve for x.

1. $5x = \dfrac{4}{5}$ **2.** $-\dfrac{1}{3}x = 2$ **3.** $\dfrac{3}{4}x - x = 2$

4. $3x - 2(x - 1) = 1$ **5.** $2(x - 1) + x = 3x$

6. $2(x + 5) - 3x = 0$ **7.** $2x - 4 + 3x - 2x = 6 + x + 7x$

8. $4x - 2 + 3x + 7 = 4 - 3x + 2 - 5x$

9. $3x + 2 = 4(x - 7) + 6$ **10.** $3(x - 5) = 2(3x - 2) - 2$

11. $4(2x - 3) = 7(x - 4 + x)$

12. $2(7 - 3x) + 10 = 5(2x + 3 - 6 + 2x)$

13. $4(2x + 5) + 3(x - 7) = x - 1$

14. $2(x - 3 + x) - x + 3 = 5(x - 2 + x - 7)$

15. $3(x + 2) - 4(x - 7) + 3(x - 12) = 2(x - 1)$

16. $\dfrac{2x - 5}{3} - \dfrac{x - 4}{2} = 1$ **17.** $\dfrac{3x - 2}{5} - \dfrac{1 + x}{10} = 1$

18. $\dfrac{x-2}{5} + \dfrac{3x-2}{6} = \dfrac{x+8}{15}$

19. $\dfrac{1}{3}(4-x) - \dfrac{3}{5}(5-2x) = \dfrac{2}{3}(x-7)$

20. $\dfrac{2}{3}(2-3x) + x - \dfrac{3}{4}(x-5) = 2 + \dfrac{1}{6}(2x+1)$

21. $m - 3(2x - 3m) = 2(3x - m)$ **22.** $a(3p + x) - am = 2(ap - ax)$

23. $6a = 4(a - 3x) + 5(a + x)$ **24.** $3a - 2x = \dfrac{x - 8a}{6}$

25. $\dfrac{5}{6}(c + 3x) = \dfrac{3}{4}(x - 2c)$ **26.** $\dfrac{c - 4x}{3} = \dfrac{2c - x}{5} + x$

27. $a^2 - 3x = ax + 9$ **28.** $px + 2 - p = p^2 + x$

29. $cx + bc + 2ad = ac + 2dx + 2bd$

30. $a(x - a^2) = b(x - b^2)$

31. $2(m + 3x) - 2(m - x) = 3(2x - 1) + 2x + 3p$

32. $\dfrac{1}{3}(x - 5a + 2) = \dfrac{3}{4}(2x - 5a + 2) + \dfrac{5}{12}(5a - 2)$

33. Solve $S = \dfrac{n}{2}(a + l)$ for l. What must be the value of l when $S = 195$, $n = 10$, and $a = 6$?

34. Solve $A = \dfrac{h}{2}(b_1 + b_2)$ (area of a trapezoid) for h; for b_2. When $A = 36$, $b_1 = 10$, and $h = 4$, what must be the value of b_2?

Solve for x.

35. $\dfrac{4}{x-3} = \dfrac{2}{x+3}$ **36.** $\dfrac{5}{4x+2} = \dfrac{1}{x+1}$

37. $\dfrac{3x-2}{2x+3} = \dfrac{9x-5}{6x+1}$ **38.** $\dfrac{1+2x}{x-4} = \dfrac{4x^2+5x}{2x^2-7x-4}$

39. $\dfrac{x+6}{6} - \dfrac{24}{9x+36} = \dfrac{3x+24}{18}$ **40.** $\dfrac{x+3}{x^2-5x+4} + \dfrac{1}{x-1} = \dfrac{2}{x-4}$

41. $\dfrac{7x}{x-3} - \dfrac{12x^2-12}{x^2-2x-3} + \dfrac{5x}{x+1} = 0$

42. $\dfrac{x}{2x+1} = \dfrac{x+1}{2x-4} + \dfrac{2x-3}{2x^2-3x-2}$

43. $\dfrac{x-2}{x+6} = \dfrac{76-14x}{x^2+2x-24} - \dfrac{2+x}{4-x}$

44. $\dfrac{x-3}{x+1} - 2 = \dfrac{x+9}{x^2-x-2} + \dfrac{x+4}{2-x}$

45. $\dfrac{2 - \dfrac{3}{x}}{3 - \dfrac{1}{x}} = \dfrac{2x + 4}{3x + 13}$

46. $\dfrac{\dfrac{2}{x} - 3}{1 - \dfrac{3}{x}} = \dfrac{3x - 2}{7 - x}$

Solve for the variable; the replacement set is the set of real numbers.

47. $x - 1 < 6$

48. $y + 5 \geq 6$

49. $-4 > t + 5$

50. $m - 4 \geq 0$

51. $3x \leq -15$

52. $100 \geq 10a$

53. $\dfrac{1}{3}p + 2 > 8$

54. $-6 + \dfrac{1}{2}x < 0$

55. $-x + 3 > 0$

56. $2m - 3 \leq 4m - 5$

57. $0 \leq 4x - 2 - 7x$

58. $3y - \dfrac{1}{2} \geq y - 4 + 3y$

59. $4(2m - 1) > 3(m + 3) + 2$

60. $2(k - 3) \leq 6 - 2(4 - k)$

61. $\dfrac{x - 3}{3} \geq x + \dfrac{2 - x}{5}$

62. $a + 3 + \dfrac{a}{2} < \dfrac{3a - 1}{3}$

63. $\dfrac{4 - 3y}{5} > 2 - \dfrac{y + 2}{15}$

64. $\dfrac{2x + 1}{3} \leq \dfrac{2x - 3}{4} - \dfrac{1}{2}$

65. $|x - 7| = 1$

66. $|y + 2| = 4$

67. $|1 - 3x| = 0$

68. $0 = |3y + 3|$

69. $2|2x - 5| = 10$

70. $3|x + 7| = 21$

71. $2|4x - 3| - 7 = -3$

72. $11 - 3|5m + 2| = 2$

73. $|m| > 8$

74. $|y| < 2$

75. $|x - 4| \leq 5$

76. $|3 - m| \geq 2$

77. $|2 - 3x| \leq 5$

78. $3 < |2x - 3|$

79. $21 < 3|3x + 7|$

80. $-4|2 - 3x| \leq -12$

81. $3 - 2|m + 3| > -9$

82. $0 \geq 4 - 2|2a - 5|$

134 Chapter 4 FIRST-DEGREE EQUATIONS AND INEQUALITIES

5 APPLICATIONS

1. INTRODUCTION

Most people study mathematics for its applications. Not only is the subject an aid to clear and incisive thinking, but it can also be used to solve a vast array of problems.

In this chapter we place special emphasis on the use of algebra to solve problems. Essentially, the approach consists of two parts.

1. Translate the problem situation into an algebraic sentence—an equation or inequality.
2. Solve the sentence by algebraic techniques.

Item 2 is readily handled, for it is simply a matter of mechanics. But item 1 is usually more difficult, and stress is given to it throughout this chapter. In order

135

to get the most from the chapter, make the development of a basic equation or inequality a part of the solution of *every* exercise, even though it may seem unnecessary in some cases.

2. SOLVING PROBLEMS BY EQUATION

That which makes a problem a problem is the existence of an unknown number whose value we seek. What makes the problem *solvable* is the availability of enough related information so that we can determine the value of the unknown.

The first step in solving a problem by algebra is to identify, from the problem, the unknown number that we wish to make known. Then *represent this unknown by a letter symbol*, which subsequently becomes a variable in an equation.

The second step is to *identify within the problem some equivalence relationship that involves known data and the unknown of the problem*. Every problem capable of solution by an equation must contain—either stated or implied—such an equivalence, as the following two problems illustrate.

problem with the equivalence relationship stated

What number when doubled and then increased by 3 is 27?

If we substitute the word "equals" for "is" and translate the phrase before it, using x to represent the unknown but sought-for number, we obtain

x represents the unknown number.

$2x$ represents the unknown number doubled.

$2x + 3$ represents the doubled unknown number increased by 3.

$2x + 3 = 27$ is the equation.

problem with the equivalence relationship implied

What is the altitude of a triangle that has an area of 36 and a base of 6?

In this problem it is necessary to know, either as a matter of fact or by formula, that the area of a triangle is numerically equal to one-half the product of its base and altitude measures. Thus if we let x represent the number associated with the unknown altitude, then

$$\underbrace{36}_{(\text{area})} = \frac{1}{2} (\underbrace{6}_{(\text{base})}) (\overset{(\text{altitude})}{\underset{\downarrow}{x}}) \left.\right\} \text{the equation}$$

The third step in problem solving by equation has already been demonstrated in the two illustrative problems: *incorporate the unknown and the knowns into an algebraic statement of equivalence.* This is the heart of the solution of the problem, for it provides an equation by which the unknown can become known.

The fourth step is purely mechanical: *solve the equation for the variable that represents the unknown.*

The final step is to check by *substituting the solution of the equation back into the problem* rather than into the equation.

Substitution of the solution back into the equation simply verifies the computational work. It will not detect any mistake in the setup of the equation, the major source of error in the solution of problems by equation. Moreover, not every solution of the equation is certain to be a solution of the problem.

PROGRAM 5.1

To solve a problem by equation:

Step 1: Determine from the statement of the problem that which is sought—the unknown—and represent it by a letter symbol.

Step 2: Identify some equivalence relationship that involves both known data and the unknown of the problem.

Step 3: State algebraically (by equation) the equivalence relationship of Step 2.

Step 4: Solve the equation of Step 3 for the variable.

Step 5: Check by substituting the solution found in Step 4 back into the problem.

The following examples can be referred to as "dimension problems." Other categories of basic problem types are discussed in the next several sections. Often a drawing or sketch of the situation posed by the problem can be useful. The drawing need not be exactly to scale, but the given as well as the unknown dimensions should be inserted.

EXAMPLES

1. A board 12 ft long is cut in two pieces so that one piece is three times the length of the shorter piece. How long is the shorter piece?

SOLUTION *Step 1:* What is sought? The length of the shorter piece. Let x represent the length of the shorter piece. (See Figure 5.1.)

Figure 5.1

Step 2: Identify an equivalence relationship:

length of short piece + length of long piece = total length

Step 3: State algebraically (by equation) the equivalence relationship of Step 2. If x represents the length of the shorter piece, then the length of the longer piece, which is three times the length of the shorter piece, can be represented as $3x$. (See Figure 5.1.) Therefore,

length of short piece + length of long piece = total length

has for a corresponding equation:

$$x \qquad + \qquad 3x \qquad = \qquad 12$$

Step 4: Solve the equation of Step 3:

$$x + 3x = 12$$
$$4x = 12$$
$$x = 3$$

Length of short piece: 3 ft

Step 5: Check by going back to the problem and reasoning: If the length of the short piece is 3 ft, then the length of the long piece, which is three times that of the short piece, must be 3×3 ft, or 9 ft; the sum of the lengths of the short and the long piece is 3 ft + 9 ft, or 12 ft, which should equal the length of the board (12 ft). Since it does, the solution of the equation also satisfies the problem.

2. A 12-ft-long board is to be cut into three pieces so that the second piece is twice the length of the first piece and the third piece is three times the length of the second piece. Find the lengths of the three pieces of board.

SOLUTION Let x represent the length of the first piece. Then $2x$ represents the length of the second piece, and $3(2x)$ represents the length of the third piece. (See Figure 5.2.)

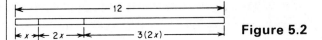

Figure 5.2

The equivalence relationship is

length of first piece + length of second piece

+ length of third piece = total length

The equation is

$$x + 2x + 3(2x) = 12$$
$$x + 2x + 6x = 12$$
$$9x = 12$$
$$x = \frac{12}{9} = 1\frac{1}{3}$$

If $x = 1\frac{1}{3}$, then $2x = 2\frac{2}{3}$ and $3(2x) = 8$. Hence the length of the first piece is 1 ft 4 in., the second piece 2 ft 8 in., and the third piece 8 ft 0 in.

Check: The lengths of the three pieces add to 12 ft: 1 ft 4 in. + 2 ft 8 in. + 8 ft 0 in. = 12 ft 0 in. So the conditions of the problem are met by this solution.

3. The perimeter of a rectangle whose length is $2\frac{1}{2}$ times its width is 21 ft. Find the dimensions of the rectangle. (See Figure 5.3.)

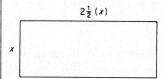

$2\frac{1}{2}(x)$

x

Figure 5.3

SOLUTION Let x represent the width of the rectangle. Then $2\frac{1}{2}(x)$, or $\frac{5}{2}x$, represents the length of the rectangle. The equivalence relationship is

$$\text{perimeter of a rectangle} = 2(\text{width} + \text{length})$$

The equation is

$$21 = 2\left(x + \frac{5}{2}x\right)$$
$$21 = 2x + 5x$$
$$3 = x$$

If $x = 3$, then $\frac{5}{2}x = 7\frac{1}{2}$. The width of the rectangle is 3 ft and its length is $7\frac{1}{2}$ ft.

Check: The distance around the rectangle, which is its perimeter, is the sum:

$$3 \text{ ft} + 7\frac{1}{2} \text{ ft} + 3 \text{ ft} + 7\frac{1}{2} \text{ ft} = 21 \text{ ft}$$

So the conditions of the problem are met by this solution.

Write an equation for each problem, then solve.

1. A piece of wire 36 ft long is cut so that one piece is 12 ft longer than the other. Find the dimensions of the two pieces.

2. A board 16 ft long is to be cut into two pieces, one $3\frac{1}{2}$ ft shorter than the other. What is the length of the longer piece?

3. A rope 6 ft long was cut into three pieces. The middle-sized piece is 6 in. shorter than one piece and 6 in. longer than the other. What is the length of the shortest piece?

4. The sum of the angles of every triangle in a plane is 180°. The largest angle of a triangle is 10° greater than one angle and 20° greater than the other. What is the measure of the largest angle?

5. A rectangle twice as long as it is wide has a perimeter of 51 in. What are its dimensions?

6. The perimeter of a rectangle is 176 ft. Its width is $\frac{3}{8}$ its length. What is its length?

7. A rectangle is 4 ft longer than it is wide. Its perimeter is 68 ft. How wide is the rectangle?

8. A pentagon (five sides) has a perimeter of 43 in. Three sides are the same length and the two remaining sides are each 6 in. shorter. How long is one of the three equal sides?

9. The perimeter of an isosceles triangle (two sides of the same length) is 44 in. One side is 8 in. shorter than the sum of the other two sides. What is the length of the longest side?

10. An airplane flew a triangular course of 70 miles (mi). The first leg was twice the second, and the third leg was 14 mi longer than the second. How long was the first leg?

11. A triangle sits on top of a square and forms a five-sided figure in which two sides are equal and the three other sides are equal. If the perimeter of the triangle is 22 in. and the perimeter of the complete five-sided figure is 38 in., find the length of the edge of the square.

12. Find the sizes of the three angles of a triangle if the smallest is half the middle-sized angle, and the largest is 27° larger than three times the size of the middle-sized angle.

13. A man paid $1044 to have his rectangular lot, perimeter 600 ft, fenced in. If fencing the 80 ft front part cost him 30¢ more per foot than that of the back and sides, how much did he pay per foot for the front fencing?

14. One dimension of a rectangular lot is $\frac{3}{4}$ the other. If the perimeter becomes 680 ft when each dimension is extended 30 ft, what are the original dimensions?

15. A bus starting from Glendale each morning makes a trip to Avalon, and from Avalon it goes 26 mi to Barret. From Barret it returns to Glendale over a route that is 9 mi longer than the first leg of the trip. If the total trip is 205 mi, what is the distance of this last leg?

16. A square and a rectangle have the same area. The length of one side of the square is 3 cm less than one side of the rectangle, and 6 cm greater than the other side of the rectangle. What is the length of the sides of the square?

3. MIXTURE PROBLEMS

Situations classified as "mixture problems" often have the key data given in terms of their component parts. For instance, 20 lb of candy at $2 a pound is worth less than 15 lb at $3 a pound because the former, although greater in weight, has a total worth of $40, while the latter, smaller in weight, is worth more at $45. This judgment could not be made on the basis of either weight or price per pound alone, but only by considering the product of the factors.

Similarly, 100 cc of 60% alcohol solution contains more pure alcohol than 150 cc of 30% alcohol solution. This is so because

$$100 \text{ cc of } 60\% \text{ alcohol solution} = 0.6 \times 100 \text{ cc}$$

or 60 cc of pure alcohol. And

$$150 \text{ cc of } 30\% \text{ alcohol solution} = 0.3 \times 150 \text{ cc}$$

or 45 cc of pure alcohol.

As usual, in setting up the equivalence relationship and equation, it is necessary that both members of the equation refer to the same thing.

EXAMPLES

1. A merchant sells a mixture of olive oil and corn oil for salad dressing. If olive oil is priced at 32¢ a unit and corn oil at 28¢ a unit, how much of each should he use to set the price for 10 units of the mixture at $2.94?

SOLUTION Let x represent the number of units of corn oil.
Then $10 - x$ will represent the number of units of olive oil.

The equivalence relationship is

value of olive oil $+$ value of corn oil $=$ value of total

$$\left[\begin{array}{c} \text{units of} \\ \text{olive oil} \end{array} \times \begin{array}{c} \text{price per} \\ \text{unit} \end{array} \right] + \left[\begin{array}{c} \text{units of} \\ \text{corn oil} \end{array} \times \begin{array}{c} \text{price per} \\ \text{unit} \end{array} \right] = \$2.94$$

$$[(10 - x)(\$0.32)] \quad + \quad [(x)(\$0.28)] \quad = \$2.94$$

The corresponding equation is

$$(10 - x)(0.32) + (x)(0.28) = 2.94$$

$$3.2 - 0.32x + 0.28x = 2.94$$

$$-0.04x = -0.26$$

$$x = \frac{-0.26}{-0.04} = \frac{26}{4}$$

$$x = 6\frac{1}{2}$$

If $x = 6\frac{1}{2}$, then

$$10 - x = 10 - 6\frac{1}{2} = 3\frac{1}{2}$$

Solution:

$$\text{Units of corn oil: } 6\frac{1}{2}$$

$$\text{Units of olive oil: } 3\frac{1}{2}$$

Check:

$$6\frac{1}{2} \text{ units corn oil @ \$0.28} = \$1.82$$

$$3\frac{1}{2} \text{ units olive oil @ \$0.32} = \$1.12$$

$$\overline{10 \text{ units}} \qquad\qquad \overline{\$2.94}$$

2. A chemist has 15 oz of 4% acid solution. How much 20% acid solution must he add to bring the total solution up to 10% strength?

Note: In Example 1 above, the "total amount" is given; in this example the "total amount" is essentially unknown.

SOLUTION Let x represent the number of ounces of 20% solution to be added. Then (see Figure 5.4),

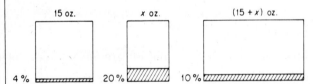

Figure 5.4

$15 + x$ represents the number of ounces in the resulting solution.
4% (or 0.04) of 15 is the number of ounces of pure acid in the original solution.
20% (or 0.2) of x is the number of ounces of pure acid that will be added when x ounces of 20% solution are added.
10% (or 0.1) of $(15 + x)$ is the number of ounces of pure acid in the resulting 10% solution.

The equivalence relationship is

pure acid in original solution + pure acid in added solution

$$= \text{pure acid in resulting solution}$$

The equation is

$$(0.04)(15) + (0.2)(x) = 0.1(15 + x)$$
$$0.6 + 0.2x = 1.5 + 0.1x$$
$$0.1x = 0.9$$
$$x = 9$$

Thus 9 oz of the 20% acid solution must be added.

Check: In the 15 oz of 4% solution there is 0.04×15 or 0.6 oz of pure acid.
In the 9 oz of 20% acid solution that is added, there is 0.20×9 or 1.8 oz of pure acid. This amounts to $0.6 + 1.8$ or 2.4 oz of pure acid in $15 + 9$ or 24 oz of solution. Then $2.4 \div 24 = 0.1 = 10\%$, the strength of the resulting solution. So the conditions of the problem are met by this solution of the equation.

3. A coin bank holds nickels, dimes, and quarters. If it contains three times as many dimes as nickels and two more quarters than dimes, and if the total value of its contents is $4.90, how many of each coin are there in the bank?

SOLUTION Let x represent the number of nickels.
Then $3x$ represents the number of dimes, and $3x + 2$ represents the number of quarters.

The equivalence relationship is

value of nickels + value of dimes + value of quarters = total value

$$\left(\begin{matrix} \text{no. of} \\ \text{nickels} \end{matrix} \times 5¢\right) + \left(\begin{matrix} \text{no. of} \\ \text{dimes} \end{matrix} \times 10¢\right) + \left(\begin{matrix} \text{no. of} \\ \text{quarters} \end{matrix} \times 25¢\right) = 490¢$$

The corresponding equation is

$$[(x)(5)] + [(3x)(10)] + [(3x + 2)(25)] = 490$$
$$5x + 30x + 75x + 50 = 490$$
$$110x = 440$$
$$x = 4$$

If $x = 4$, then $3x = 12$ and $3x + 2 = 14$; therefore,

nickels: 4
dimes: 12
quarters: 14

Check:

$$4 \text{ nickels } = \$0.20$$
$$12 \text{ dimes } = 1.20$$
$$14 \text{ quarters } = \underline{3.50}$$
$$\$4.90$$

Write an equation for each problem and solve.

1. A grocer has some cereal that sells for 50¢ a pound and some that sells for 80¢ a pound. To make a 45-lb blend of the two that will sell for 60¢ a pound, what proportions should he use?

2. The manager of a candy store wishes to put a mixture of hard candy on sale for 72¢ a pound. He prepared 60 lb, mixing candy that usually sells for 60¢ a pound with another kind that usually sells for 90¢ a pound. How many pounds of the cheaper candy did he use?

3. Two metals are used in an alloy. One costs 40¢ a gram and the other 65¢ a gram. How many grams of each should be used if 10 g of the alloy is to cost $5?

4. A vendor wishes to prepare 12 quarts of a blend of two juices for $1 a quart. One juice costs him 80¢ a quart and the other $1.40 a quart. How many quarts of each juice should he use?

5. Meat in a dog food preparation costs $1 a kilogram (kg), and cereal added to the meat costs 30¢ a kilogram. How much cereal should be added to 8 kg of meat to produce a mixture costing 70¢ a kilogram?

6. A candy dealer mixes chocolates that normally sell for $1.80 a pound with others that sell for $1.20 a pound. If he has 40 lb of the cheaper candy that he wants to use up in an assortment that will sell for $1.60 a pound, how many pounds of the expensive candy should he use?

7. A caterer has some dried fruit in her inventory that cost her $1 a pound. She decides to mix 9 lb of the fruit with some nuts for a snack. How many pounds of the nuts should she buy, at $1.60 per pound, if the total mixture is to cost $1.20 per pound?

8. A nut mixture of peanuts and cashews costs a merchant $1.60 a pound. If he pays $1.10 a pound for the peanuts and $1.80 a pound for the cashews, what percent peanuts must be in the mixture?

9. A druggist must make 20 oz of 12% argyrol from his supply of 5% and 15% solutions. How much of the 5% solution should he use?

10. Two alloys, one 1 part silver and 5 parts copper and the other 3 parts silver and 1 part copper, are mixed to form 350 lb of an alloy that is equal in silver and copper content. How many pounds of the first alloy should be used?

11. How much 24% butterfat cream must be mixed with 4% butterfat milk to make 10 gal of a lighter cream that is 20% butterfat?

12. A can of evaporated milk contains 0.2% fat, and a certain brand of whole milk has 3.9% fat. How much whole milk must be mixed with the contents of a 13-oz can of evaporated milk in order to make a 2.0% fat mixture?

13. How much water may be added to 1 liter of glycerin without reducing its strength by more than 20%? [*Hint:* Consider the strength of the glycerin as 100% and the strength of the added water as 0%.]

14. A nurse must make a 20% alcohol solution by diluting 10 oz of 25% solution. How much water must she add?

15. A 24-qt automobile radiator contains a mixture that is 25% alcohol and 75% water. How much must be drained off and replaced with an 85% alcohol solution so as to bring the alcoholic content in the radiator up to 30%?

16. A chemistry student has 50 cc of 30% sulfuric acid. How much of this solution must he remove and replace with pure sulfuric acid to bring the acid strength up to 58%?

17. A jar contains nickels and quarters that are worth $3.90. There are six less nickels than quarters. How many of each coin are in the jar?

18. The worth of a collection of ten coins—quarters and dimes—is $1.75. How many of each are there?

19. A boy had 12 coins, all quarters and dimes. If his dimes were quarters and his quarters were dimes, he would be 30¢ richer. How many of each coin has he?

20. Postage on a parcel amounted to $1.68. The sender used 14 stamps, all 8¢ and 15¢ types. How many of each stamp was used?

21. Sideline seats at a football game cost $4, and end-zone seats cost $1.50. One-third of the paying spectators bought end-zone seats and the rest sideline seats. If the total paid admission was $2090, how many of each kind of seat was sold?

22. Marie put two more 10¢ stamps on a package than 25¢ stamps and twice as many 25¢ stamps as 5¢ stamps. If the total postage was $2.45, how many of each kind of stamp did she use?

4. INVESTMENT PROBLEMS

When a person borrows money, he pays a rental called *interest* for the use of the money he has borrowed, just as one pays rent for the use of a house or apartment. When someone lends money, he receives this interest as compensation.

In general, interest is determined by the product of three factors:

$$\text{Interest} = \begin{bmatrix} \text{principal} \\ \text{(money invested} \\ \text{or borrowed)} \end{bmatrix} \times \begin{bmatrix} \text{rate} \\ \text{(\%, expressed} \\ \text{as a decimal)} \end{bmatrix} \times \begin{bmatrix} \text{time} \\ \text{(duration of} \\ \text{investment or loan)} \end{bmatrix}$$

By formula,

$$I = Prt$$

Thus the (simple*) interest on a loan of $600, at 8% interest rate, for $2\frac{1}{2}$ years, is

$$I = (600)(0.08)(2.5)$$

$$= 120 \text{ (dollars)}$$

* Another form of interest is compound interest, which includes interest on interest.

Countless business transactions each day revolve around the borrowing and lending of money, each with an agreed-upon rate of interest. Just as the comparison of distance to time produces a *rate* of speed—a measure of motion—the comparison of the rental or interest to the amount of money borrowed produces a *rate* of interest—a measure of the cost of borrowing money. From the lender's point of view, the rate of interest is a measure of the profit that his loaned money is producing.

Note: Although the situations are financial, problems of the type below are basically "mixture" problems, discussed in the previous section.

EXAMPLES

1. A man invested part of $4000 at 6% and the rest at 5%. If he expects a return of $210 a year on his investment, how much has he invested at each rate?

SOLUTION Let x represent the amount of money invested at 6%.
Then $4000 - x$ represents the amount of money invested at 5%.
6% of x is the return each year on 6% investments.
5% of $(4000 - x)$ is the return each year on 5% investments.

The equivalence relationship is

$$\begin{bmatrix} \text{interest on} \\ 6\% \text{ investment} \end{bmatrix} + \begin{bmatrix} \text{interest on} \\ 5\% \text{ investment} \end{bmatrix} = \begin{bmatrix} \text{total interest} \\ \text{return} \end{bmatrix}$$

The equation is

$$(0.06)(x) + (0.05)(4000 - x) = 210$$
$$0.06x + 200 - 0.05x = 210$$
$$0.01x = 10$$
$$x = 1000$$

If $x = 1000$, then $4000 - x = 3000$. The man has $1000 invested at 6% and $3000 invested at 5%.

Check: $1000 invested at 6% will produce an annual return of 0.06×1000, or $60; $3000 invested at 5% will produce an annual return of 0.05×3000, or $150. Together the two investments will produce an annual return of $60 + $150, or $210. So the conditions of the problem are met by this solution.

2. A woman pays $1440 interest annually for a sum of borrowed money. Part she borrows at 6%, and the balance, which exceeds the other part by $4000, costs her 8%. How much money has she borrowed at the 6% rate?

SOLUTION Let x represent the number of dollars borrowed at 6%.
Then $x + 4000$ represents the number of dollars borrowed at 8%.
6% of x is the cost of that borrowed at 6%.
8% of $(x + 4000)$ is the cost of that borrowed at 8%.

The equivalence relationship is

   cost of part at 6% + cost of part at 8% = total cost of borrowed money

The equation is

$$(0.06)(x) + (0.08)(x + 4000) = 1440$$
$$0.06x + 0.08x + 320 = 1440$$
$$0.14x = 1120$$
$$x = 8000$$

The woman has borrowed $8000 at the 6% rate.

Check: $8000 borrowed at 6% costs 0.06 × 8000, or $480, per year; $4000 more, or $12,000, borrowed at 8% costs 0.08 × 12,000, or $960, per year. Together these two loans cost a total of $480 + $960, or $1440. So the conditions of the problem are met by this solution.

EXERCISE 5-3

Write an equation for each problem and solve.

1. Mr. Wallace has part of his $7000 in a bank at 4% and the rest in bonds at 7%. If his annual return is $400, how much has he invested at each rate?

2. An investor can borrow money on her credit rating at 6% and can invest it at $7\frac{3}{4}$ %. How much should she borrow to realize a clear profit of $875 a year?

3. Mr. Walton has three times as much money in a savings account at 4% as he does in a stock that pays 7%. If his annual return from both investments is $456, how much has he invested in the savings account?

4. A man had invested $3 in a safe 4% government bond for every $1 he had invested in a speculative stock that paid him 7%. If his annual return on these investments was $209, how much must he have had invested in the stock?

5. Part of a $10,000 trust fund is invested at 6% and the balance at 10%. The return from these two investments is the equivalent of $8\frac{1}{2}$ % on the total investment. What amount is invested at 6%?

6. There is $2800 invested at one rate and $4600 at another rate that is 1% more than the former. If the total return for both investments is $342, what are the two rates?

7. A widow invested $7000, part at 6% and part at 8%. The income from both investments was the same. How much had she invested at each rate?

8. An investor has $8000 invested at 8%, and has an opportunity to invest more at 10%. How much must she invest at the 10% rate to bring her overall return up to $9\frac{1}{2}$%?

9. Mr. Miller has invested $6000 at one rate of interest and another $4000 at half that rate. If his annual return is $360, what are the two rates?

10. Mrs. Carey has $9000 invested in various bonds. She has twice as much invested in a 9% issue as she does in a 6% issue; the balance is invested in a 10% issue. How much money has she invested in each of these bond issues if her annual return is $840?

11. A man pays $955 annually for borrowed money for which his house is used as security. He borrowed part through a first mortgage at $5\frac{1}{2}$% and the rest, which is $6000 less than the first part, by means of a second mortgage at 7%. How much did he borrow on the first mortgage?

12. For every two dollars invested in a company at 5%, the investor is allowed to invest another dollar in a 8% investment. If a person desires a yield of $900 per year from such an investment arrangement, how much should be invested altogether?

13. Mr. Jones, years ago, had invested part of $6000 at 3% and the remainder at 5%. A reversal of these investments would have resulted in a return of $48 less. How much did he invest at 3%?

14. A man invested $8000, part at 4% and the rest at 6%. Had he switched these investments, he would have earned $60 more. What amounts did he have invested at these two rates?

15. A borrower receives a loan of $6000 at 9% annual simple interest. However, the lender immediately subtracts 2% from the funds as service charge. If the borrower pays the loan off in one year, what true annual rate of interest did she pay for the cash she actually received?

16. A total of $7000 is invested and yields $430 per year. Some is invested at 5%, and the rest is invested in equal parts at 6% and 8%. How much is invested at each rate?

5. UNIFORM-MOTION PROBLEMS

Basic to all uniform-motion problems is the relationship

$$\text{distance} = (\text{rate of speed}) \cdot (\text{time}) \quad \text{or} \quad d = rt$$

From it we can derive the associated relationships

$$\text{rate} = \text{distance} \div \text{time}, \quad r = \frac{d}{t}$$

and

$$\text{time} = \text{distance} \div \text{rate}, \quad t = \frac{d}{r}$$

When seeking an equivalence relationship that will lead to an equation for solving a uniform-motion problem, keep in mind that the members of the equation must represent equal measures of the same kind, such as equal distances, equal times, or equal rates.

EXAMPLES

1. An airplane flies 980 mi in $3\frac{1}{2}$ hr against a steady headwind blowing at 20 mph. How fast would the airplane be flying in still air?

SOLUTION Let x represent the rate of the airplane in still air; then $x - 20$ represents its rate against the wind; and $980 \div 3\frac{1}{2}$ $\left(\text{i.e., } \frac{d}{t} = r\right)$ represents its rate during flight. The equivalence relationship is:

$$\text{rate} = \text{rate}$$

The equation is:

$$x - 20 = \frac{980}{3\frac{1}{2}}$$

$$x - 20 = 280$$

$$x = 300$$

The airplane's rate of speed in still air would be 300 mph.

Check: The rate in still air (300 mph) reduced by the rate of the headwind (20 mph) implies the airplane can make good only 280 mph. Flying at that speed for $3\frac{1}{2}$ hr, it would be able to travel $3\frac{1}{2} \times 280$, or 980 mi. So the conditions of the problem are met by this solution.

2. A stream flows at the rate of 4 mph. A boat takes 3 hr to make a trip to a certain town downstream, but it takes 7 hr to return because of the adverse current. What would be the speed of the boat in still water?

SOLUTION Let x represent the speed of the boat in still water.
Then $x + 4$ represents the speed of the boat going downstream, and $x - 4$ represents the speed of the boat going upstream.

The basic equivalence for this problem will be in terms of distance:

$$\text{distance down} = \text{distance back}$$

$$(\text{rate down}) \cdot (\text{time down}) = (\text{rate back}) \cdot (\text{time back})$$

$$(x + 4) \quad \cdot \quad (3 \text{ hr}) \quad = \quad (x - 4) \quad \cdot \quad (7 \text{ hr})$$

Chapter 5 APPLICATIONS 149

The corresponding equation is:

$$3(x + 4) = 7(x - 4)$$
$$3x + 12 = 7x - 28$$
$$40 = 4x$$
$$10 = x$$

The speed of the boat in still water would be 10 mph.

Check: If the boat moves at the rate of 10 mph in still water, with the current it travels at $10 + 4$, or 14 mph. A 3-hr trip at this rate must be $3 \cdot 14$, or 42 mi long. Coming back, against the current, the boat travels at only $10 - 4$, or 6 mph. At this rate, a 7-hr trip must be $7 \cdot 6$, or 42 mi long. Hence

$$\text{distance down (42 mi)} = \text{distance back (42 mi)}$$

and the solution of the equation satisfies the problem.

3. A truck leaves the highway terminal and heads for New York, averaging 40 mph. A mistake in shipping orders is noted, and 24 min after the truck leaves, a car is dispatched to overtake the truck. If the car averages 50 mph, how long will it take it to catch up with the truck?

SOLUTION Let x represent the time in hours that the car travels.

Then $x + \dfrac{24}{60}$ represents the time in hours that the truck travels.

$40\left(x + \dfrac{24}{60}\right)$ represents the distance $(r \cdot t)$ that the truck travels.

$50(x)$ represents the distance $(r \cdot t)$ that the car travels.

	r $\cdot$	t $=$	d
Truck	40	$x + \dfrac{24}{60}$	$40\left(x + \dfrac{24}{60}\right)$
Car	50	x	$50(x)$

distance for car = distance for truck

The equation is

$$50x = 40\left(x + \frac{24}{60}\right)$$
$$50x = 40x + 16$$
$$10x = 16$$
$$x = 1.6$$

It will take the car 1.6 hr, or 1 hr 36 min, to overtake the truck.

Check: When the car overtakes the truck, the truck will have been traveling for 1 hr 36 min + 24 min, or 2 hr, exactly. In that time, at 40 mph, the truck would have traveled 80 mi. To travel those 80 mi at 50 mph, the car would have to travel for 80 ÷ 50, or 1.6 hr. So the conditions of the problem are met by this solution.

EXERCISE 5-4

Write an equation for each problem and solve.

1. A boat travels 63 mi upstream in $4\frac{1}{2}$ hr against a steady current of 4 mph. How fast would the boat be moving if there were no current?

2. An airplane flies 560 mi in $2\frac{1}{2}$ hr, aided by a tailwind of 30 mph. How fast would the plane have flown in still air?

3. One cyclist pedals at a rate of 6 mph faster than another. At these speeds, one covers a certain distance in 3 hr and the other in 4 hr. What is the average rate of speed of the faster cyclist?

4. Two men are jogging. The faster man jogs a distance of 1000 yd in the same time that the slower one jogs 900 yd at a pace 20 yd per minute slower than that of the faster man. At what rate is the slower man jogging?

5. Two automobiles leave towns 470 mi apart at the same time. They travel toward each other, one at a rate that is 20 mph faster than the other. If they meet in 5 hr, what is the speed of the faster automobile?

6. Two airplanes pass each other, going in opposite directions, at 0602 on the clock. One is traveling at a speed of 270 mph and the other at 210 mph. At what time will they be 100 mi apart?

7. A speeding automobile is going 55 mph, and a highway trooper, 2 mi behind, is chasing it at a speed of 70 mph. How long will it take the trooper to overtake the speeding car if both maintain these speeds?

8. A boy rides his bike 6 mi from his home to the bus stop at the rate of 9 mph. He arrives just in time to catch the city bus, which averages 30 mph. If the boy spent 1 hr and 28 min traveling from home to city, how far did he travel by bus?

9. Abel ran a 440-yd race in 59.2 sec, and Baker came in second with a time of 60.4 sec. Assuming that each runner ran his race at a uniform speed throughout the race, how far back was Baker when Abel crossed the finish line?

10. A pair of hikers, 18 mi apart, begin at the same time to hike toward each other. If one walks at a rate that is 1 mph faster than the other and if they meet 2 hr later, how fast is the slower hiker traveling?

11. A driver averaged 50 mph on the turnpike and 30 mph on secondary roads. If a trip of 185 mi over turnpike and secondary roads took 4 hr and 30 min, how many miles were over the turnpike?

12. An automobile 19 ft long overtakes a 25-ft-long truck traveling at 35 mph. At what rate must the automobile move in order to pass the truck completely in 3 sec?

13. A man drives the first mile of a 2-mi stretch at an average speed of 30 mph. What speed must he average in the second mile in order to average 40 mph overall?

14. At how many minutes after 4 P.M. will the minute hand of the clock be even with the hour hand?

15. How long after 12:30 will it take the two hands of a clock to overlap?

16. A motorist travels for 2 hr at a constant rate of speed, then averages 10 mph more for the next 3 hr. If in the 5 hr she drives a distance of 245 mi, what must have been her rate of speed during the first 2 hr?

17. A family cruises upstream against the current for 3 hr to a picnic spot. After 2 hr of the return trip, with the current, they pass a landmark 5 mi above their point of departure. If the boat's speed throughout the trip would have been at the rate of 15 mph in still water, what must have been the speed of the current?

18. Two cyclists start at 2 P.M. from points 28 mi apart, and pedal toward each other. If they meet at 4 P.M., and one traveled at $\frac{3}{4}$ the speed of the other, what were their respective speeds?

6. WORK PROBLEMS

Fractions play an important part in a category of problems called work problems. They are similar to motion problems in that they involve rates:

$$\text{rate of work} = \frac{\text{amount of work}}{\text{time}}$$

or

$$(\text{rate of work}) \cdot (\text{time}) = \text{amount of work}$$

For example, if a mechanic can complete a "job" in 4 hr, then his rate of work is $\frac{1}{4}$ "job" per hour.

Often the problems involve two or more workers, each having different rates of work, who combine their efforts to complete a job. In such cases, the basic equivalence relationship is

$$\begin{bmatrix} \text{amount of} \\ \text{work by A} \end{bmatrix} + \begin{bmatrix} \text{amount of} \\ \text{work by B} \end{bmatrix} = \begin{bmatrix} \text{amount of work} \\ \text{done together} \end{bmatrix}$$

1. A can paint a billboard in 4 hr, but B can do it in 3 hr. How long will it take them, working together, to paint the billboard?

Let x = time it takes to paint the billboard together.
Then in 1 hr:

A can paint $\dfrac{1}{4}$ of the billboard.

B can paint $\dfrac{1}{3}$ of the billboard.

Together they will get $\dfrac{1}{x}$ of the billboard painted each hour.

The equation and its solution:

$$\frac{1}{4} + \frac{1}{3} = \frac{1}{x}$$

LCD: $12x$

$$3x + 4x = 12$$

$$7x = 12$$

$$x = \frac{12}{7}$$

$$x = 1\frac{5}{7}$$

Together A and B can paint the billboard in $1\dfrac{5}{7}$ hr.

2. It takes 12 hr to fill a swimming pool with a garden hose but less time with a pump. If it takes 4 hr to fill the pool using both the garden hose and the pump, how long would it take to fill the pool by using the pump alone?

Let x = time to fill the pool using the pump alone.
Then, in 1 hr, this much of the pool will be filled:

By hose: $\dfrac{1}{12}$

By pump: $\dfrac{1}{x}$

By both: $\dfrac{1}{4}$

Equation:

$$\frac{1}{12} + \frac{1}{x} = \frac{1}{4}$$

LCD: $12x$

$$x + 12 = 3x$$

$$2x = 12$$

$$x = 6 \quad \text{(hours)}$$

3. Water from one of two taps will fill a tub in 12 min, the other in 8 min. The drain will empty the full tub in 6 min. If both taps are turned on and the drain is left open, how long will it take before the tub overflows?

SOLUTION   Let x = time to fill the tub to overflowing. In 1 min one tap will fill $\frac{1}{12}$ of the tub, the other $\frac{1}{8}$, and the open drain will let out $\frac{1}{6}$ tub; the net gain will be $\frac{1}{x}$ tubful. Thus

$$\underbrace{\frac{1}{12} + \frac{1}{8}}_{\text{in}} \underbrace{- \frac{1}{6}}_{\text{out}} = \underbrace{\frac{1}{x}}_{\text{net}}$$

LCD: $24x$

$$2x + 3x - 4x = 24$$

$$x = 24 \quad \text{(minutes)}$$

EXERCISE 5-5

Write an equation for each problem and solve.

1. How long will it take two boys to wax a car if one can do the job alone in 3 hr and the other in 2 hr?

2. A laborer can paint 100 yd of fence in 7 hr, but a painter can do it in 5 hr. How long will it take them, working together, to paint the fence?

3. A young woman can mow the lawn in 30 min, but it takes her little brother 45 min. If they use two mowers and work together, how quickly can they mow the lawn?

4. A full tank can be drained by means of a small pipe in 20 min, and through a larger pipe in 16 min. How long will it take to drain the tank with both pipes opened?

5. Two roofers, working together, finish a roof in 12 hr. Had one done the job alone it would have taken him 18 hr. How long would it have taken the other roofer, working alone?

6. A new copier can produce a set of tests in 8 min. An old copier takes longer. If both copiers are put to work, the job can be done in 5 min. How long would it have taken the old copier to produce the tests?

7. Three typists work on a manuscript. If each were to type it alone, they would take 10, 12, and 15 hr. How long should it take them to complete the manuscript if they work together?

8. Pipe A can fill a tank in 32 min, pipe B in 36 min, and pipe C in 16 min. How long will it take to fill the tank if all three pipes are used?

9. Tom can line the football field in 40 min. Bill helps Tom and they complete the job in 25 min. How long would it have taken Bill to do the job alone?

10. Two mechanics complete a motor job in 4 hr. Working alone, it would have taken one mechanic 6 hr to do the job. How long would it have taken the other one if working alone?

11. When the outflow pipe is open, a full tank will drain in $1\frac{1}{2}$ hr. That same tank can be filled in 2 hr through one pipe and in 3 hr through another. An employee opens up both input pipes but forgets to close the outflow. Will the tank ever fill? If so, in how many hours?

12. Same situation as Exercise 11 except that the input pipes will fill the tank in 3 hr and 4 hr, respectively. Will the tank ever fill? If so, in how many hours?

13. Ann can sort a bag of mail in 36 min and Betty in 24 min. Ann starts sorting and works for 9 min before Betty begins to help. How long will it take them to complete the sorting? $\left[Hint: \frac{9}{36}, \text{ or } \frac{1}{4}, \text{ of the job is completed when Betty starts; so } \frac{3}{4} \text{ of the job is left.} \right]$

14. A tank can be filled in 18 min through pipe A and in 15 min through pipe B. If a man starts filling the tank through pipe A and after 6 min turns on pipe B, how long will it take the two pipes to fill the tank completely?

15. A press, which can print 300 copies a minute, starts on a run of 22,500 copies. After 5 min a second press is turned on, and together they complete the run in another 50 min. At what speed must the second press have run?

16. After 2 hr of typing, a job is half done. A second typist is brought in, and together the two complete the job in another 45 min. How long would it have taken the second typist to do the entire job?

17. Two clerks begin to process 200 pieces of mail. One can do 15 pieces a minute, and the other 10 pieces a minute. After 5 min the slower clerk quits. How long will it take the faster clerk to complete the job?

18. Alone, machine F can complete a job in 24 min, and machine S in 36 min. Both machines start to work on a run together, but after 8 min, machine F breaks down, and the job is completed by machine S. How long did it take to complete the whole job?

7. INTEGER AND DIGIT PROBLEMS★

One common category of algebra problems relates to integer sequences and to the digits of a standard Hindu–Arabic numeral. Although the relevance of these problems is more historical than practical, they do provide a useful experience in translating worded statements into mathematical symbolism.

By consecutive integers we mean a set of integers arranged in the same order as would occur in counting. Examples of sets of consecutive integers are 5, 6, 7, 8; or 18, 19; or 126, 127, 128, 129, 130. Since each integer in sequence differs from the previous integer by 1, such a sequence can be represented algebraically as

$$n, n + 1, n + 2, n + 3, \ldots$$

where n is the smallest of the integers under consideration.

Each integer can be further classified as being odd or even. Even integers are those that are exactly divisible by 2, whereas odd integers always have a remainder of 1 when divided by 2. The nonnegative even integers are 0, 2, 4, 6, $\ldots$, and the nonnegative odd integers are 1, 3, 5, 7, $\ldots$. In both cases, one integer differs from the next by 2; so both sequences can be expressed algebraically as

$$n, n + 2, n + 4, n + 6, \text{etc.}$$

where n represents the smallest odd or even integer, as the case may be.

The numerical aspect of integer problems usually refers to the way we express the integers in the decimal (tens) system of number notation. For instance, the numeral for eighty-six is 86, which implies 8 tens + 6 ones, or $(8 \cdot 10) + (6 \cdot 1)$.

If we reverse the order of the two digits to 68, we have a numeral that represents a number 18 less than the other, for 68 implies 6 tens + 8 ones, or $(6 \cdot 10) + (8 \cdot 1)$.

EXAMPLES

1. Given three consecutive integers. If the sum of the two smaller integers is decreased by three times the largest, the result is -37. What are the integers?

SOLUTION   Following the pattern of Program **5.1**:

Step 1: Three consecutive integers are sought. Let x represent the smallest of them.

Step 2: Equivalence relationship:

$$\text{integer}_1 + \text{integer}_2 - 3(\text{integer}_3) = -37$$

Step 3: If the smallest integer is represented by x, the next consecutive integer will be represented by one greater than x, or

$$x + 1$$

★ The content of this section, in more basic courses of study, may be postponed without significant effect upon the remaining parts of the book.

The integer after the second will be two greater than the smallest, or one greater than the second integer. So $(x) + 2$, or $(x + 1) + 1$, or $(x + 2)$ represents the third integer:

$$\text{integer}_1 + \text{integer}_2 - 3(\text{integer}_3) = -37$$
$$x \quad + (x + 1) - \quad 3(x + 2) \ = -37$$

Step 4: Solve the equation of Step 3:

$$x + (x + 1) - 3(x + 2) = -37$$
$$x + x + 1 - 3x - 6 = -37$$
$$x + x - 3x = -37 + 6 - 1$$
$$x = 32$$

If $x = 32$, then $x + 1 = 33$ and $x + 2 = 34$. Therefore, integer$_1$ is 32, integer$_2$ is 33, and integer$_3$ is 34.

Step 5: The sum of the two smaller integers is $32 + 33$ or 65; if 65 is diminished by three times the larger ($3 \times 34 = 102$), the result should be -37: $65 - 102 = -37$. The solution checks in the problem.

2. There are three consecutive odd integers whose sum is five more than twice the largest. What are the integers?

SOLUTION Let x represent the smallest odd integer.
Then
$x + 2$ represents the middle odd integer,
$x + 4$ represents the largest odd integer, and
$2(x + 4) + 5$ represents five more than twice the largest odd integer.

The equivalence relationship is

$$\text{integer}_1 + \text{integer}_2 + \text{integer}_3 = \text{twice the largest, increased by 5}$$

The equation is

$$(x) + (x + 2) + (x + 4) = 2(x + 4) + 5$$
$$x + x + 2 + x + 4 = 2x + 8 + 5$$
$$3x + 6 = 2x + 13$$
$$3x - 2x = 13 - 6$$
$$x = 7$$

If $x = 7$, then $x + 2 = 9$ and $x + 4 = 11$. The integers are 7, 9, 11.

Check: 7, 9, 11 are three consecutive odd integers. Their sum is 27, which *is* 5 more than twice the largest, 2×11, or 22. So the conditions of the problem are met by this solution.

3. The digits of a two-digit numeral are consecutive integers. The number expressed when these two digits are reversed differs from the original by an amount equal to the sum of the digits. What are the digits?

SOLUTION Let x represent the smaller of the two consecutive digits.

Then $x + 1$ represents the larger of the two consecutive digits,

$(x + 1)$ tens $+ (x)$ ones represents the larger of the two numbers expressed by these digits,

(x) tens $+ (x + 1)$ ones represents the smaller of the two numbers expressed by these digits, and

$(x + 1) + x$ represents the sum of the two digits.

The equivalence relationship is

$$\text{larger number} - \text{smaller number} = \text{sum of digits of either}$$

$$[(x + 1) \text{ tens} + (x) \text{ ones}] - [(x) \text{ tens} + (x + 1) \text{ ones}] = (x + 1) + x$$

The equation is

$$[10(x + 1) + 1(x)] - [10(x) + 1(x + 1)] = (x + 1) + x$$
$$10x + 10 + x - 10x - x - 1 = x + 1 + x$$
$$10 - 1 = 2x + 1$$
$$8 = 2x$$
$$4 = x$$

If $x = 4$, then $x + 1 = 5$. The smaller digit is 4 and the larger is 5.

Check: The two integers that have the digits 4 and 5 in their standard numeral are 45 and 54. Their difference is $54 - 45$, or 9, which is also the sum of the digits. So the conditions of the problem are met by this solution.

Here are the data of this problem in table form.

Smaller digit	Larger digit	In words	In symbols
		Larger as tens digit	$10(x + 1) + 1(x)$
x	$x + 1$	Smaller as tens digit	$10(x) + 1(x + 1)$
		Sum of the two digits	$(x) + (x + 1)$

EXERCISE 5-6

Write an equation for each problem and solve.

1. The sum of three consecutive integers is 39. What are the integers?

2. What four consecutive integers have a sum of 2?

3. The sum of three consecutive even integers is 54. What are the integers?

4. The sum of eight consecutive odd integers is 112. What are the integers?

5. The two digits of a numeral differ by 4. The sum of the digits is 10. What number is represented by the numeral?

6. In a three-digit numeral, two of the digits are the same and the third is 3 larger. If the sum of the digits is 15, what is the largest number that can be expressed by the digits?

7. The sum of the digits of a two-digit numeral is 6. When the larger digit is used as the tens digit, the number represented is 18 more than when it is used as the ones digit. What are the digits?

8. The tens digit of a two-digit numeral is 2 more than twice the other digit. If the sum of these digits is 8, what is the two-digit numeral?

9. The sum of the digits of a three-digit Hindu–Arabic numeral is 12. Its hundreds digit is 3 greater than the tens digit and 3 less than the ones digit. What is the numeral?

10. An integer has three consecutive digits in its numeral, in ascending order. Eight times the hundreds digit added to seven times the ones digit yields a sum that is 190 less than the original integer. What is the integer?

11. What four consecutive multiples of 3 have a sum of 102?

12. Find four consecutive odd integers such that the sum of the first three, when subtracted from three times the fourth, leaves a remainder of 12.

13. Four consecutive integers differ, one from the next, by 3. Their sum is 10. Name the integers.

14. Name four consecutive integers that differ one from the next by 4 if their sum is -8.

15. Three consecutive even integers are such that three times the middle one, less twice the smallest, is 2 more than the largest. Name the integers.

16. If 3 is added to the smallest of a set of three consecutive odd integers and 8 is subtracted from the largest, their sum changes to 34. Name the original integers.

17. An even integer and the previous three consecutive odd integers add to 47. Which are they?

18. An odd integer and the next three consecutive even integers have 29 for a sum. Which are they?

8. PROBLEMS INVOLVING INEQUALITIES

For some problems, the appropriate mathematical model is not an equation but an inequality. As with equations, the approach consists of two main parts.

1. Translate the problem situation into an algebraic sentence—an inequality.
2. Solve the inequality by algebraic techniques (e.g., the methods discussed in Chapter 4).

The normal procedure is essentially that of Program **5.1** except that the mathematical sentence expressing the relationship between knowns and unknown is an inequality rather than an equation.

PROGRAM 5.2

To solve a problem expressed by an inequality:

Step 1:   Determine from the statement of the problem that which is sought—the unknown—and represent it by a letter symbol.

Step 2:   Involve the known and the unknown of the problem in an inequality.

Step 3:   Solve the inequality of Step 2 for the variable.

Step 4:   Check by interpreting the solutions obtained in Step 3 against the statement of the problem.

EXAMPLES

1.   A board is 12 ft long. It is to be cut into two pieces so that the longer piece is always greater than twice the shorter piece. How long should the longer piece be?

SOLUTION   *Step 1:* Associate a letter (variable) with the unknown. Let $x =$ length of the longer piece.

Step 2: The length of the shorter piece must be $12 - x$. Since the longer piece must always be more than twice the shorter piece, the following inequality represents the problem situation:

$$x > 2(12 - x)$$

Step 3: Solve the inequality:

$$x > 2(12 - x)$$
$$x > 24 - 2x$$
$$3x > 24$$
$$x > 8$$

Step 4: The solution of Step 3 implies that the longer piece must always be greater than 8 ft. Some sample situations are shown in tabular form.

Longer (ft)	Shorter (ft)	Twice shorter (ft)	Is longer > 2 × shorter?
8	4	8	No; =
$8\frac{1}{4}$	$3\frac{3}{4}$	$7\frac{1}{2}$	Yes
9	3	6	Yes
10	2	4	Yes
11	1	2	Yes
12	0	0	Yes

2. The sum of four consecutive integers must be 34 or less. Describe the sets of numbers that meet this condition.

SOLUTION *Step 1:* Let x = largest possible integer.

Step 2: The next three smaller consecutive integers must be $x - 1$, $x - 2$, and $x - 3$. The inequality:

$$(x) + (x - 1) + (x - 2) + (x - 3) \leq 34$$

Note: " $\leq$ " allows for the sum to be 34 or less.

Step 3: Solving,

$$x + x - 1 + x - 2 + x - 3 \leq 34$$
$$4x - 6 \leq 34$$
$$4x \leq 40$$
$$x \leq 10$$

Step 4: Check solutions.
Four consecutive integers, the largest of which is 10 or less:

$$10, 9, 8, 7; \text{ sum } 34$$
$$9, 8, 7, 6; \text{ sum } 30$$
$$8, 7, 6, 5; \text{ sum } 26$$
$$\text{etc.}$$

3. A son, 8 years old, and his father, 28 years old, have birthdays on the same day of the year. Beyond what age will the son be more than half his father's age?

SOLUTION Let x = the number of years from now.
Then
$8 + x$ = the son's age x years from now; and
$28 + x$ = the father's age x years from now.
The inequality:

$$8 + x > \frac{1}{2}(28 + x)$$

The solution:

$$2(8 + x) > 28 + x$$
$$16 + 2x > 28 + x$$
$$x > 12$$

Interpretation: After 12 years from now (when he is 20) the son will be more than half the age of his father (who will then be 40).

Write an inequality for each problem and then use it to solve the problem.

1. Two integers differ by 4, and their sum is less than 16. Describe the set of possible pairs of numbers.

2. Two integers differ by 3, and their sum is greater than 21. Describe the set of possible pairs.

3. One number is twice the other number when the other number is increased by 4. The sum of the two numbers is always less than −5. Describe the set of possible larger numbers.

4. One number is twice the other, and their sum is never less than 32. Describe the set of possible smaller numbers.

5. If three times a number is diminished by 3, the result is greater than 15. Describe the set of numbers that meet this condition.

6. The sum of three consecutive odd integers is to be less than 21. What is the largest possible integer?

7. The sum of three consecutive odd integers is to be held between, but not including, 3 and 27. Which sets of integers qualify? [*Hint:* Work as two inequalities; then combine solutions to meet the requirements of the problem.]

8. Three consecutive integers differ one from the other by 3. Their sum is to be at least 9 but not greater than 15. Name the sets that qualify.

9. A father is now 30 years old. In 10 years he will still be more than twice his daughter's age. How old might the daughter be now?

10. According to a prior agreement, a broker is to receive at least $50 but not more than half of what his client receives of a forfeited deposit, depending on negotiations. How much might the broker receive of a forfeited deposit of $270?

11. Two sides of a triangle are to be equal and the third side is 17 meters (m) long. If the perimeter of the triangle is to be no greater than 33 m: (a) How long might each of the two equal sides be? (b) Is the triangle possible?

12. Ann was 6 years old when her brother was born. For how long will the brother be less than $\frac{2}{3}$ Ann's age?

13. Some tourists decide to drive up to a scenic point 8 mi off the highway. If they average 24 mph going up and wish to keep the driving time up and back to under 32 min, what rates of speed may they use coming down?

14. An investor has $12,000 to invest. He can invest $8000 of it at 7%. At what rate must he invest the rest if his annual income from the two investments is to exceed $760?

15. Two children are on a seesaw. One weighs 80 lb and sits 4 ft from the fulcrum. The other weighs 60 lb. How far from the fulcrum must the lighter child sit so as to keep the heavier child up in the air?

16. Two campers plan to start out at opposite ends of a 14-mi trail at the same time and hike toward each other. One will hike at a steady pace and the other at 1 mph, or more, faster. What is the greatest distance that the slower hiker will have to cover if they plan to meet 2 hr after starting?

17. A trustee has $12,000 to invest. She invests $5000 at a 7% return. If she wants to average more than $8\frac{1}{2}$% on the total, what interest rate must she consider before investing the rest?

18. A caterer decides to mix some raisins and nuts for snacks. Raisins will cost her $1.20 a pound and nuts 75¢ a pound. How many pounds of nuts should she use to keep the cost of the final mix below 90¢ a pound?

REVIEW

SOLVE

1. The perimeter of a rectangle is 148 in., and the length of one side is 1 in. short of being $1\frac{1}{2}$ times the length of the other side. What are the dimensions of the rectangle?

2. A circular "pie" graph is to be divided into three sectors. One sector is to have twice the central angle of the first and 40° less than the central angle of the third. Find the measures of the three central angles.

3. Find the sizes of the angles of a triangle if one is 19° larger than the middle-sized angle, and the third is 19° less than that angle.

4. A line 7 ft long is to be cut into three pieces. If one piece is to be twice the length of another and 4 in. shorter than the third, what should be the lengths of the three pieces of line?

5. An isosceles (two legs equal) triangle has a base which is $\frac{3}{4}$ the length of one of its legs. If each of the legs is increased by 4 in., and the base by 4 in., the perimeter of the new triangle is 56 in. Find the dimensions of the original triangle.

6. When opposite sides of a square are increased by 7 ft, and the other pair of opposite sides are decreased by 2 ft, the result is a rectangle whose area is 31 sq ft more than that of the square. What is the length of side of the square?

7. Two kinds of nuts are in a mixture. One costs $2 a kilogram and the other $3 a kilogram. Seven kilograms of the cheaper kind is stirred in with the more expensive kind to make up a mixture at a cost of $2.30 a kilogram. How much of the more expensive nut was used?

8. A confectioner has some candy that normally sells for $1.90 a pound and some that sells for 70¢ a pound. How much of the expensive candy must he use to produce 20 lb of a mixture to sell for $1 a pound?

9. What volumes of 10% and 4% solutions must be mixed together to yield 81 cc of 6% solution?

10. A basketball team scored 86 points in a game in which they scored 16 more field goals than foul shots. (Field goal = 2 points; foul shot = 1 point.) How many field goals were scored?

11. An 8-gal tank contains a 40% salt solution. How much of the solution must be drawn off and replaced with pure water in order to bring the salt concentration down to 25%?

12. Thirty-six ounces of 1 part gold, 8 parts copper alloy are to be mixed with an alloy containing 5 parts gold, 3 parts copper to make an alloy containing 7 parts gold and 5 parts copper. How much of the second alloy must be used?

13. An investor can borrow money on her good credit at $4\frac{1}{4}$% and can invest these same funds at $5\frac{3}{4}$% elsewhere. If she earned $600 that way last year, how much did she borrow?

14. A man has $10,000 to invest. He invests $3000 at 6% and $2500 at $4\frac{1}{2}$%. What rate should he seek in investing the balance so that his total return is $585 annually?

15. Mr. Wilkins has $42,000 invested in two real estate holdings which pay him 8% and 11%. What must be the amount invested in the 11% property if his total yield is $3690?

16. Several years ago, a man borrowed money on his house at 5% and a quarter of that amount on his furniture at 6%. If his total interest charge was $234 a year, how much must he have borrowed against the furniture?

17. Mr. Allen had part of $40,000 invested at 4% and the rest at 6%. Had he switched these investments, his yield would have been $400 more. How much must he have had invested at 6%?

18. Two men each invested $8000. The first invested part at 4% and the balance at 5%. The second invested $1000 more at $2\frac{1}{2}$% than the first did at 4%, and the rest at 10%. If both realized the same return, how much must the first man have had invested at 4%?

19. Two hikers, Al and Pete, are carrying loads of 60 lb and 70 lb, respectively. How many pounds must be removed from Al's load and added to Pete's so that Al will then be carrying a load that is just $\frac{3}{4}$ of Pete's?

20. Two automobiles leave towns 333 mi apart and travel toward each other, one averaging 10 mph faster than the other. If they meet in $4\frac{1}{2}$ hr, what must have been the average speed of the slower automobile?

21. A driver averaged 52 mph on one leg of a 206-mi trip, and 40 mph on the second. If the whole trip took $4\frac{1}{4}$ hr, how many miles did she cover at the faster speed?

22. A room can be painted by a fast workman in 12 hr, and by a slow workman in 16 hr. How long would it take a slow and a fast workman to paint the room working together?

23. A little girl runs to the store at the rate of 9 mph and walks back at the rate of 3 mph. If her total traveling time is 10 min, how far away is the store?

24. A passenger train 325 ft long overtakes a freight train 555 ft long and traveling at a speed of 32 mph on a parallel track. If it takes exactly 1 min for the passenger train to pass the freight train completely, how fast must the passenger train be moving?

25. When the hands of a clock register 5:15, how long will it be before the hands are together?

26. Promptly at 8 A.M., two trucks leave from terminals that are 175 mi apart, and drive toward each other. If the average speed of one is $\frac{2}{3}$ the other, and they meet at 10:30 A.M., what must have been the average speeds?

27. Alf can paint a barn in 10 hr and Mabel can do it in 14 hr. How long will it take them to paint the barn if they work together?

28. Two typists can finish a job in 36 min. The faster one could have done it herself in 60 min. How long would it have taken the slower one to do the complete job?

29. A tank can be drained by pipe A in 20 min, by pipe B in 30 min, and filled by pipe C in 12 min. If all three pipes are opened, will the tank overflow? If so, in how long?

30. Same situation as Problem 29, except that pipe B is shut. Will the tank overflow? If so, in how long?

31. One roofer can complete a roof in a line of townhouses in 16 hr. It takes his helper 20 hr to do the same job. The roofer starts the job, works for 4 hr, and is joined by the helper. How long should it take them to finish the roof?

32. Same situation as Problem 31, except that after the roofer is joined by the slower helper it takes the two of them 8 hr to complete the roof. How long would it have taken the helper to do the whole roof himself?

33. Name three consecutive integers that differ one from the next by 4, and have a sum of 9.

34. Seven consecutive integers add to 728. Which are they?

35. A numeral has two digits. The tens digit is $2\frac{1}{2}$ times the size of the ones digit, and the sum of the digits is 7. What is the numeral?

36. Reversing the digits, which differ by 2, of a two-digit numeral and subtracting the number it represents from that of the first yields a difference of 18. What are the digits?

37. Five integers, in ordered sequence, differ one from the next by 7. The sum of the four smallest is 10 less than the largest. What are the integers?

38. Two different numbers can be represented by a pair of digits that differ by 3, and the difference between these two numbers is 3 less than twice the sum of the digits. Which is the larger of the two numbers in question?

39. A father, who is 36, has two sons, one twice the age of the other. In 3 years, the sum of all three ages will be 63. How old are the sons now?

40. An uncle is 3 times the age of his nephew, and in 8 years he will be twice the nephew's age. How old is the nephew now?

41. A woman has 3 times as many quarters as dimes in her purse. If the quarters were dimes and the dimes were dollars, she would be richer by $1.80. How much money does she have in the purse?

42. A man spent $3.20 for 4¢, 7¢, and 30¢ stamps. If he bought $2\frac{1}{2}$ times as many 4¢ stamps as 7¢ stamps, and his 30¢ stamps numbered $\frac{1}{7}$ of the combined number of the other two types, how many 7¢ stamps did he buy?

43. Two integers in a pair differ by 3. Their sum is greater than 15 but less than 21. Name the possible pairs.

44. Three consecutive odd positive integers have a sum that is less than or equal to 21. Name the possible sets of three integers.

45. Mother and daughter have the same birthday each year. The mother is now 32 years old. In 6 years she will still be more than 3 times the age of her daughter. How old might the daughter be now?

46. How many milliliters (ml) of water can be added to 30 ml of 5% acid and not reduce the concentration of the acid solution below 2%?

47. Two ships pass each other going in opposite directions at 0600, one at 11 knots and the other at 13 knots. If they hold to their respective courses and speeds, how long will it take them to be more than 60 mi apart?

48. Two cyclists start out at the same time from locations 56 mi apart, and ride toward each other. The speed of one of them is 12 mph. If they are to meet in 2 hr or less, what must be the speed of the other cyclist?

49. An investor has $10,000 to invest. If he invests $4000 at 8%, at what rate must he invest the balance if he is to have better than 10% return on the total investment?

50. Two children are on a seesaw. Sue sits 6 ft from the fulcrum and Barbara 5 ft from it. If Barbara can keep Sue up in the air, and Sue weighs 75 lb, how heavy might Barbara be?

6 EXPONENTS AND RADICALS

1. POSITIVE INTEGRAL EXPONENTS

A *power*, we recall, is a product of equal factors. The number of times the equal factor (*base*) occurs in the product is indicated by the *exponent*:

$$\underbrace{a \cdot a \cdot a \cdot a \cdot a}_{5 \text{ factors of } a} = a^{5} \begin{array}{l} \leftarrow \text{exponent} \\ \leftarrow \text{base} \end{array} \Big\} \text{power}$$

From this definition we develop the basic **Laws of Exponents**.

If a and b are real numbers and m and n are positive integers, then

$$\textbf{I.} \quad a^m \cdot a^n = a^{m+n}$$

$$\textbf{II.} \quad a^m \div a^n = a^{m-n} \text{ if } m > n \qquad (a \neq 0)$$

$$= \frac{1}{a^{n-m}} \text{ if } m < n \qquad (a \neq 0)$$

$$\textbf{III.} \quad (a^n)^m = a^{m \times n}$$

$$\textbf{IV.} \quad (ab)^n = a^n b^n$$

$$\textbf{V.} \quad \left(\frac{a}{b}\right)^n = \frac{a^n}{b^n} \qquad (b \neq 0)$$

EXAMPLES

I. $\quad a^m \cdot a^n = a^{m+n}$

"When multiplying factors having the same base, add the exponents."

1. $a^2 \cdot a^5 = a^{2+5} = a^7$

2. $b^4 \cdot b = b^{4+1} = b^5$

II. $\quad a^m \div a^n = a^{m-n} \text{ if } m > n \qquad (a \neq 0)$

$$= \frac{1}{a^{n-m}} \text{ if } m < n \qquad (a \neq 0)$$

"When dividing terms having the same base, subtract the exponents."

1. $a^5 \div a^2 = a^{5-2} = a^3$

2. $\dfrac{a^7}{a^3} = a^7 \div a^3 = a^{7-3} = a^4$

3. $\dfrac{a^3}{a^7} = a^3 \div a^7 = \dfrac{1}{a^{7-3}} = \dfrac{1}{a^4}$

III. $\quad (a^n)^m = a^{m \times n}$

"When raising a power to a given power, multiply the exponents."

1. $(b^5)^3 = b^{(3)(5)} = b^{15}$
This result can be verified by Exponent Law I:
$(b^5)^3 = b^5 \cdot b^5 \cdot b^5 = b^{5+5+5} = b^{15}$.

2. $(x^3)^5 = x^{(5)(3)} = x^{15}$

3. $(x)^4 = x^{(4)(1)} = x^4$

IV. $(ab)^n = a^n b^n$

"When raising a product to a given power, raise each factor of the product to the given power."

1. $(bc^2)^3 = b^{(3)(1)}c^{(3)(2)} = b^3 c^6$

This result can be verified by Exponent Law I:

$$(bc^2)^3 = bc^2 \cdot bc^2 \cdot bc^2 = b \cdot c^2 \cdot b \cdot c^2 \cdot b \cdot c^2$$
$$= b \cdot b \cdot b \cdot c^2 \cdot c^2 \cdot c^2$$
$$= b^{1+1+1} c^{2+2+2}$$
$$= b^3 c^6$$

2. $(x^2 y^3)^4 = x^{(4)(2)} y^{(4)(3)} = x^8 y^{12}$

3. $(3xy^2)^3 = 3^{(3)(1)} x^{(3)(1)} y^{(3)(2)} = 3^3 x^3 y^6 = 27 x^3 y^6$

V. $\left(\dfrac{a}{b}\right)^n = \dfrac{a^n}{b^n} \qquad (b \neq 0)$

"When raising a fraction to a given power, raise both the numerator and the denominator to the given power."

1. $\left(\dfrac{p^2}{q^4}\right)^3 = \dfrac{p^{(3)(2)}}{q^{(3)(4)}} = \dfrac{p^6}{q^{12}} \qquad (q \neq 0)$

This result can be verified by Exponent Law I:

$$\left(\frac{p^2}{q^4}\right)^3 = \left(\frac{p^2}{q^4}\right)\left(\frac{p^2}{q^4}\right)\left(\frac{p^2}{q^4}\right)$$
$$= \frac{p^2 \cdot p^2 \cdot p^2}{q^4 \cdot q^4 \cdot q^4} = \frac{p^{2+2+2}}{q^{4+4+4}} = \frac{p^6}{q^{12}}$$

2. $\left(\dfrac{2x}{y^2}\right)^4 = \dfrac{2^{(4)(1)} \cdot x^{(4)(1)}}{y^{(4)(2)}} = \dfrac{2^4 x^4}{y^8} = \dfrac{16x^4}{y^8} \qquad (y \neq 0)$

3. $\left(\dfrac{3x}{5ab^2}\right)^3 = \dfrac{3^{(3)(1)} \cdot x^{(3)(1)}}{5^{(3)(1)} \cdot a^{(3)(1)} \cdot b^{(3)(2)}} = \dfrac{27x^3}{125a^3 b^6} \qquad (a, b \neq 0)$

EXERCISE 6-1

Simplify; assume that literal exponents represent positive integers.

1. 2^3 **2.** 2^5 **3.** $-(3)^2$

4. -3^3 **5.** $(-3)^3$ **6.** $(-2)^4$

7. -4^3 **8.** $-(-3)^3$ **9.** $\left(-\dfrac{2}{3}\right)^3$

10. $-\left(-\dfrac{3}{4}\right)^2$ **11.** $(-0.3)^4$ **12.** $-(0.2)^5$

13. $a^3 \cdot a^5$ 14. $a^3 \cdot a^2 \cdot a$ 15. $x^3 \cdot x^2 \cdot x^4$

16. $(a^2y^3)(a^2y)$ 17. $x^4 \div x$ 18. $y^6 \div y^6$

19. $\dfrac{a^2b^3}{a^3b}$ 20. $\dfrac{x^3y^3}{x^3y}$ 21. $\dfrac{a^2b^2c^4}{a^3bc^3}$

22. $(x^2)^3$ 23. $-(b^2)^2$ 24. $(-m^3)^4$

25. $2^2 \cdot 2^4$ 26. $(3^2)^3$ 27. $(2)^2(3)^3$

28. $-(a^2b^3)^3$ 29. $(-a^5b^4)^3$ 30. $(2xy^2)^3$

31. $(3a^2bc)^4$ 32. $(-2x^2yz^3)^3$ 33. $(4-6)^3$

34. $\left(\dfrac{3+4}{4}\right)^2$ 35. $\left(-\dfrac{a^2}{4}\right)^2$ 36. $\left(\dfrac{6}{m}\right)^3$

37. $2^m \cdot 2^n$ 38. $6^a \cdot 2^a$ 39. $(7^m)^n$

40. $\left(\dfrac{2^x}{3^x}\right)^y$ 41. $\left(\dfrac{1}{3^x}\right)^y$ 42. $2^n \div 3^n$

43. $\left(\dfrac{x^3}{y^5}\right)^4$ 44. $\left(\dfrac{3a^2}{4b^3}\right)^2$ 45. $\left(-\dfrac{3}{v^2}\right)^4$

46. $(a^{2m})^p$ 47. $\left(\dfrac{3^n}{2}\right)^3$ 48. $\left(-\dfrac{x^2}{y^n}\right)^4$

49. $x^{4n} \div x^n$ 50. $x^{5n} \div x$ 51. $x^{2n} \div x^{3n}$

52. $a^{2+n} \cdot a^n$ 53. $(3a+x)^2$ 54. $r^{xy} \cdot r^{x+1}$

55. $3(x^n)^2$ 56. $(m^a)^{p+2}$ 57. $(a^2b^3)^r$

58. $\left(-\dfrac{a^2b^3}{2b}\right)^7$ 59. $\left(\dfrac{a^2b^3}{c}\right)\left(\dfrac{ac^3}{b}\right)$ 60. $\left(-\dfrac{x^3y}{a}\right)^2\left(\dfrac{x^4y^2}{a^2}\right)$

61. $\left(\dfrac{x^2y}{a}\right) \div \left(\dfrac{xy^5}{a^2}\right)$ 62. $\left(-\dfrac{ma}{b}\right)^3 \div \left(\dfrac{mc^4}{b^3}\right)$ 63. $\left[\left(\dfrac{3a^2}{b}\right)\left(-\dfrac{b^3}{9ac}\right)\right]^4$

2. ZERO AND NEGATIVE INTEGRAL EXPONENTS

In the previous section we limited exponents to the positive integers: 1, 2, 3, 4, ... This restriction was in keeping with our interpretation of the exponent as indicator of the number of times that a factor is repeated in a product.

What about the other integers—zero and the negative integers? How might we interpret those numbers as exponents?

Let us consider the number 0 first. In order for 0 to be acceptable as an exponent, by Exponent Law I,

$$a^0 \cdot a^n = a^{0+n} = a^n$$

This implies that a^0 acts as though it were 1, since only

$$1 \cdot a^n = a^n$$

Such an interpretation also fits well with Exponent Law II, since

$$a^n \div a^n = \frac{a^n}{a^n} = \boxed{1}$$

and

$$a^n \div a^n = a^{n-n} = \boxed{a^0}$$

Consequently, by definition,

Any power having a nonzero base and an exponent of zero is equal to 1.

EXAMPLES

1. $3^0 = 1$

2. $b^0 = 1$ $(b \neq 0)$

3. $3x^0 = 3$ $(x \neq 0)$
   The exponent is considered related only to the factor x and not to the 3.
   So $3x^0 = (3)(x^0) = (3)(1) = 3$.

4. $(3x)^0 = 1$ $(x \neq 0)$
   By Exponent Law IV, $(3x)^0 = (3)^0(x)^0 = (1)(1) = 1$.

5. $\left(\dfrac{p}{q}\right)^0 = 1$ $(p, q \neq 0)$

   By Exponent Law V, $\left(\dfrac{p}{q}\right)^0 = \dfrac{p^0}{q^0} = \dfrac{1}{1} = 1$.

6. $(a + 3b)^0 = 1$ $(a + 3b \neq 0)$

7. $(-6m)^0 = (-6)^0(m)^0 = (1)(1) = 1$ $(m \neq 0)$

Let us consider next an interpretation for negative integers as exponents. Suppose that we had a product in which one factor is a^n (n any positive integer, $a \neq 0$) and the other factor is the as-yet-uninterpreted a^{-n}.

By Exponent Law I and the preceding definition for 0 exponents,

$$a^n \cdot a^{-n} = a^{n+(-n)} = a^{n-n} = a^0 = 1$$

Since the only pair of factors that has 1 for a product is a factor and its *reciprocal*, it is evident that a^{-n} acts as the reciprocal of a^n. That is, $a^{-n} = \dfrac{1}{a^n}$.

By definition, then,

Any power having a nonzero base a and negative exponent $-n$ is equivalent to $\dfrac{1}{a^n}$.

A useful consequence of this definition is the following generalization: An equivalent fraction results when any nonzero *factor* in the numerator of a fraction is placed in the denominator of that fraction or when any nonzero *factor* in the denominator of a fraction is placed in the numerator of that fraction, *provided that the sign of the factor's exponent is reversed at the same time.*

$$\text{Caution:} \quad \frac{1}{3^{-1} + 2^{-1}} \neq 3 + 2$$

$$\text{but} \quad = \frac{1}{\frac{1}{3} + \frac{1}{2}} = \frac{1}{\frac{5}{6}} = \frac{6}{5}$$

$$\text{In general,} \quad \frac{1}{a^{-1} + b^{-1}} = \frac{ab}{a + b} \text{ and not } a + b.$$

EXAMPLES

1. $a^{-4} = \dfrac{1}{a^4}$

   (a^{-4} may be read "a to the negative 4 power.")

2. $8^{-1} = \dfrac{1}{8^1} = \dfrac{1}{8}$

3. $a^3 b^{-2} = a^3 \left(\dfrac{1}{b^2}\right) = \dfrac{a^3}{b^2}$

4. $\dfrac{1}{a^{-2}} = \dfrac{1}{\dfrac{1}{a^2}} = 1 \div \dfrac{1}{a^2} = 1 \cdot a^2 = a^2$

5. $\dfrac{a^{-2}}{b^{-3}} = \dfrac{b^3}{a^2}$

6. $\left(\dfrac{p}{q}\right)^{-1} = \dfrac{q}{p}$

7. $\dfrac{1}{x^{-2} + y^{-2}} = \dfrac{1}{\dfrac{1}{x^2} + \dfrac{1}{y^2}} = \dfrac{1}{\dfrac{y^2 + x^2}{x^2 y^2}} = \dfrac{x^2 y^2}{y^2 + x^2}$

With the introduction of negative exponents, the two parts of Exponent Law II can be collapsed into one:

$$a^m \div a^n = a^{m-n}$$

When $m > n$, then $m - n > 0$ and the quotient has a positive exponent.

When $m < n$, then $m - n < 0$ and the quotient has a negative exponent.

When $m = n$, then $m - n = 0$ and the quotient has 0 as an exponent.

EXERCISE 6-2

Simplify.

1. $(-3x)^0$

2. $-3x^0$

3. $m^2 \cdot m^3 \cdot m^0$

4. $3 \cdot 10^0$

5. $\left(\dfrac{3p}{q}\right)^0$

6. $\left(\dfrac{3p^0}{q}\right)^2$

7. $\left(\dfrac{5+4}{5}\right)^0$

8. $\dfrac{3x^0y^2z^0}{3xy^0z^0}$

9. $(2^0)(4^0)$

10. $(3^2)(2)^0$

11. $\dfrac{6^2}{2^0}$

12. $\left(\dfrac{3^0}{6}\right)^0$

Write an equivalent expression using a negative exponent.

13. x^2

14. $\dfrac{1}{m^3}$

15. $-\dfrac{1}{n^4}$

16. $-d^2$

Eliminate negative exponents; simplify.

17. 2^{-3}

18. 4^{-2}

19. 6^{-1}

20. $(3)^2\left(\dfrac{1}{2^{-1}}\right)$

21. $8^{-2} \div 2^{-1}$

22. $4^{-1} \div 2^{-3}$

23. $2^{-2} - 2^{-1}$

24. $3^{-2} + 3^0$

25. $\dfrac{1}{3^{-2}} + 3^{-2}$

26. $2^{-3} - 3^{-2}$

27. $(2^{-3})^{-2}$

28. $\left[\left(\dfrac{1}{2}\right)^{-3}\right]^{-1}$

29. $(a^{-2}b)(ab^2)$

30. $(x^2y^{-1})(x^{-1}y^2)$

31. $a^{-2}b^{-2}(a^2b^2 - 1)$

32. $x^{-2}y(1 - x^2y)$

33. $ab^{-2}(ab^{-1} + b)$

34. $(x^{-1}y^2 - 2)(xy^{-1})$

Simplify the following to equivalent expressions containing only positive exponents.

35. $(3x)^{-3}$

36. $\left(\dfrac{2}{x}\right)^{-2}$

37. $(x^{-2})^{-2}$

38. $a^5 \cdot a^{-3} \cdot a^{-2}$

39. $(xy^{-1})^{-2}$

40. $(3x^0y)^{-4}$

41. $(2^3x^{-3})^{-2}$

42. $\left(\dfrac{a^{-3}}{c^{-4}}\right)^3$

43. $\left(-\dfrac{a^{-1}}{b}\right)^{-2}$

44. $\left(\dfrac{ax^4}{b^3y}\right)^3\left(\dfrac{by^3}{a^2x^2}\right)^2$

45. $(x^2y^3)^3\left(\dfrac{a^2x^3}{y^2}\right)^2$

46. $x^{-1} - y^{-1}$

47. $(x^{-2})^{-2} \cdot (y^{-3})^2$

48. $2x^{-1} - y^{-2}$

49. $3x^{-2} + 3y^{-2}$

50. $x^{-3} \div y^{-2}$

51. $a^{-2} \div \left(\dfrac{a}{b}\right)^{-3}$

52. $\dfrac{a}{b^{-1}} + \dfrac{b}{a^{-1}} + ab$

53. $\dfrac{x}{x^{-1} + y^{-1}}$

54. $\dfrac{a^{-1}}{a^{-1} - b^{-1}}$

55. $\dfrac{x^{-2}}{\dfrac{1}{x^2} + y^{-2}}$

56. $\dfrac{(x - y)^{-1}}{x^{-1} - y^{-1}}$

57. $\dfrac{x^2 - 2xy + y^2}{x^{-2} - y^{-2}}$

58. $\dfrac{m^2 + n}{m^{-2} + n^{-1}}$

59. $\dfrac{ab}{2b^{-2} - 3a^{-2}}$

60. $\dfrac{3x + 2}{3x^{-1} + 2x^{-2}}$

61. $(x^{-1} - y^{-1})(x^{-1} + y^{-1})$

62. $(x^{-1} - y^{-1})^2$

63. $(2a^{-1} + b^{-1})^2$

64. $(3m^{-1} - 2n^{-1})(3m^{-1} + 2n^{-1})$

3. SCIENTIFIC NOTATION

Some applications of mathematics involve very large and very small numbers. For instance, the average distance from earth to Jupiter is approximately

$$480{,}000{,}000 \text{ miles}$$

The approximate length of a certain bacterium is

$$0.00006 \text{ inch}$$

A more convenient notation to express numbers of this sort is an exponential form called **scientific notation**:

$$4.8 \times 10^8 \text{ miles}$$

$$6 \times 10^{-5} \text{ inch}$$

With scientific notation we can express any decimal as a product of two factors in which one factor is a number between 1 and 10 and the other is a power of 10. To illustrate,

Decimal	Scientific Notation
6,200,000	6.2×10^6
0.000004	4×10^{-6}
3,240,000,000	3.24×10^9
0.000623	6.23×10^{-4}

PROGRAM 6.1

To express a decimal in scientific notation:

Step 1: Move the point in the decimal, right or left, to a new position to the right of the first significant digit.*

Step 2: Count the number of digits or places in the decimal through which the decimal point was moved in Step 1.

Step 3: Express as the equivalent of the given decimal a pair of factors in which

(a) One factor is the number developed in Step 1.
(b) The other factor is a power of 10 having as exponent
—the number of Step 2 if the move of Step 1 was to the left; or
—the negative of the number of Step 2 if the move of Step 1 was to the right.

EXAMPLES

1. Express 13,200. in scientific notation.

SOLUTION *Step 1:* Move the decimal point (left) so that it is to the right of the first significant digit:

$$13{,}200.$$

Step 2: Count the digits, or places, moved in Step 1:

$$1.3200$$

left, 4 digits

Step 3: Write the pair of factors:

from Step 1

$$13{,}200 = \mathbf{1.32} \times 10^{4}$$

from Step 2

2. Express 0.000067 in scientific notation.

SOLUTION *Step 1:* Move the decimal point (right) so that it is to the right of the first significant digit:

$$0.000067$$

---

* All digits in a decimal are considered significant except those zeros that are used principally to locate the decimal point. For example, 206 has three significant digits; 0.026 has two significant digits; 0.004 has one significant digit. Numerals such as 200 may have three, two, or one significant digits. If 200 means 200 ones to the nearest one—three significant digits; if 20 tens to the nearest ten—two significant digits; if 2 hundreds to the nearest hundred—one significant digit.

Step 2: Count the places moved in Step 1:

$$00006.7$$

$$\xrightarrow{\hspace{2cm}}$$

right, 5 digits

Step 3: $0.000067 = 6.7 \times 10^{-5}$

3. Expressed in scientific notation,

(a) $0.000000007594 = 7.594 \times 10^{-9}$
(b) $3,000,000,000 = 3 \times 10^{9}$ (or 3.0×10^{9})

Computations with very large or very small numbers can often be simplified by using scientific notation.

EXAMPLES

1. Multiply $0.0000031 \times 60,000$.

SOLUTION Express the factors in scientific notation:

$$0.0000031 = 3.1 \times 10^{-6}$$

$$60,000 = 6 \times 10^{4}$$

Then

$$0.0000031 \times 60,000 = (3.1 \times 10^{-6})(6 \times 10^{4})$$

$$= (3.1 \times 6)(10^{-6} \times 10^{4})$$

$$= (18.6)(10^{-6+4})$$

$$= 18.6 \times 10^{-2} \left(\text{i.e., } \frac{18.6}{100} \right)$$

$$= 0.186$$

2. Divide 0.000042 by 0.014.

SOLUTION $$0.000042 \div 0.014 = \frac{0.000042}{0.014}$$

$$= \frac{4.2 \times 10^{-5}}{1.4 \times 10^{-2}}$$

$$= \frac{4.2}{1.4} \times 10^{-5} \times 10^{2}$$

$$= 3 \times 10^{-3}$$

$$= 0.003$$

3. Simplify $\dfrac{0.0081 \times 30{,}000}{9{,}000{,}000}$.

SOLUTION

$$\frac{0.0081 \times 30{,}000}{9{,}000{,}000} = \frac{(8.1 \times 10^{-3})(3 \times 10^{4})}{9 \times 10^{6}}$$

$$= \frac{(8.1 \times 3)(10^{-3} \times 10^{4})}{9 \times 10^{6}}$$

$$= \left(\frac{8.1 \times 3}{9}\right) \times 10^{-3} \times 10^{4} \times 10^{-6}$$

$$= 2.7 \times 10^{-5}$$

$$= 0.000027$$

EXERCISE 6-3

Use scientific notation to express each number.

1. 36,000   **2.** 143,000   **3.** 62,800

4. 850   **5.** 0.000037   **6.** 0.0000004

7. 0.0000621   **8.** 0.00073   **9.** 14

10. 0.83   **11.** 0.61   **12.** 13

13. 0.7   **14.** 6.0   **15.** 1.3

Express as a standard decimal.

16. 3.2×10^{2} **17.** 4.5×10^{3} **18.** 3.41×10^{2}

19. 6.02×10^{5} **20.** 3.5×10^{-3} **21.** 6.7×10^{-2}

22. 8.34×10^{-5} **23.** 6.71×10^{-6} **24.** 3.0×10^{-4}

Simplify.

25. $(4 \times 10^{6})(3 \times 10^{-2})$ **26.** $(8 \times 10^{4})(6 \times 10^{-1})$

27. $\dfrac{8 \times 10^{8}}{4 \times 10^{2}}$ **28.** $\dfrac{9 \times 10^{6}}{3 \times 10^{2}}$

29. $(9 \times 10^{-3})(4 \times 10^{6})(3 \times 10^{-3})$

30. $(7 \times 10^{2})(5 \times 10^{-8})(3 \times 10^{4})$

31. $\dfrac{0.0063 \times 14{,}000}{0.007}$ **32.** $\dfrac{1600 \times 0.04}{2{,}000{,}000}$ **33.** $\dfrac{0.000042}{80{,}000 \times 0.0007}$

34. $\dfrac{42{,}000}{0.00006 \times 0.007}$ **35.** $\dfrac{3.2 \times 860{,}000}{0.0064 \times 0.00043}$ **36.** $\dfrac{100{,}000 \times 0.0075}{0.0006 \times 25{,}000}$

4. ROOTS

The inverse operation of *raising to a power* is that of *extracting a root*. Because these two operations apply to single numbers, they are called **unary** operations—as opposed to *binary* operations which involve two numbers. Thus, in raising to a power:

$$2^2 = 2 \cdot 2 = 4 \quad \text{(or "two squared} = 4\text{")}$$

$$2^3 = 2 \cdot 2 \cdot 2 = 8 \quad \text{(or "two cubed} = 8\text{")}$$

$$2^4 = 2 \cdot 2 \cdot 2 \cdot 2 = 16 \quad \text{(or "two to the fourth} = 16\text{")}$$

When the circumstances are reversed and we ask "What number squared is 4?" or "What number cubed is 8?" and so on, we are seeking a **root**. Thus

2 is a *square root* of 4

2 is a *cube root* of 8

2 is a *fourth root* of 16

Although powers of rational numbers are always rational numbers, roots of rational numbers are not always rational numbers. For instance, the rational number $+9$ has two rational numbers as square roots, $+3$ and -3, because

$$(+3)^2 = +9 \quad \text{and} \quad (-3)^2 = +9$$

On the other hand, the square roots of 5 or 2 or 6 are not to be found among the rational numbers. They are in another set called the irrational numbers. The irrational numbers include all the nonrational roots of rational numbers with the exception of the even roots of negative numbers, such as

$$\sqrt{-4}, \quad \sqrt[4]{-8}, \quad \sqrt[10]{-1}$$

These roots are available only in a still larger set, the **complex numbers**, to be discussed in Section 12 of this chapter.

As noted in Chapter 1, the rational and irrational numbers together make up the set of numbers called the real numbers:

$$\text{real numbers} \begin{cases} \text{rational numbers} \begin{cases} \text{integers} \\ \text{nonintegers} \end{cases} \\ \text{irrational numbers} \end{cases}$$

Irrational numbers that are the roots of rational numbers are usually expressed with a symbol called a **radical**: $\sqrt{}$. A numeral appearing in the crook of the radical to denote the root is called the **index**. The expression appearing within the radical is called the **radicand**. Thus

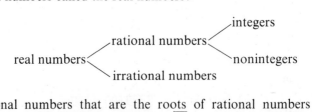

cube root of 7:

square root of 6: $\sqrt[2]{6}$ or $\sqrt{6}$

Note: The index is normally omitted from the square root radical.

fourth root of 5: $\sqrt[4]{5}$

nth root of a: $\sqrt[n]{a}$

In general, the nth root of a given number is one of n equal factors whose product is the given number. For instance,

$$\sqrt{3} \cdot \sqrt{3} = 3$$
$$\sqrt[3]{4} \cdot \sqrt[3]{4} \cdot \sqrt[3]{4} = 4$$

Some numbers have more than one real root; for example, the square roots of $+4$ are the two numbers $+2$ and -2, since $(+2)(+2) = +4$ and $(-2)(-2) = +4$. This discussion leads us to define the **principal root** of various numbers.

The principal nth root of a positive number is the positive root.
The principal nth root of 0 is 0.
The principal nth root of a negative number is the negative root when n is odd.

Note: The principal nth root of a negative number when n is even does not exist among the real numbers.

EXAMPLES

1. The principal square root of 9 is $\sqrt{9}$ or 3; $-\sqrt{9}$ or -3 is also a square root of 9, since $(-3)(-3) = 9$, but -3 is not the principal square root.

2. The principal cube root of 8 is $\sqrt[3]{8}$ or 2.

3. The principal cube root of -8 is $\sqrt[3]{-8}$ or -2.

4. The principal square root of 5 is $\sqrt{5}$; $-\sqrt{5}$ is also a square root but not the principal square root.

5. The principal square root of -9 does not exist among the real numbers—that is, $\sqrt{-9}$ is not a real number.

5. FRACTIONAL EXPONENTS

Up to this point we have developed definitions for integral exponents—that is, zero and the positive and negative integers. We now develop a useful and consistent interpretation for exponents in the form of unit fractions, such as $\frac{1}{3}$.

If we were to cube an expression like $a^{1/3}$ (which suggests a power) according to Exponent Law III, the resulting power would have an exponent of 1:

$$(a^{1/3})^3 = a^{3 \times 1/3} = a^1$$

It is reasonable, then, to interpret $a^{1/3}$ as *cube root of a*, or

$$a^{1/3} = \sqrt[3]{a}$$

Similarly,

$$a^{1/2} = \sqrt{a} \qquad 8^{1/3} = \sqrt[3]{8} = 2$$

$$a^{1/7} = \sqrt[7]{a} \qquad -8^{1/3} = \sqrt[3]{-8} = -2$$

$$x^{1/17} = \sqrt[17]{x} \qquad 9^{1/2} = \sqrt{9} = 3$$

By definition, then,

$a^{1/n} = \sqrt[n]{a}$, where $\sqrt[n]{a}$ represents the principal nth root of a, provided that a is nonnegative when n denotes an even integer.

The qualifying condition in the above definition—that the radicand must be nonnegative when the index is an even integer—is intended to rule out such expressions as $\sqrt{-2}$, which have no meaning among the real numbers.

Now that we have an interpretation for exponents of the form $1/n$, all we need is an interpretation for exponents of the form m/n and our discussion is complete. Consider the product

$$a^p \cdot a^p \cdot a^p = a^{p+p+p} = a^{3p}$$

If we let p represent an exponent of the form $1/n$, say 1/4, then by Exponent Law I

$$a^{1/4} \cdot a^{1/4} \cdot a^{1/4} = a^{1/4 + 1/4 + 1/4} = a^{3/4} \cdot$$

Expressing the product in another way, as the cube of the factor $a^{1/4}$, we have

$$a^{1/4} \cdot a^{1/4} \cdot a^{1/4} = (a^{1/4})^3 = (\sqrt[4]{a})^3 = a^{3/4}$$

Consider now the related example

$$\sqrt[4]{a \cdot a \cdot a} = \sqrt[4]{a^3} = (a^3)^{1/4}$$

To be consistent with Exponent Law III,

$$(a^3)^{1/4} = a^{(1/4)(3)} = a^{3/4}$$

So we see that

$$a^{3/4} = \sqrt[4]{a^3} = (\sqrt[4]{a})^3$$

which illustrates the generalization: The numerator of a fractional exponent can be interpreted as the power to which the base is to be raised and the denominator of a fractional exponent as the root to be extracted.

1. Simplify $8^{2/3}$.

SOLUTION

$$8^{2/3} = \begin{cases} \sqrt[3]{8^2} = \sqrt{64} = 4 \\ (\sqrt[3]{8})^2 = (2)^2 = 4 \end{cases}$$

2. Simplify $9^{3/2}$.

SOLUTION

$$9^{3/2} = \begin{cases} \sqrt{9^3} = \sqrt{729} = 27 \\ (\sqrt{9})^3 = (3)^3 = 27 \end{cases}$$

Note: When the radicand yields a rational root, it is generally simpler to extract the root first and then raise to the power.

6. SUMMARY—LAWS OF EXPONENTS

We now have developed plausible and consistent interpretations for exponents that range from the original positive integers to the complete set of rational numbers.
In general,

If a and b represent real numbers and m and n represent rational numbers, then

  I. $a^m \cdot a^n = a^{m+n}$
  II. $a^m \div a^n = a^{m-n}$ $(a \neq 0)$
 III. $(a^n)^m = a^{mn}$
  IV. $(ab)^n = a^n b^n$
  V. $\left(\dfrac{a}{b}\right)^n = \dfrac{a^n}{b^n}$ $(b \neq 0)$

To *simplify* an exponential expression normally means to carry out all operations possible by the foregoing laws of exponents (I to V), to eliminate all negative exponents from the final expression, and to expand all numerical coefficients having integral exponents.

1. $(x^{3/4})^8 = x^{8(3/4)} = x^6$

2. $x^{1/2}x^{2/3} = x^{1/2 + 2/3} = x^{3/6 + 4/6} = x^{7/6}$
$$= x^{1 + 1/6} = x^1 x^{1/6} = x\sqrt[6]{x}$$

3. $(9x^{-3})^{2/3} = 9^{2/3}x^{(2/3)(-3)} = (3^2)^{2/3}(x^{-2})$

$$= 3^{4/3}x^{-2} = \frac{3\sqrt[3]{3}}{x^2}$$

4. $\left(-\dfrac{1}{27}\right)^{-2/3} = \left[\left(-\dfrac{1}{3}\right)^3\right]^{-2/3} = \left(-\dfrac{1}{3}\right)^{(-2/3)(3)}$

$$= \left(-\dfrac{1}{3}\right)^{-2} = \dfrac{1}{\left(-\dfrac{1}{3}\right)^2} = \dfrac{1}{\dfrac{1}{9}} = 9$$

5. $\dfrac{9x^{-2}y^3}{3x^4y^{-5}} = 3^2 \cdot x^{-2} \cdot y^3 \cdot 3^{-1} \cdot x^{-4} \cdot y^5$

$$= 3^{2-1}x^{-2-4}y^{3+5}$$

$$= 3^1x^{-6}y^8 = \dfrac{3y^8}{x^6}$$

6. $(a^2 - b^2)\sqrt{(a-b)^{-2}} = (a^2 - b^2)[(a-b)^{-2}]^{1/2}$

$$= (a^2 - b^2)(a-b)^{-1}$$

$$= \dfrac{(a^2 - b^2)}{(a-b)} = -\dfrac{\cancel{(a-b)}(a+b)}{\cancel{(a-b)}}$$

$$= a + b$$

7. $\dfrac{(3x^2y)^{1/3}}{9x^{-2/3}y^2} = \dfrac{3^{1/3}x^{2/3}y^{1/3}}{3^2x^{-2/3}y^2}$

$$= 3^{1/3} \cdot x^{2/3} \cdot y^{1/3} \cdot 3^{-2} \cdot x^{2/3} \cdot y^{-2}$$

$$= 3^{-5/3}x^{4/3}y^{-5/3}$$

$$= \dfrac{x \cdot x^{1/3}}{3 \cdot 3^{2/3} \cdot y \cdot y^{2/3}} = \dfrac{x\sqrt[3]{x}}{3y\sqrt[3]{9y^2}}$$

EXERCISE 6-4

Write the real number principal root for each.

1. $\sqrt{4}$ **2.** $\sqrt{36}$ **3.** $\sqrt[3]{8}$

4. $\sqrt[3]{-125}$ **5.** $\sqrt[3]{-8}$ **6.** $\sqrt{0.01}$

7. $\sqrt{\dfrac{1}{4}}$ **8.** $\sqrt[3]{-\dfrac{1}{8}}$ **9.** $\sqrt{-36}$

10. $\sqrt{\dfrac{16}{25}}$ **11.** $\sqrt{\dfrac{16}{100}}$ **12.** $\sqrt[3]{-\dfrac{1}{1000}}$

13. $\sqrt[3]{-\dfrac{8}{27}}$ **14.** $\sqrt[6]{-64}$ **15.** $\sqrt[5]{-32}$

16. $\sqrt[3]{0.008}$

Express in exponential form.

17. $\sqrt{3}$ **18.** $\sqrt[3]{5}$ **19.** $\sqrt[4]{11}$ **20.** $\sqrt{10}$

21. $\sqrt{8}$ **22.** $\sqrt[3]{100}$ **23.** $\sqrt{\dfrac{1}{4}}$ **24.** $\sqrt[3]{\dfrac{1}{27}}$

25. $\sqrt{2^{-2}}$ **26.** $\sqrt{10^{-4}}$ **27.** $\sqrt{\left(\dfrac{1}{2}\right)^{-1}}$ **28.** $\sqrt[3]{3^{-2}}$

29. $\sqrt{p}$ **30.** $\sqrt[3]{s}$ **31.** $\sqrt[10]{t}$ **32.** $\sqrt[7]{-x}$

33. $\sqrt[3]{p^2}$ **34.** $\sqrt[5]{x^2y^2}$ **35.** $\sqrt[n]{y^2}$ **36.** $\sqrt[r]{s^2t^3}$

37. $\dfrac{1}{\sqrt[r]{p^s}}$ **38.** $\sqrt[k]{p^{3m}}$ **39.** $\sqrt{c-6}$ **40.** $\sqrt[3]{(a+b)^2}$

41. $\sqrt[4]{\dfrac{c^3}{b}}$ **42.** $\sqrt[4]{x^2y^{1/2}}$

Simplify, if possible; express results in radical form.

43. $3^{1/3}$ **44.** $2^{1/5}$ **45.** $4^{-1/2}$ **46.** $3^{-1/3}$

47. $6^{2/5}$ **48.** $6^{5/2}$ **49.** $a^{1/5}$ **50.** $b^{1/3}c^{2/3}$

51. $a^{5/8}$ **52.** $3x^{1/2}$ **53.** $(3x)^{1/2}$ **54.** $(8x)^{1/3}$

55. $\left(\dfrac{2}{3}\right)^{-1/3}$ **56.** $\left(\dfrac{1}{3^{-3}}\right)^{1/2}$ **57.** $\left(\dfrac{3}{2^{-5}}\right)^{-1/2}$ **58.** $\left(\dfrac{2}{3^{-3}}\right)^{-1/3}$

59. $\left(\dfrac{1}{2^{-1}+1}\right)^{1/2}$ **60.** $\dfrac{1}{(3^{-1}+2)^{-1/2}}$ **61.** $(2x^2)^0$

62. $a^{1/3}b^{2/3}c$ **63.** $(b-c)^{1/2}$ **64.** $\dfrac{1}{(a+2)^{-1/3}}$

65. $[(x+y)^2]^{1/3}$ **66.** $m^0n^{-1/6}$ **67.** $(s^2t^{1/4})^{1/n}$

68. $\dfrac{x^{-1/2}}{y^{1/2}}$ **69.** $\left(p-\dfrac{pr}{r}\right)^{1/3}$ **70.** $(a^km^{-p})^{-1/p}$

71. $x^{2/3} \div x^{1/2}$ **72.** $m^{2/5} \div m^{2/3}$ **73.** $\dfrac{a^{7/2}}{a^{1/2}}$

74. $\dfrac{b^{5/2}}{\sqrt[3]{b^{10}}}$ **75.** $\dfrac{a^{0.8}}{a^{0.2}}$ **76.** $\dfrac{a^{0.12}}{a^{0.6}}$

77. $\dfrac{m^{2/3}n^{1/3}}{m^0n^{1/4}}$ **78.** $\left(\dfrac{4x^{-1}y}{x^{2/3}y^{-1}}\right)^{-3}$ **79.** $\left(-\dfrac{8}{125}\right)^{2/3}$

80. $\left(\dfrac{1}{16}\right)^{-3/4}$ **81.** $\sqrt{(x+y)^{-2}}$ **82.** $\sqrt{x^{-1}y^{-1}}$

83. $(-2^{-6})^{2/3}$

84. $-(2^6)^{-2/3}$

85. $\dfrac{(2x^2y^3)^{1/4}}{16x^{-2}y}$

86. $\dfrac{x^{(2m+3)/c}}{x^{(m-4)/c}}$

87. $\sqrt[5]{s^{1/2}}$

88. $(2^0 + 3^0 + 4^0)^{-1/2}$

89. $\sqrt[4]{\sqrt[3]{\sqrt{x}}}$

90. $\sqrt[a]{\sqrt[b]{\sqrt[c]{x^d}}}$

7. SIMPLIFYING RADICAL EXPRESSIONS

Radical notation, we have observed, is an alternative to fractional exponents. In many instances it is a more convenient symbolism. Because symbolism is a matter of expression, and not substance, the various properties of exponents are equally applicable to radicals. In particular:

If a and b represent real numbers, and n represents a positive integer (and the roots exist), then

1. $\sqrt[n]{ab} = \sqrt[n]{a} \cdot \sqrt[n]{b}$ [by Exponent Law IV: $(ab)^{1/n} = a^{1/n}b^{1/n}$]

2. $\sqrt[n]{\dfrac{a}{b}} = \dfrac{\sqrt[n]{a}}{\sqrt[n]{b}} \ (b \neq 0)$ $\left[\text{by Exponent Law V:} \left(\dfrac{a}{b}\right)^{1/n} = \dfrac{a^{1/n}}{b^{1/n}} \ (b \neq 0) \right]$

Note: In the case of $\sqrt{a^2}$, where a may represent either a positive or a negative number, we define $\sqrt{a^2} = |a|$. This is to eliminate a potential conflict with the prior definition that the principal nth root of a positive number is the positive root (a^2 is positive, even though a may be negative). In general:

$$\sqrt[n]{a^n} = a \text{ (when } n \text{ is odd)}$$

$$\sqrt[n]{a^n} = |a| \text{ (when } n \text{ is even)}$$

Hereafter, to avoid repeating this distinction, unless indicated otherwise, we assume that variables in the radicand represent only nonnegative numbers.

To *simplify* a radical expression usually means to express it in an equivalent form in which

- the radicand contains no fractions,
- the index of the radical and the exponents of the radicand are positive integers, and
- the power of any factor of the radicand is less than the index of the radical.

1. Simplify $\sqrt{18}$.

SOLUTION
The model: $\sqrt[n]{ab} = \sqrt[n]{a}\,\sqrt[n]{b}$

$$\sqrt{18} = \sqrt{9} \cdot \sqrt{2}$$
$$= 3\sqrt{2}$$

2. Simplify $\sqrt{\dfrac{36}{49}}$.

SOLUTION
The model: $\sqrt[n]{\dfrac{a}{b}} = \dfrac{\sqrt[n]{a}}{\sqrt[n]{b}}$

$$\sqrt{\frac{36}{49}} = \frac{\sqrt{36}}{\sqrt{49}} = \frac{6}{7}$$

3. Simplify $\sqrt[3]{-24}$.

SOLUTION

$$\sqrt[3]{-24} = \sqrt[3]{-8 \cdot 3}$$
$$= \sqrt[3]{-8} \cdot \sqrt[3]{3}$$
$$= -2\sqrt[3]{3}$$

4. Simplify $\sqrt[4]{9a^2b^4}$.

SOLUTION

$$\sqrt[4]{9a^2b^4} = \sqrt[4]{3^2} \cdot \sqrt[4]{a^2} \cdot \sqrt[4]{b^4}$$
$$= 3^{2/4} \cdot a^{2/4} \cdot b^{4/4}$$
$$= 3^{1/2} \cdot a^{1/2} \cdot b^1$$
$$= \sqrt{3} \cdot \sqrt{a} \cdot b$$
$$= b\sqrt{3a}$$

5. Simplify $\sqrt[3]{-\dfrac{256a^2b^{-1}}{8b^2}}$.

SOLUTION

$$\sqrt[3]{-\frac{256a^2b^{-1}}{8b^2}} = \sqrt[3]{-\frac{32a^2}{b^3}} = \frac{\sqrt[3]{-32} \cdot \sqrt[3]{a^2}}{\sqrt[3]{b^3}}$$

$$= \frac{\sqrt[3]{(-8)(4)} \cdot \sqrt[3]{a^2}}{b} = \frac{-2\sqrt[3]{4} \cdot \sqrt[3]{a^2}}{b}$$

$$= -\frac{2}{b}\sqrt[3]{4a^2}$$

6. Simplify $\sqrt[4]{\dfrac{36x^6y^{-3}}{3xy^2}}$.

SOLUTION

$$\sqrt[4]{\frac{36x^6y^{-3}}{3xy^2}} = \sqrt[4]{\frac{12x^5}{y^5}} = \frac{\sqrt[4]{2^2 \cdot 3 \cdot x^5}}{\sqrt[4]{y^5}}$$

$$= \frac{x\sqrt[4]{12x}}{y\sqrt[4]{y}}$$

EXERCISE 6-5

Simplify. (Table I, page 429, may be useful in some instances.)

1. $\sqrt{8}$ **2.** $\sqrt{45}$ **3.** $\sqrt{84}$ **4.** $\sqrt{96}$

5. $\sqrt[3]{125}$ **6.** $\sqrt[3]{250}$ **7.** $\sqrt[3]{-81}$ **8.** $\sqrt[3]{-1000}$

9. $\sqrt{\dfrac{9}{25}}$ **10.** $\sqrt{\dfrac{42}{7}}$ **11.** $\sqrt{\dfrac{368}{4}}$ **12.** $\sqrt{32}$

13. $\sqrt[3]{x^4}$ **14.** $\sqrt[3]{a^5}$ **15.** $\sqrt{3a^2}$ **16.** $\sqrt{8b^2}$

17. $\sqrt{x^{-2}}$ **18.** $\sqrt[3]{-27x^6}$ **19.** $\sqrt[5]{x^7}$ **20.** $\sqrt[10]{y^{12}}$

Reduce the order (i.e., index), and simplify.

21. $\sqrt[4]{9a^2}$ **22.** $\sqrt[4]{64a^2b^2}$ **23.** $\sqrt[6]{4m^2}$ **24.** $\sqrt[6]{-8x^3}$

25. $\sqrt[9]{27m^6}$ **26.** $\sqrt[12]{36s^8}$ **27.** $\sqrt[9]{(x^2 + y^2)^3}$

28. $\sqrt[6]{\dfrac{27x^3}{64y^3}}$

Simplify.

29. $\sqrt[3]{27x^4y^2}$ **30.** $\sqrt{45m^2}$ **31.** $\sqrt{18x^2y^3z^4}$ **32.** $\sqrt[3]{kx^5y^4}$

33. $3\sqrt{12x^3}$ **34.** $x\sqrt{3xy^3}$ **35.** $\sqrt{\dfrac{x}{16}}$ **36.** $\sqrt{\dfrac{5s^2}{t^6}}$

37. $4\sqrt[3]{\dfrac{m^2}{8}}$ **38.** $3\sqrt[4]{\dfrac{m^6}{81}}$ **39.** $\sqrt{36x^2y^{-4}}$ **40.** $\sqrt{5x^{-6}y^3}$

41. $\sqrt[3]{8x^{-2}y^6z^{-1}}$ **42.** $\sqrt{25^{-1}p^3y}$ **43.** $\sqrt{(a + b)^{-4}}$

44. $\sqrt[3]{(p + q)^{-3}}$ **45.** $\sqrt{x + y^{-1}}$ **46.** $\sqrt[3]{a^{-1} + b}$

47. $\sqrt{4x^2 + 12x + 9}$ **48.** $\sqrt{a^2 - 4ab + 4b^2}$ **49.** $\sqrt{x^2 - y^2}$

50. $\sqrt{x^{-2} - y^{-2}}$ **51.** $\sqrt{\dfrac{x - a}{x^2 - a^2}}$ **52.** $\sqrt[3]{\dfrac{a + b}{a^2 + 2ab + b^2}}$

8. ADDITION AND SUBTRACTION WITH RADICAL EXPRESSIONS

Radical expressions are said to be *like* when they have the same index and radicand. Thus

$$3\sqrt[3]{4}, \quad 6\sqrt[3]{4}, \quad \text{and} \quad x\sqrt[3]{4}$$

are like because each possesses the same index (3) and the same radicand (4).

On the other hand,

$$2\sqrt{6} \quad \text{and} \quad 2\sqrt{5}$$

are not like because they differ in radicand. And

$$3\sqrt[3]{2} \quad \text{and} \quad 3\sqrt{2}$$

are not like because they differ in index.

Factors appearing outside the radical, such as the 3 in $3\sqrt{2}$, or the $2x$ in $2x\sqrt[3]{xy}$, are referred to as the *coefficient* of the radical expression.

Computing sums and differences of like radical expressions amounts to addition or subtraction of their coefficients, in accordance with the distributive property:

$$3\sqrt{5} + 4\sqrt{5} = (3 + 4)\sqrt{5} = 7\sqrt{5}$$

When the terms are unlike radical expressions, sums and differences are expressed as a polynomial. For instance, the sum of $2\sqrt{6}$ and $3\sqrt[3]{6}$ would be written simply as

$$2\sqrt{6} + 3\sqrt[3]{6}$$

Similarly, the difference of $4\sqrt{7}$ and $3\sqrt{5}$ would be expressed as the polynomial

$$4\sqrt{7} - 3\sqrt{5}$$

PROGRAM 6.2

To compute the sum (difference) of radical expressions:

Step 1: Simplify each term involving a radical expression.

Step 2: Add (subtract) the like radical expressions by adding (subtracting) their coefficients; express the sum (difference) of the unlike radical expressions as a polynomial.

EXAMPLES

1. Add $5\sqrt{27} + 7\sqrt{12} + 3\sqrt{3}$.

SOLUTION   *Step 1:* Simplify each addend:

$$5\sqrt{27} = 5\sqrt{9\cdot 3} = 5\cdot\sqrt{9}\cdot\sqrt{3} = 5\cdot 3\cdot\sqrt{3} = 15\sqrt{3}$$
$$7\sqrt{12} = 7\sqrt{4\cdot 3} = 7\cdot\sqrt{4}\cdot\sqrt{3} = 7\cdot 2\cdot\sqrt{3} = 14\sqrt{3}$$
$$3\sqrt{3} = 3\sqrt{3}$$

Step 2: Add the coefficients of the like radical expressions:

$$15\sqrt{3} + 14\sqrt{3} + 3\sqrt{3} = (15 + 14 + 3)\sqrt{3}$$
$$= 32\sqrt{3}$$

2. Subtract $9\sqrt{2} - \sqrt{18}$.

SOLUTION   *Step 1:* Simplify:

$$9\sqrt{2} = 9\sqrt{2}$$
$$\sqrt{18} = \sqrt{9\cdot 2} = \sqrt{9}\cdot\sqrt{2} = 3\sqrt{2}$$

Step 2: Subtract coefficients of like radicals:

$$9\sqrt{2} - 3\sqrt{2} = (9 - 3)\sqrt{2} = 6\sqrt{2}$$

3. Simplify $3\sqrt{125} - 4\sqrt[3]{5} - \sqrt{45} + 3\sqrt[3]{625}$.

SOLUTION   *Step 1:* $3\sqrt{125} = 3\sqrt{25\cdot 5} = 3\cdot 5\cdot\sqrt{5} = 15\sqrt{5}$

$$4\sqrt[3]{5} = 4\sqrt[3]{5}$$
$$\sqrt{45} = \sqrt{9\cdot 5} = 3\sqrt{5}$$
$$3\sqrt[3]{625} = 3\sqrt[3]{125\cdot 5} = 3\cdot 5\cdot\sqrt[3]{5} = 15\sqrt[3]{5}$$

Step 2: $15\sqrt{5} - 4\sqrt[3]{5} - 3\sqrt{5} + 15\sqrt[3]{5}$

$$= (15\sqrt{5} - 3\sqrt{5}) + (15\sqrt[3]{5} - 4\sqrt[3]{5})$$
$$= 12\sqrt{5} + 11\sqrt[3]{5}$$

4. Simplify $\sqrt{3x^3} - 2\sqrt{3x} + \sqrt{12x^5}$.

SOLUTION

$$\sqrt{3x^3} - 2\sqrt{3x} + \sqrt{12x^5} = x\sqrt{3x} - 2\sqrt{3x} + 2x^2\sqrt{3x}$$
$$= (2x^2 + x - 2)\sqrt{3x}$$

5. Simplify $\sqrt{3a^{-1}} + \sqrt{2a^{-1}}$.

SOLUTION

$$\sqrt{3a^{-1}} + \sqrt{2a^{-1}} = \frac{\sqrt{3}}{\sqrt{a}} + \frac{\sqrt{2}}{\sqrt{a}}$$

$$= \frac{\sqrt{3} + \sqrt{2}}{\sqrt{a}}$$

EXERCISE 6-6

Simplify.

1. $4\sqrt{a} + 3\sqrt{a} - 2\sqrt{a}$
2. $\sqrt{18} + \sqrt{50} - \sqrt{72}$
3. $3\sqrt{5} + \sqrt{20} - \sqrt{45} + \sqrt{5}$
4. $\sqrt{4a} + \sqrt{16a} - \sqrt{36a}$
5. $\sqrt[3]{81} - 2\sqrt[3]{3} + 3\sqrt[3]{24}$
6. $3\sqrt{45} - 2\sqrt{25} + \sqrt{5}$
7. $\sqrt{8} + \sqrt{18} - \sqrt{32}$
8. $\sqrt[3]{2} + \sqrt[3]{-54} + \sqrt[3]{16}$
9. $\sqrt[3]{6} - 3\sqrt[3]{48} + 2\sqrt[3]{162}$
10. $\sqrt{8} - \sqrt[3]{2} + \sqrt{18}$
11. $\sqrt{6x^2} - x\sqrt{54}$
12. $2\sqrt{24} + 3\sqrt{150} - 3\sqrt{96}$
13. $3\sqrt{72} - 7\sqrt{18} + 2\sqrt[3]{54}$
14. $\sqrt[3]{a^4} + \sqrt[3]{27a^4} - \sqrt[6]{a^8}$
15. $\sqrt{8} + \sqrt[3]{16} + \sqrt{50} - \sqrt[3]{54}$
16. $\sqrt{72} + \sqrt[3]{16} - \sqrt{50} - \sqrt[3]{128}$
17. $\sqrt{6x^2} - \sqrt{24x^2} + \sqrt{9x^0}$
18. $\sqrt[m]{a^m x} + \sqrt[m]{b^m x} - \sqrt[2m]{a^{2m} x^2}$
19. $\sqrt[3]{(a - b)^2} + \sqrt[3]{(a - b)^5}$
20. $\sqrt{a^2 b} - \sqrt{9b} + a\sqrt{b}$
21. $\frac{3}{4}\sqrt{a^2 b} - \frac{2}{3}a\sqrt{b} + \frac{1}{6}\sqrt{ab}$
22. $\frac{1}{2}\sqrt{x^3 y} - \frac{1}{3}\sqrt{xy^3} - \frac{1}{2}\sqrt{xy}$
23. $\sqrt{a^2 b} - \sqrt{b^5} - \sqrt{a^4 b^3}$
24. $\sqrt{cx^2} - x\sqrt{4c} + 3c\sqrt{25cx^2}$
25. $\sqrt{a^3} - 3a\sqrt[4]{a^2} + 2a\sqrt[6]{a^3}$
26. $\sqrt[3]{m^2} - 3\sqrt[6]{m^4} + 2m\sqrt[4]{m^2}$
27. $\sqrt{ab^{-1}} + 2\sqrt{b^{-1}}$
28. $\sqrt{a^{-1}} + \sqrt{2a^{-1}b}$
29. $\sqrt{x^{-1}y} - \sqrt{2x^{-1}y^2}$
30. $\sqrt[3]{ab^{-2}} - 2\sqrt[3]{b^{-2}}$

9. MULTIPLICATION WITH RADICAL EXPRESSIONS

A common method for computing the products of radical expressions (Program **6.4**, following) is based upon one of the properties discussed in Section 7, namely

$$\sqrt[n]{a} \cdot \sqrt[n]{b} = \sqrt[n]{ab}$$

For instance, when the factors have the same index:

$$\sqrt[3]{3} \cdot \sqrt[3]{7} = \sqrt[3]{(3)(7)} = \sqrt[3]{21}$$

$$\sqrt[5]{a^2} \cdot \sqrt[5]{a^4} = \sqrt[5]{a^2 \cdot a^4} = \sqrt[5]{a^6} = a\sqrt[5]{a}$$

However, when the factors involve radicals of different order (i.e., index), a preliminary step is needed to change one or more of the radicals to a common order. In general, for c a positive integer

$$\sqrt[n]{a^m} = \sqrt[cn]{a^{cm}}$$

because

$$\sqrt[n]{a^m} = a^{m/n} = a^{c \cdot m/c \cdot n} = a^{cm/cn} = \sqrt[cn]{a^{cm}}$$

and conversely.

The following program translates the generalization above into practice.

PROGRAM 6.3

To produce an equivalent radical expression of specified order (index):

Step 1: Express the radicand exponentially with all numerical coefficients in prime factor form.

Step 2: Replace each of the exponents of Step 1 with an equivalent fractional exponent in which the denominator is the specified index.

Step 3: Write the equivalent of the expression of Step 2 in radical notation.

EXAMPLES

1. Produce a radical expression with an index of 8 that is the equivalent of $\sqrt{3}$.

SOLUTION *Step 1:* Express the radicand, 3, exponentially:

$$\sqrt{3} = 3^{1/2}$$

Step 2: Replace the exponent with an equivalent fraction having the specified index, 8, as denominator:

$$3^{1/2} = 3^{4/8}$$

Step 3: Express the result of Step 2 in radical notation:

$$3^{4/8} = \sqrt[8]{3^4} = \sqrt[8]{81}$$

2. Produce an equivalent of $\sqrt[3]{9xy^2}$ with an index of 6.

SOLUTION *Step 1:* Express the radicand exponentially:

$$\sqrt[3]{9xy^2} = (3^2xy^2)^{1/3} = 3^{2/3} \cdot x^{1/3} \cdot y^{2/3}$$

Step 2: Replace the fractional exponents with equivalent fractions having denominators of 6:

$$3^{2/3}x^{1/3}y^{2/3} = 3^{4/6}x^{2/6}y^{4/6}$$

Step 3: Express the result of Step 2 in radical notation:

$$\sqrt[6]{3^4 x^2 y^4} = \sqrt[6]{81x^2 y^4}$$

3. Given $\sqrt[4]{7a^2 b^3}$; write an equivalent radical expression having an index of 12.

SOLUTION *Step 1:* $\sqrt[4]{7a^2 b^3} = 7^{1/4}a^{2/4}b^{3/4}$
Step 2: $7^{1/4}a^{2/4}b^{3/4} = 7^{3/12}a^{6/12}b^{9/12}$
Step 3: $\sqrt[12]{7^3 a^6 b^9} = \sqrt[12]{343a^6 b^9}$

The usual procedure for multiplying two or more factors involving radical expressions, as given in the following program, is similar to that for multiplying polynomials. Recall that the "coefficient" of the radical expression, mentioned in the program, is the factor outside the radical symbol.

PROGRAM 6.4

To compute the product of radical expressions:

Step 1: Express each of the radical factors as an equivalent radical expression of a common order.

Step 2: Write the product of the coefficients as the coefficient of the product and the product of the radicands as the radicand of the product (expressed under the radical of the common order).

Step 3: (Optional but usual.) Simplify if possible.

EXAMPLES

1. Multiply $2\sqrt{3x} \cdot 4\sqrt{6xy} \cdot \sqrt{2py}$.

SOLUTION *Step 1:* All factors are of a common order.
Step 2: Multiply the coefficients:

$$2 \cdot 4 \cdot 1 = 8$$

Multiply the radicands:

$$3x \cdot 6xy \cdot 2py = 36x^2 y^2 p$$

Express as a product:

$$8\sqrt{36x^2 y^2 p}$$

Step 3: Simplify:

$$8\sqrt{36x^2y^2p} = 8 \cdot 6 \cdot x \cdot y \cdot \sqrt{p} = 48xy\sqrt{p}$$

2. Multiply $4x\sqrt[3]{2x^2y} \cdot y\sqrt{2xy}$.

SOLUTION *Step 1:* A common order is 6. (Recall Program **6.3**.)

$$\sqrt[3]{2x^2y} = (2x^2y)^{1/3} = (2x^2y)^{2/6} = \sqrt[6]{(2x^2y)^2} = \sqrt[6]{4x^4y^2}$$
$$\sqrt{2xy} = (2xy)^{1/2} = (2xy)^{3/6} = \sqrt[6]{(2xy)^3} = \sqrt[6]{8x^3y^3}$$

Step 2: Product of coefficients:

$$4x \cdot y = 4xy$$

Product of radicands:

$$\sqrt[6]{4x^4y^2} \cdot \sqrt[6]{8x^3y^3} = \sqrt[6]{4x^4y^2 \cdot 8x^3y^3} = \sqrt[6]{4 \cdot 8 \cdot x^{4+3}y^{2+3}} = \sqrt[6]{32x^7y^5}$$

Product of the given factors:

$$4xy\sqrt[6]{32x^7y^5}$$

Step 3: Simplify:

$$4xy\sqrt[6]{32x^7y^5} = 4x^2y\sqrt[6]{32xy^5}$$

3. Multiply $(\sqrt{3} - 2\sqrt{2})(\sqrt{3} + 2\sqrt{2})$.

SOLUTION The radical expressions are of a common order. Treat the factors as polynomials and use Program **2.4** for computing the product of the sum and difference of two terms:

$$(\sqrt{3} - 2\sqrt{2})(\sqrt{3} + 2\sqrt{2}) = (\sqrt{3})^2 - (2\sqrt{2})^2$$
$$= 3 - 8$$
$$= -5$$

4. Multiply $(3\sqrt{x} + 2\sqrt{y})(2\sqrt{x} - 5\sqrt{y})$.

SOLUTION Radical expressions are of common order. Compute the product by Program **2.6**:

$$6\sqrt{x^2} - 11\sqrt{xy} - 10\sqrt{y^2} = 6x - 11\sqrt{xy} - 10y$$

Complete the equivalent.

1. $\sqrt{3} = \sqrt[4]{}$
2. $\sqrt{2xy} = \sqrt[6]{}$
3. $\sqrt[3]{4xy} = \sqrt[6]{}$
4. $\sqrt[3]{6x^2y} = \sqrt[9]{}$
5. $\sqrt{16} = \sqrt[3]{}$
6. $\sqrt[4]{13x^2y^3z} = \sqrt[8]{}$
7. $\sqrt{x^2 - y^2} = \sqrt[6]{}$
8. $\sqrt[3]{(a - b)^2} = \sqrt[6]{}$

Compute the products.

9. $\sqrt[3]{2a} \cdot \sqrt[3]{4a^2}$
10. $(4x\sqrt{3x})^2$
11. $\sqrt{3a^3} \cdot \sqrt{6}$
12. $\sqrt{2xy} \cdot \sqrt{6y}$
13. $\sqrt{2} \cdot \sqrt[3]{3}$
14. $\sqrt[3]{4} \cdot \sqrt{6}$
15. $\sqrt[3]{2xy} \cdot \sqrt[6]{4y^2}$
16. $\sqrt{3a} \cdot \sqrt[4]{5a^3}$
17. $(2a\sqrt{3a})(b\sqrt[3]{ab})$
18. $(3x\sqrt[3]{2xy^2})(x\sqrt{2x})$
19. $\sqrt{3}(\sqrt{3} - 2\sqrt{5} + \sqrt{6})$
20. $\sqrt{6}(\sqrt{2} - 3\sqrt{3} + 2\sqrt{6})$
21. $\sqrt{x}(\sqrt{x} - 1)$
22. $\sqrt{b}(3 + \sqrt{b})$
23. $\sqrt{a}(\sqrt{2a} + \sqrt{3})$
24. $\sqrt{3}(\sqrt{3b} - \sqrt{a})$
25. $\sqrt[3]{a}(\sqrt{a} - \sqrt[3]{a})$
26. $\sqrt{x}(3\sqrt{x} - \sqrt[4]{x^3})$
27. $(\sqrt{3} - 4)^2$
28. $(\sqrt{5} + 1)^2$
29. $(\sqrt{3} - 2)(\sqrt{3} + 2)$
30. $(2 + \sqrt{7})(2 - \sqrt{7})$
31. $(\sqrt{5} + \sqrt{3})(\sqrt{5} - \sqrt{3})$
32. $(\sqrt{2} - \sqrt{11})(\sqrt{2} + \sqrt{11})$
33. $(3\sqrt{5} - 2\sqrt{3})^2$
34. $(2\sqrt{6} + 3\sqrt{2})^2$
35. $(\sqrt{a} - \sqrt{b})(\sqrt{a} + \sqrt{b})$
36. $(\sqrt{a} - \sqrt{b})^2$
37. $(\sqrt{3} - 2\sqrt{5})(2\sqrt{3} + \sqrt{5})$
38. $(\sqrt{2} - \sqrt{3})(2\sqrt{2} - \sqrt{3})$
39. $(\sqrt{3} + \sqrt{a})(2\sqrt{3} + 3\sqrt{a})$
40. $(\sqrt{2} - \sqrt{x})(3\sqrt{2} + 4\sqrt{x})$
41. $(2\sqrt{x} - 3\sqrt{y})(2\sqrt{x} + 5\sqrt{y})$
42. $(\sqrt{x} - 4\sqrt{y})(5\sqrt{x} - 3\sqrt{y})$
43. $(\sqrt{2x} - \sqrt{3y})(\sqrt{2x} - 3\sqrt{3y})$
44. $(\sqrt{3a} - \sqrt{b})(2\sqrt{3a} - 3\sqrt{b})$
45. $(2\sqrt{3a} - 2\sqrt{2b})(3\sqrt{3a} + \sqrt{2b})$
46. $(3\sqrt{2x} - 4\sqrt{5y})(\sqrt{2x} + 3\sqrt{5y})$
47. $(\sqrt[3]{x} - \sqrt[3]{y})^2$
48. $(\sqrt[3]{a} - \sqrt[3]{b})(\sqrt[3]{a} + \sqrt[3]{b})$
49. $(\sqrt{a + 3} + 4)^2$
50. $(\sqrt{x - 2} + \sqrt{2})^2$
51. $(\sqrt{a + b} - \sqrt{a - b})^2$
52. $(\sqrt[3]{a^2b} - \sqrt{ab})^2$

Find the value of

53. $x^2 + x - 1$ when $x = 2 - \sqrt{2}$
54. $2x^2 - x + 1$ when $x = \sqrt{3} - 1$
55. $3x^2 + 2x - 4$ when $x = \sqrt{3} + 2$
56. $3x^2 - 3x - 5$ when $x = 2\sqrt{3} - \sqrt{2}$

10. RATIONALIZING A DENOMINATOR

To **rationalize the denominator** of a fraction means to express the fraction equivalently, but without any radical expression in the denominator. The following program for doing so is based on the fact that an equivalent fraction results when the numerator and denominator of a fraction are both multiplied by the same non-zero number. A suitable multiplier is one that yields a denominator whose terms have only positive integral exponents and rational coefficients.

Note: The resulting equivalent fraction with rational denominator may or may not have a rational numerator.

PROGRAM 6.5

To rationalize a denominator:

Step 1:   Express the denominator exponentially.

Step 2:   Determine a factor that, when multiplied with the given denominator, yields a product containing only positive integral exponents and rational coefficients.

Step 3:   Multiply the numerator and denominator by the factor of Step 2.

EXAMPLES

1.   Express an equivalent of $\dfrac{5}{\sqrt[3]{a}}$ with a rational denominator.

SOLUTION   *Step 1:* Express the denominator, $\sqrt[3]{a}$, exponentially:

$$\sqrt[3]{a} = a^{1/3}$$

Step 2: A suitable factor is $a^{2/3}$, since

$$a^{1/3} \cdot a^{2/3} = a^{1/3 + 2/3} = a^{3/3} = a^1$$

Step 3: Multiply numerator and denominator by $a^{2/3}$:

$$\frac{5 \cdot a^{2/3}}{a^{1/3} \cdot a^{2/3}} = \frac{5a^{2/3}}{a^1} = \frac{5\sqrt[3]{a^2}}{a}$$

2.   Rationalize the denominator: $\dfrac{p}{\sqrt[3]{x^2 y}}$.

SOLUTION   *Step 1:* $\sqrt[3]{x^2 y} = x^{2/3} y^{1/3}$

Step 2: A suitable factor is $x^{2/3}y^{1/3}$, since

$$(x^{2/3}y^{1/3})(x^{1/3}y^{2/3}) = x^{2/3+1/3}y^{1/3+2/3} = x^1y^1$$

Step 3: $\dfrac{(p)(x^{1/3}y^{2/3})}{(x^{2/3}y^{1/3})(x^{1/3}y^{2/3})} = \dfrac{p\sqrt[3]{xy^2}}{xy}$

3. Express an equivalent of $\sqrt{\dfrac{3}{5}}$ with a rational denominator.

Note: $\sqrt{\dfrac{3}{5}} = \dfrac{\sqrt{3}}{\sqrt{5}} = \dfrac{3^{1/2}}{5^{1/2}}$.

SOLUTION | *Step 1:* Express the denominator $\sqrt{5}$ exponentially: $5^{1/2}$.

Step 2: A suitable factor is $5^{1/2}$, since $5^{1/2} \cdot 5^{1/2} = 5^{1/2+1/2} = 5^1$.

Step 3: $\dfrac{3^{1/2} \cdot 5^{1/2}}{5^{1/2} \cdot 5^{1/2}} = \dfrac{3^{1/2} \cdot 5^{1/2}}{5^1} = \dfrac{(3 \cdot 5)^{1/2}}{5} = \dfrac{(15)^{1/2}}{5} = \dfrac{\sqrt{15}}{5}$.

When the denominator of a radical expression is a binomial containing one or two square root radicals, it can be rationalized by a special technique. It makes use of the fact that the product of the sum and difference of two numbers is the difference of their squares (Program **2.4**). For instance,

$$(\sqrt{7} + \sqrt{2})(\sqrt{7} - \sqrt{2}) = (\sqrt{7})^2 - (\sqrt{2})^2 = 7 - 2 = 5$$

PROGRAM 6.6

To rationalize a binomial denominator involving square roots:

Step 1:   Multiply both numerator and denominator by the denominator with the sign of the second term reversed.

Step 2:   (Optional, but usual.) Simplify if possible.

Note: This program, as stated, is valid only for square roots; it is not valid when the index of the radical is other than 2.

EXAMPLES

1. Express equivalently with a rational denominator: $\dfrac{1}{\sqrt{5} - \sqrt{3}}$

SOLUTION | *Step 1:* The denominator is $\sqrt{5} - \sqrt{3}$; multiply numerator and denominator by $\sqrt{5} + \sqrt{3}$:

$$\frac{1}{(\sqrt{5} - \sqrt{3})} \cdot \frac{(\sqrt{5} + \sqrt{3})}{(\sqrt{5} + \sqrt{3})} = \frac{\sqrt{5} + \sqrt{3}}{(\sqrt{5})^2 - (\sqrt{3})^2}$$

Step 2: Simplify:

$$\frac{\sqrt{5} + \sqrt{3}}{(\sqrt{5})^2 - (\sqrt{3})^2} = \frac{\sqrt{5} + \sqrt{3}}{5 - 3} = \frac{\sqrt{5} + \sqrt{3}}{2}$$

2. Rationalize the denominator: $\dfrac{8}{\sqrt{7} + \sqrt{3}}$.

SOLUTION Multiply numerator and denominator by $\sqrt{7} - \sqrt{3}$; then simplify:

$$\frac{8}{(\sqrt{7} + \sqrt{3})} \cdot \frac{(\sqrt{7} - \sqrt{3})}{(\sqrt{7} - \sqrt{3})} = \frac{8(\sqrt{7} - \sqrt{3})}{7 - 3}$$

$$= \frac{\overset{2}{\cancel{8}}(\sqrt{7} - \sqrt{3})}{\underset{}{\cancel{4}}}$$

$$= 2(\sqrt{7} - \sqrt{3}) = 2\sqrt{7} - 2\sqrt{3}$$

3. Rationalize the denominator: $\dfrac{y}{1 - \sqrt{y}}$.

SOLUTION

$$\frac{(y) \cdot (1 + \sqrt{y})}{(1 - \sqrt{y}) \cdot (1 + \sqrt{y})} = \frac{y(1 + \sqrt{y})}{1 - y}$$

EXERCISE 6-8

Write equivalents for each in simplest terms with rational denominators.

1. $\dfrac{1}{\sqrt{3}}$ **2.** $\dfrac{2a}{\sqrt{2a}}$ **3.** $\dfrac{3x}{\sqrt{5}}$

4. $\dfrac{7a}{2\sqrt{3}}$ **5.** $\dfrac{3}{\sqrt[3]{4}}$ **6.** $\dfrac{2p}{\sqrt[3]{p^2}}$

7. $\dfrac{12x}{\sqrt[3]{x}}$ **8.** $\sqrt{\dfrac{6a}{5b}}$ **9.** $\sqrt{7m^{-1}}$

10. $\dfrac{\sqrt[3]{9}}{\sqrt[3]{3}}$ **11.** $\sqrt[3]{\dfrac{-3a}{4b^2c^7}}$ **12.** $\sqrt{\dfrac{a - b}{a + b}}$

13. $\dfrac{1}{\sqrt{3} - \sqrt{2}}$ **14.** $\dfrac{2}{\sqrt{3} + \sqrt{5}}$ **15.** $\dfrac{\sqrt{2}}{\sqrt{2} - \sqrt{3}}$

16. $\dfrac{\sqrt{3}}{3 - \sqrt{3}}$ **17.** $\dfrac{x}{1 - \sqrt{x}}$ **18.** $\dfrac{\sqrt{a}}{\sqrt{a} - \sqrt{b}}$

19. $\dfrac{\sqrt{2}+\sqrt{3}}{\sqrt{2}-\sqrt{3}}$ **20.** $\dfrac{\sqrt{5}-\sqrt{3}}{\sqrt{5}+\sqrt{3}}$ **21.** $\dfrac{\sqrt{3}}{2\sqrt{3}-3\sqrt{2}}$

22. $\dfrac{\sqrt{a}-\sqrt{b}}{\sqrt{a}+\sqrt{b}}$ **23.** $\dfrac{2\sqrt{2}-\sqrt{3}}{3\sqrt{2}+\sqrt{3}}$ **24.** $\dfrac{3\sqrt{x}+2\sqrt{y}}{\sqrt{x}-\sqrt{y}}$

Use the techniques of rationalizing denominators to rationalize the numerator of

25. Exercise 19   **26.** Exercise 20   **27.** Exercise 23   **28.** Exercise 24

11. DIVISION WITH RADICAL EXPRESSIONS

The quotient of radical expressions is most readily found by expressing the two terms in fraction form: the dividend as the numerator and the divisor as the denominator. If the divisor-denominator contains a radical expression, it is ordinarily rationalized. The resulting equivalent fraction (preferably simplified) is the desired quotient.

PROGRAM 6.7

To compute the quotient of two radical expressions:

Step 1: Express the dividend as the numerator and the divisor as the denominator of a fraction.

Step 2: Rationalize the denominator.

Step 3: Simplify the rationalized fraction.

EXAMPLES

1. Divide $12\sqrt{5}$ by $\sqrt{3}$.

SOLUTION *Step 1:* Express as a fraction:

$$12\sqrt{5} \div \sqrt{3} = \frac{12\sqrt{5}}{\sqrt{3}}$$

Step 2: Rationalize the denominator by multiplying by $\dfrac{\sqrt{3}}{\sqrt{3}}$:

$$\frac{12\sqrt{5}\cdot\sqrt{3}}{\sqrt{3}\cdot\sqrt{3}} = \frac{12\sqrt{15}}{3}$$

Step 3: Simplify:

$$\frac{12\sqrt{15}}{3} = 4\sqrt{15} \quad \text{(quotient)}$$

2.   Divide $(4x + \sqrt{6})$ by $3\sqrt{2}$.

SOLUTION   *Step 1:* Express as a fraction:

$$(4x + \sqrt{6}) \div 3\sqrt{2} = \frac{4x + \sqrt{6}}{3\sqrt{2}}$$

Step 2: Rationalize the denominator:

$$\frac{(4x + \sqrt{6}) \cdot (\sqrt{2})}{(3\sqrt{2}) \cdot (\sqrt{2})} = \frac{4x\sqrt{2} + \sqrt{12}}{3 \cdot 2}$$

Step 3: Simplify:

$$\frac{4x\sqrt{2} + \sqrt{12}}{6} = \frac{4x\sqrt{2}}{6} + \frac{2\sqrt{3}}{6}$$

$$= \frac{2x\sqrt{2}}{3} + \frac{\sqrt{3}}{3} \left(\text{or } \frac{2}{3}x\sqrt{2} + \frac{1}{3}\sqrt{3} \right)$$

3.   Divide $1 - \sqrt{x}$ by $1 + \sqrt{x}$.

SOLUTION $(1 - \sqrt{x}) \div (1 + \sqrt{x}) = \dfrac{1 - \sqrt{x}}{1 + \sqrt{x}}$

$$= \frac{(1 - \sqrt{x})(1 - \sqrt{x})}{(1 + \sqrt{x})(1 - \sqrt{x})}$$

$$= \frac{1 - 2\sqrt{x} + x}{1 - x}$$

4.   Divide $4\sqrt{5} - 2$ by $3\sqrt{2} - 2\sqrt{5}$.

SOLUTION   *Step 1:* $(4\sqrt{5} - 2) \div (3\sqrt{2} - 2\sqrt{5}) = \dfrac{4\sqrt{5} - 2}{3\sqrt{2} - 2\sqrt{5}}$

Step 2: $\dfrac{(4\sqrt{5} - 2) \cdot (3\sqrt{2} + 2\sqrt{5})}{(3\sqrt{2} - 2\sqrt{5}) \cdot (3\sqrt{2} + 2\sqrt{5})}$

$$= \frac{12\sqrt{10} - 6\sqrt{2} + 8\sqrt{25} - 4\sqrt{5}}{(3\sqrt{2})^2 - (2\sqrt{5})^2}$$

Step 3:
$$= \frac{12\sqrt{10} - 6\sqrt{2} + (8 \cdot 5) - 4\sqrt{5}}{18 - 20}$$

$$= \frac{12\sqrt{10} - 6\sqrt{2} + 40 - 4\sqrt{5}}{-2}$$

$$= -6\sqrt{10} + 3\sqrt{2} - 20 + 2\sqrt{5}$$

5.   Divide $\sqrt{x} - 2\sqrt{y}$ by $\sqrt{x} - \sqrt{y}$.

SOLUTION
$$\frac{\sqrt{x} - 2\sqrt{y}}{\sqrt{x} - \sqrt{y}} = \frac{(\sqrt{x} - 2\sqrt{y})(\sqrt{x} + \sqrt{y})}{(\sqrt{x} - \sqrt{y})(\sqrt{x} + \sqrt{y})}$$

$$= \frac{x - \sqrt{xy} - 2y}{x - y}$$

| **EXERCISE 6-9**

Compute the quotients.

1. $8 \div 2\sqrt{2}$ **2.** $3 \div \sqrt[3]{9}$

3. $(\sqrt{6} - 2\sqrt{15}) \div \sqrt{3}$ **4.** $(3\sqrt{2} - \sqrt{3} + 2\sqrt{6}) \div \sqrt{6}$

5. $(2\sqrt{6} - 3\sqrt{7} + 2\sqrt{3}) \div 3\sqrt{42}$

6. $(4\sqrt{3} + 3\sqrt{2} + \sqrt{8} - 3\sqrt{6}) \div 2\sqrt{6}$

7. $(2\sqrt{3} + 3\sqrt[3]{2}) \div \sqrt[3]{3}$ **8.** $(3\sqrt[3]{4} - 2\sqrt[3]{3} + 2\sqrt{6}) \div \sqrt[3]{6}$

9. $3 \div (1 + \sqrt{10})$ **10.** $a \div (1 + \sqrt{a})$

11. $(2\sqrt{5} - 7\sqrt{3}) \div (\sqrt{5} - 2\sqrt{3})$ **12.** $(3\sqrt{6} - 2\sqrt{3}) \div (\sqrt{6} + 2\sqrt{3})$

13. $(5 + 3\sqrt{2}) \div (\sqrt{6} - \sqrt{3})$ **14.** $(\sqrt{6} - 3\sqrt{2}) \div (4\sqrt{2} - 3\sqrt{6})$

15. $(3 - \sqrt{a}) \div (3 + \sqrt{a})$ **16.** $(\sqrt{x} - 2) \div (\sqrt{x} - 5)$

17. $(\sqrt{x} - y) \div (y - \sqrt{x})$ **18.** $(7 - \sqrt{x}) \div (\sqrt{x} + 4)$

19. $(\sqrt{a} - 2\sqrt{b}) \div (\sqrt{a} + 3\sqrt{b})$ **20.** $(2\sqrt{x} - \sqrt{y}) \div (\sqrt{x} - 2\sqrt{y})$

21. $(3\sqrt{m} - 2\sqrt{n}) \div (2\sqrt{m} + 3\sqrt{n})$ **22.** $(2\sqrt{x} + 3\sqrt{y}) \div (\sqrt{x} - 2\sqrt{y})$

23. $(\sqrt{2x} - 2\sqrt{y}) \div (\sqrt{x} - 3\sqrt{y})$ **24.** $(2\sqrt{a} - \sqrt{3b}) \div (\sqrt{2a} - 3\sqrt{b})$

12. COMPLEX NUMBERS

The real numbers are sufficient for the operations of addition, subtraction, multiplication, division (except by zero), raising to a power, and extracting most roots. The exception to the last operation is extracting even roots of negative numbers. In that case, we need to draw on a larger set of numbers—a set that includes a number called the **imaginary unit**, which is symbolized by i and defined as having the property $i^2 = -1$.*

Thus

$$\sqrt{-1} = \sqrt{i^2} = i$$

$$\sqrt{-9} = \sqrt{(9)(-1)} = \sqrt{9i^2} = \sqrt{9} \cdot \sqrt{i^2} = 3 \cdot i = 3i$$

$$\sqrt{-5} = \sqrt{(5)(-1)} = \sqrt{5i^2} = \sqrt{5} \cdot \sqrt{i^2} = i\sqrt{5}$$

Such numbers, of the form bi, where b denotes a real number and i the imaginary unit, are called **pure imaginary numbers**. They are part of a larger set of numbers called the **imaginary numbers**, which are sums and differences of real and pure imaginary numbers. For instance,

$$3 + 2i, \quad 7 - i, \quad 21 + i\sqrt{3}, \quad \sqrt{5} + i, \quad \sqrt{7} - i\sqrt{3}$$

Together the real and the imaginary numbers constitute a more comprehensive set called the **complex numbers**. Within the set of complex numbers we can carry out all the operations of the algebra of polynomials: addition, subtraction, multiplication, division (except by zero), raising to a power, and, without exception, extracting roots.

Standard numerals for complex numbers are expressed in binomial form, $a + bi$, in which a and b represent real numbers and i the imaginary unit. The first term, a, is referred to as the *real part* of the complex number. The second term, bi, is referred to as the *imaginary part* of the complex number.

Note that real numbers are complex numbers in which the coefficient of the imaginary part, b, is 0:

$$a + 0i = a$$

On the other hand, pure imaginary numbers are complex numbers in which the real part, a, is 0:

$$0 + bi = bi$$

The complete number system for the algebra of polynomials can be diagrammed as follows.

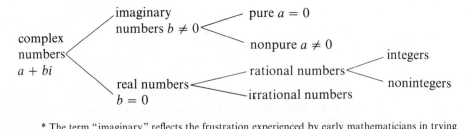

* The term "imaginary" reflects the frustration experienced by early mathematicians in trying to explain the square root of negative numbers; ironically, the term has endured.

The number whose numeral is i has an important cyclic property:

$$i^1 = i$$
$$i^2 = -1$$
$$i^3 = i^2 \times i = (-1)(i) = -i$$
$$i^4 = i^2 \times i^2 = (-1)(-1) = +1$$
$$i^5 = i^4 \times i = (+1)(i) = i$$
$$i^6 = i^4 \times i^2 = (+1)(-1) = -1$$
$$i^7 = i^4 \times i^3 = (+1)(i^3) = i^3 = -i$$
$$i^8 = i^4 \times i^4 = (+1)(+1) = +1$$
$$i^9 = i^8 \times i = (+1)(i) = i$$

etc.

In other words,

$$i^1 = i^5 = i^9 = i^{13} = \cdots = i$$
$$i^2 = i^6 = i^{10} = i^{14} = \cdots = -1$$
$$i^3 = i^7 = i^{11} = \cdots = -i$$
$$i^4 = i^8 = i^{12} = \cdots = +1$$

Thus, every integral power of i can be expressed equivalently by $i, -1, -i,$ or 1. To illustrate,

$$i^{15} = i^4 \cdot i^4 \cdot i^4 \cdot i^3 = (1)(1)(1)(-i) = -i$$
$$i^{26} = (i^4)^6 \cdot (i^2) = (1)^6(-1) = -1$$
$$i^{100} = (i^4)^{25} = (1)^{25} = 1$$

13. COMPUTING WITH COMPLEX NUMBERS

The operations of addition, subtraction, multiplication, and division (except by zero), as well as the operations of raising to a power and extracting a root, are all possible within the set of complex numbers. This fact we noted in the previous section. Moreover, the complex numbers are commutative and associative under addition and multiplication and distributive for multiplication over addition.

A complex number, $a + bi$, can be treated under the various operations as though it were an algebraic binomial. The equivalencies among the various powers of i, also noted in the previous section, simplify matters of expression and computation.

PROGRAM 6.8

To add, subtract, or multiply complex numbers:

Step 1: Replace the i expression in each term by its equivalent, i, -1, $-i$, or $+1$.

Step 2: Treat i as though it were an algebraic variable and compute the desired sum, difference, or product.

Step 3: Replace all expressions involving i by its equivalent—i, -1, $-i$, or $+1$—and simplify to the form of $a + bi$.

EXAMPLES

1. Add $(6 + 3i^3) + (2 - i) + (4 - 3i^7)$.

SOLUTION *Step 1:* Simplify the i-expressions:

$$6 + 3i^3 = 6 - 3i$$
$$2 - i = 2 - i$$
$$4 - 3i^7 = 4 + 3i$$

SOLUTION *Steps 2 and 3:* Treat i as an algebraic variable, simplify, and express the result in the form, $a + bi$:

$$(6 + 3i^3) + (2 - i) + (4 - 3i^7) = (6 - 3i) + (2 - i) + (4 + 3i)$$
$$= 6 - 3i + 2 - i + 4 + 3i$$
$$= (6 + 2 + 4) + (-3i - i + 3i)$$
$$= 12 - i$$

2. Subtract $(4 - 8i^3) - (6 - 2i^{12})$.

SOLUTION *Step 1:* $\quad 4 - 8i^3 = 4 + 8i$

$\qquad\qquad 6 - 2i^{12} = 6 - (2)(+1) = 4$

Steps 2 and 3:

$$(4 - 8i^3) - (6 - 2i^{12}) = (4 + 8i) - 4$$
$$= (4 - 4) + 8i$$
$$= 0 + 8i$$

3. Multiply $(4 - 3i^3)(6 + 7i^7)$.

SOLUTION *Step 1:* $4 - 3i^3 = 4 + 3i$

$\qquad\qquad\quad 6 + 7i^7 = 6 - 7i$

Step 2: $(4 - 3i^3)(6 + 7i^7) = (4 + 3i)(6 - 7i)$

$\qquad\qquad\qquad\qquad\quad = 24 - 10i - 21i^2$

Step 3: $24 - 10i - 21i^2 = 24 - 10i + 21 = 45 - 10i$

4.   Multiply $(-6 - 2i^2)(3 - 5i^3)(1 + 3i^{13})$.

SOLUTION   *Step 1:* $-6 - 2i^2 = -6 + 2 = -4$
$$3 - 5i^3 = 3 + 5i$$
$$1 + 3i^{13} = 1 + 3i$$

Step 2: $(-4)(3 + 5i)(1 + 3i) = (-4)(3 + 14i + 15i^2)$
$$= -12 - 56i - 60i^2$$

Step 3: $-12 - 56i - 60i^2 = -12 - 56i + 60$
$$= 48 - 56i$$

The **conjugate of a complex number**, $a + bi$, is the number $a - bi$. That is, two complex numbers are conjugates of one another if they have the same real part, a, and if their imaginary parts, bi, are negatives of one another.

For instance,

$$3 - 2i \quad \text{and} \quad 3 + 2i$$

are conjugate complex numbers. So are

$$5 + \frac{1}{2}i \quad \text{and} \quad 5 - \frac{1}{2}i$$

Note, however, that

$$-3 + 5i \quad \text{and} \quad 3 - 5i$$

are *not* conjugates but the negatives of one another, since their sum is zero:

$$(-3 + 5i) + (3 - 5i) = -3 + 5i + 3 - 5i = 0$$

Computing the quotient of two complex numbers involves a procedure similar to Program **6.7** and the use of the conjugate of the denominator as multiplier.

PROGRAM 6.9

To divide one complex number by another:

Step 1:   Replace the i expression in each term by its equivalent, i, -1, $-i$, or $+1$.

Step 2:   Express the terms of Step 1—dividend and divisor—as numerator and denominator of a fraction.

Step 3:   Multiply numerator and denominator of the fraction of Step 2 by the conjugate of the denominator.

Step 4:   Simplify and express the result of Step 3 in the form $a + bi$.

EXAMPLES

1. Divide $3 + i^5$ by $1 - i^3$.

SOLUTION *Step 1:* Simplify the *i*-expressions:

$$3 + i^5 = 3 + i$$
$$1 - i^3 = 1 + i$$

Step 2: Express the division as a fraction:

$$(3 + i^5) \div (1 - i^3) = (3 + i) \div (1 + i) = \frac{3 + i}{1 + i}$$

Step 3: The conjugate of $1 + i$ is $1 - i$; use it to multiply numerator and denominator:

$$\frac{(3 + i) \cdot (1 - i)}{(1 + i) \cdot (1 - i)} = \frac{3 - 2i - i^2}{1 - i^2}$$

Step 4: Simplify to the form, $a + bi$:

$$= \frac{3 - 2i - (-1)}{1 - (-1)} = \frac{4 - 2i}{2} = 2 - i$$

2. Divide $3 - 2i$ by $4 - 3i^7$.

SOLUTION *Step 1:* $\quad 3 - 2i = 3 - 2i$
$\qquad\qquad 4 - 3i^7 = 4 + 3i$

Step 2: $(3 - 2i) \div (4 - 3i^7) = \dfrac{3 - 2i}{4 + 3i}$

Step 3: The conjugate of $4 + 3i$ is $\quad 4 - 3i$:

$$\frac{(3 - 2i)(4 - 3i)}{(4 + 3i)(4 - 3i)} = \frac{12 - 17i + 6i^2}{16 - 9i^2}$$

Step 4: $\qquad\qquad\qquad = \dfrac{12 - 17i - 6}{16 + 9}$

$$= \frac{6 - 17i}{25} = \frac{6}{25} - \frac{17}{25}i$$

3. Divide $(4 - 2i^6) \div (i^5 - 3)$.

SOLUTION

$$4 - 2i^6 = 4 - (2)(-1) = 4 + 2 = 6$$
$$i^5 - 3 = -3 + i^5 = -3 + i$$

$$(4 - 2i^6) \div (i^5 - 3) = \frac{6}{-3 + i} = \frac{(6)(-3 - i)}{(-3 + i)(-3 - i)}$$

$$= \frac{-18 - 6i}{9 - i^2}$$

$$= \frac{-18 - 6i}{9 - (-1)} = \frac{-18 - 6i}{10} = -\frac{9}{5} - \frac{3}{5}i$$

Replace each term by its equivalent: i, -1, $-i$, or $+1$.

1. i^4 2. i^6 3. i^{12}

4. i^9 5. i^{13} 6. i^{19}

7. i^{72} 8. i^{103} 9. i^{-2}

10. i^{-3} 11. i^{-5} 12. i^{-8}

Simplify the following sums.

13. $2i + 3i^2 + 2i^3$ 14. $i^4 - 3i + 2i^2$

15. $i^5 - 2i^2 - 3$ 16. $3i^5 + 2i^6 + 3i^8 - 3$

17. $i\sqrt{3} + i - 2i\sqrt{4} + 3i^2 + i^{17}$ 18. $i^5 - 3i^6 + 2i^7 + 8i^2$

19. $i^6 + i^{-2} + i^{-3} - 3i^{-5}$ 20. $i^{-8} + i^{-17} + i^{-18} + i^{-23}$

Express in the form $a + bi$.

21. $(3 + i) + (2 - 3i)$ 22. $(2 + 3i) - (4 + 6i)$

23. $(1 - 6i) + (3 + 2i) + (2 - i)$ 24. $(4 - i\sqrt{3}) + (2 - i\sqrt{8})$

25. $(3 + 6i) + (5 - 3i) + (4 + 2i^3)$ 26. $(4 + 3i - 2i^2) + (3 + 5i^2 - 4i^5)$

27. $\left(\dfrac{1}{2} + \dfrac{1}{\sqrt{3}}i\right) + \left(\dfrac{3}{4} - i\sqrt{3}\right) - \left(\dfrac{1}{8} - \dfrac{2}{3}i\sqrt{3}\right)$

28. $[(3 + i^2) + (4 - i^3)] + [4 - i^8]$

29. $(2 + 3i)(3 - 4i)$ 30. $(3 - 5i)(2 + 3i)$

31. $(5 - 3i)(2 - 3i)$ 32. $(6 - 3i^3)(2 - i^5)$

33. $(3 - 2i^5)(4 - 3i^4)$ 34. $(2 - 3i)^2$

35. $(-3 - 2i)^3$ 36. $(3 - 2i)(-4 - 3i)(2 + 2i)$

37. $(3 - 2i^3)(4 - 3i^2)(5 + i^5)$ 38. $(3 - 2i^{-2})(3 + 4i^3)^2(-1 + 3i^{-8})$

39. $(3 + 2i) \div (4 - i)$ 40. $(2 - 5i) \div (3 - 5i)$

41. $(6 - i^4) \div (2 - 3i^2)$ 42. $(3 - 2i^3) \div (-2 - i^7)$

43. $(-3 + 2i^{-3}) \div (3 + 2i^{-4})$ 44. $(2 + 3i)^2 \div (3 - 2i)$

45. $(2 + 3i - 6i^2) \div (3 - 2i - i^5)$ 46. $(\sqrt{3} - i\sqrt{2}) \div (\sqrt{3} + 2i\sqrt{2})$

REVIEW

PART A

Answer True or False.

1. $3.264^0 = 3.264$.
2. 6.23×10^{-2} is the number 623 expressed in scientific notation.
3. a^{-n} and a^n are reciprocals of one another.

4. $\sqrt{-4}$ does not exist among the real numbers.

5. $-3^4 = (-3)^4$.

6. $\sqrt[3]{n^2} = n^{3/2}$.

7. When radical expressions have the same radicand and index, they are said to be *like*; for example, $2\sqrt{5}$ and $5\sqrt{5}$ are *like*.

8. The result of a computation is incorrect if a radical expression appears in the denominator.

9. The set of whole numbers, 0, 1, 2, 3, ... are included in the set of complex numbers.

10. The conjugate of $4 - 7i$ is $-4 + 7i$.

PART B

Simplify; assume that literal exponents represent positive integers.

1. $(-3)^4$

2. $-(5)^4$

3. $(-0.3)^3$

4. $-(3g)^2$

5. $\left(-\dfrac{2}{5}\right)^3$

6. $a^4 \cdot a^2 \cdot a$

7. $m^8 \div m^{10}$

8. $\dfrac{a^3b^2c^4}{a^2b^5c}$

9. $(-b)^{23}$

10. $(-a^6b^4c)^3$

11. $\left(\dfrac{6+3}{6}\right)^3$

12. $\left(\dfrac{x^7y^2}{xy^3}\right)^4$

13. $\left(-\dfrac{3p}{q^4}\right)^5$

14. $-\left(\dfrac{2m}{y^2}\right)^2$

15. $a^{5n} \div a^4$

16. $p^{st} \cdot p^{s+2}$

17. $(a^2b^m)^a$

18. $\left(-\dfrac{3a^3}{b^2}\right)^3 \div \left(\dfrac{2a}{b}\right)^2$

19. $\left[-\left(\dfrac{xy}{b^2}\right)\left(\dfrac{b^2c}{xy}\right)^3\right]^2$

20. $\left[-\left(\dfrac{3xy}{2cd}\right)^3 \div \left(\dfrac{3x^2y}{2cd^2}\right)^2\right]$

21. $(-3a^2b^0)^0$

22. $3 \times 10^0 \times 2^2 \times 5^0$

23. $\left(\dfrac{5+4}{4}\right)^0$

24. $\dfrac{2x^0y^2z}{(3xy^0)^0}$

Write an equivalent expression using a negative exponent.

25. $\dfrac{1}{x^2}$

26. $\dfrac{1}{a^5}$

27. m^3

28. t^4

29. $-\dfrac{1}{n^3}$

30. $-a^2$

Simplify each to an expression involving only positive exponents.

31. $\left(\dfrac{x}{2y}\right)^{-2}$

32. $\left(\dfrac{2}{xy}\right)^{-3}$

33. $a^4 \cdot a^{-3} \cdot a^m$

34. $(-2x^0y^{-3})^{-3}$

35. $\left(\dfrac{a^{-4}b^{-2}}{b^{-3}c^2}\right)^3$

36. $\left(\dfrac{ax^5}{m^2}\right)^{-2}\left(\dfrac{ab}{m^2n}\right)^4$

37. $a^{-1} - y^{-1} + z^{-1}$

38. $3x^{-2} + y^{-2}$

39. $(s^{-2}t^{-2}) \div (st)^3$

40. $\left(\dfrac{m}{n} - \dfrac{n}{m}\right)^{-1}$

41. $(x^{-2} - y^{-2})^{-2}$

42. $\dfrac{mn^{-2}}{n^{-2} + m^2}$

43. $\dfrac{a^2 + ab - 2b^2}{a^{-1} - b^{-1}}$

44. $\dfrac{b^4 - 4a^4}{a^{-2} - 2b^{-2}}$

Use scientific notation to express each number.

45. 280,000

46. 1,350,000

47. 0.000321

48. 0.000067

49. 3

50. 0.1

Express in the usual decimal notation.

51. 6.32×10^4

52. 5.7×10^3

53. 3.4×10^{-3}

54. 6.2×10^{-5}

55. 4.0×10^{-1}

56. 7.0×10^0

Simplify; express results in scientific notation.

57. $(3 \times 10^4)(2 \times 10^{-2})(6 \times 10^5)$

58. $(2.1 \times 10^{-2})(2.0 \times 10^{-6})(1 \times 10^{-3})$

59. $\dfrac{(6 \times 10^{-6})(2 \times 10^4)}{3 \times 10^{-3}}$

60. $\dfrac{(5 \times 10^3)(2 \times 10^{-6})}{1 \times 10^3}$

Write the real-number principal root for each.

61. $\sqrt{10,000}$

62. $\sqrt[3]{-1000}$

63. $\sqrt{-0.04}$

64. $\sqrt[5]{243}$

65. $\sqrt[3]{-\dfrac{8}{27}}$

66. $\sqrt[4]{0.0016}$

Express each in exponential form.

67. $\sqrt[5]{-y^3}$

68. $\sqrt[7]{a^2b^{1/2}}$

69. $\sqrt[r]{s^2t^4}$

70. $\sqrt[k]{p^{4k}}$

71. $\sqrt[5]{\dfrac{a^p}{b^q}}$

72. $\sqrt[4]{m^3n^2s^{1/2}}$

Express each in radical form.

73. $(27a)^{1/3}$

74. $(2x^3)^0$

75. $(a^2b^{1/2})^{1/p}$

76. $\left(\dfrac{1}{p + q^{-1}}\right)^{-1/2}$

77. $\left(a - \dfrac{ab}{b}\right)^{-1/5}$

78. $(a^rm^{-b}s)^{-1/b}$

Simplify.

79. $a^{1/2} \div a^{2/3}$

80. $\left(\dfrac{8}{27}\right)^{-2/3}$

81. $\dfrac{a^{3/4}b^0}{a^{1/3}b^{1/2}}$

82. $\left(-\dfrac{32}{243}\right)^{1/5}$ **83.** $[(-3)^{-4}]^{-1/2}$ **84.** $\dfrac{2x^3y^2z^{1/2}}{(2xy)^3}$

85. $(x^{2/3}y^{1/2})^6$ **86.** $(a^{-3/4}b^{1/2})^8$

87. $\left(\dfrac{a^{4/3}b^{-1/2}}{c^{1/3}}\right)^{-6}$ **88.** $\left(\dfrac{x^{-10}y^5}{z^0}\right)^{-3/5}$

Reduce the order of each.

89. $\sqrt[4]{64a^6b^2}$ **90.** $\sqrt[6]{144x^4y^2}$

91. $\sqrt[2m]{36a^4b^2}$ **92.** $\sqrt[6]{\dfrac{16x^4y^2}{25a^4b^2}}$

Simplify.

93. $\sqrt{50x^2y^7}$ **94.** $\sqrt{18a^3b^2c}$

95. $\sqrt[3]{-16}$ **96.** $x\sqrt[3]{3x^5y^4z^0}$

97. $\sqrt[3]{ax^4m^{-2}}$ **98.** $x\sqrt[6]{81x^8y^2}$

99. $\sqrt[3]{27x^{-3}y^2z^{-2}}$ **100.** $\sqrt{8^{-1}x^{-3}y^3}$

101. $\sqrt{16^{-1}m^3n^{-6}}$ **102.** $x\sqrt[4]{(x+y)^6}$

Complete the equivalent radical expression.

103. $\sqrt{3} = \sqrt[6]{}$ **104.** $\sqrt[3]{3xy^2} = \sqrt[9]{}$

105. $\sqrt[3]{2x^2y^{-1}} = \sqrt[12]{}$ **106.** $\sqrt{4x^2y^4} = \sqrt[3]{}$

Express each in simplest terms with rational denominators.

107. $\dfrac{3x}{\sqrt{p}}$ **108.** $\dfrac{3a}{\sqrt[4]{p^3}}$

109. $\sqrt[3]{8b^{-2}}$ **110.** $\sqrt{\dfrac{a^2 - 2ab + b^2}{a^2 - b^2}}$

111. $\dfrac{3}{\sqrt{2} - \sqrt{7}}$ **112.** $\dfrac{\sqrt{7}}{2 + \sqrt{3}}$

113. $\dfrac{\sqrt{x}}{1 - \sqrt{x}}$ **114.** $\dfrac{\sqrt{5} - \sqrt{3}}{2\sqrt{3} + \sqrt{5}}$

115. $\dfrac{\sqrt[3]{a}}{\sqrt{a} - \sqrt{b}}$ **116.** $\dfrac{\sqrt{76} - \sqrt{102}}{\sqrt{102} - \sqrt{76}}$

117. $\dfrac{3\sqrt{a} + 4\sqrt{b}}{2\sqrt{b} - \sqrt{a}}$ **118.** $\sqrt{\dfrac{a}{b} + 1}$

Simplify.

119. $3\sqrt{12} + \sqrt{3} - \sqrt{12} + \sqrt{48}$

120. $\sqrt{45} - \sqrt{60} + \sqrt{15} - \sqrt{135} + \sqrt{20}$

121. $\sqrt[3]{3a^3} + \sqrt[3]{48} - \sqrt[3]{24a^3} - a\sqrt[3]{-3}$

122. $\sqrt{2} - \sqrt[4]{324} + \sqrt[6]{512} + \sqrt[8]{16}$

123. $\sqrt{\dfrac{5}{27}} - \sqrt{\dfrac{12}{125}}$ **124.** $4\sqrt{5} - \dfrac{2}{\sqrt{5}} + \dfrac{7}{3\sqrt{5}} + \dfrac{\sqrt{5}}{2}$

125. $\sqrt{12x} - \sqrt{48x} - \sqrt{3x^3} + \sqrt{12x^3}$

126. $\sqrt{(a+b)^3} - \sqrt{a+b}$

127. $\dfrac{\sqrt{x^3y}}{3} - x\sqrt{\dfrac{xy}{25}} + \dfrac{\sqrt{4x^3y}}{5}$

128. $\sqrt{a^{-3}} + 2a\sqrt[3]{b^{-1}} + 3ab\sqrt{b^{-2}}$

129. $\dfrac{3}{\sqrt{2}} + \dfrac{5\sqrt{2}}{2}$ **130.** $\dfrac{2}{\sqrt[n]{a}} + \dfrac{\sqrt[n]{a}}{a}$

Multiply, and simplify.

131. $\sqrt[4]{27x^3} \cdot \sqrt{3x}$ **132.** $\sqrt[4]{36}(\sqrt{3} - 2\sqrt[3]{2} + \sqrt{6})$

133. $(3m\sqrt[3]{4m^2n})(m\sqrt{n})(\sqrt[6]{mn^4})$ **134.** $(8 - \sqrt{3})(2 + \sqrt[3]{9})$

135. $(2\sqrt{10} - 3\sqrt{5})(\sqrt{10} + 4\sqrt{5})$ **136.** $(\sqrt{316} - \sqrt{214})(\sqrt{316} + \sqrt{214})$

137. $(3 - 2\sqrt{5})(\sqrt{3} - \sqrt{5})$ **138.** $(\sqrt{x+y} - \sqrt{y})^2$

139. $(\sqrt{a-2} + \sqrt{a+3})^2$ **140.** $(\sqrt[3]{9x^2y} - \sqrt{6xy})^2$

141. If $x = 2\sqrt{5} - 2$, then $x^2 - 4x + 7 = ?$

142. If $x = 3\sqrt{2} - \sqrt{3}$, then $2x^2 - 3x + 1 = ?$

Divide, and simplify.

143. $(4\sqrt{3} - 2\sqrt{6} + 3\sqrt{2}) \div \sqrt{18}$

144. $(3\sqrt{110} - 4\sqrt{15} + 2\sqrt{21} - 2\sqrt{11}) \div 2\sqrt{30}$

145. $(3\sqrt[3]{2} - 2\sqrt{3}) \div \sqrt[3]{3}$ **146.** $(\sqrt{3} - \sqrt{7}) \div (\sqrt{3} - 4\sqrt{7})$

147. $(3\sqrt{2} - 2\sqrt{3}) \div (\sqrt{2} + 3\sqrt{3})$ **148.** $(2\sqrt{3} - 3\sqrt{2}) \div (3 - \sqrt{2})$

149. $(4 - \sqrt{6}) \div (2\sqrt{3} + 3\sqrt{6})$ **150.** $(\sqrt{172} - \sqrt{326}) \div (\sqrt{326} - \sqrt{172})$

151. $(\sqrt{3x} - \sqrt{2y}) \div (\sqrt{2x} + \sqrt{3y})$

152. $\dfrac{\sqrt{3} - \sqrt{2}}{\sqrt{3} + 2\sqrt{2}} \div \dfrac{3\sqrt{3} - 2\sqrt{2}}{2\sqrt{3} - \sqrt{2}}$

Simplify.

153. $3i - 4i^2 + 3i^3 - 7$

154. $4i^4 - 3i^3 + 2i^7 + i^8$

155. $4i^{17} + 3i^{12} + 2i^{13} - 2i^{19}$

156. $i\sqrt{3} + 2i - 3\sqrt{3} + 4i^5$

157. $i^{-4} + i^{-3} + 2i^{-8} + i^{-6}$

158. $i^{-8} - i^{-18} + i^{-13} - 2i^{-9} + i^{-38}$

Express in the form a + bi.

159. $(1 - 3i) + (2 - 3i) - (4 + 2i)$

160. $(3 + 2i - 6i^2) + (3i - 2 + 4i^5)$

161. $\left(\dfrac{1}{3} + \dfrac{i}{\sqrt{2}}\right) + \left(\dfrac{5}{6} - i\sqrt{2}\right) - \left(\dfrac{2}{3} + \dfrac{3i}{\sqrt{2}}\right)$

162. $(3 - 2i)(4 - 3i)$

163. $(3 - 2i)(4 - 3i)(1 - i)$

164. $(3 - 2i + 5i^6 - 3i^7)^2$

165. $(4 - 3i) \div (-2 + 6i)$

166. $(4 - i^3) \div (2 - 3i^{-6} + i^{-5})$

167. $(\sqrt{3} - i) \div (2\sqrt{3} + 3i)$

168. $(3 - 2i + 4i^2 - 3i^3)^2 \div (2 - 3i + 5i^3)$

169. $\dfrac{(3 - 2i)}{(4 - 3i)} \div \dfrac{(2 - i)}{(2 + 3i)}$

170. $\dfrac{2 - 3i}{3 + 2i^2} \div (4 + 3i^3)$

7 QUADRATIC EQUATIONS AND INEQUALITIES

1. POLYNOMIAL EQUATIONS

A polynomial equation in a single variable is one in which the variable carries only positive integral exponents. The **degree** of a polynomial equation is that of the variable's greatest exponent.

When the greatest exponent present is 1, the degree is the *first*, and the equation is called **linear**. When the greatest exponent present is 2, the degree is the *second*, and the equation is called **quadratic**. When the greatest exponents are 3, 4, 5, the degrees are *third*, *fourth*, and *fifth*, respectively, and the equations are referred to as **cubic**, **quartic** (or **biquadratic**), and **quintic**, respectively.

Polynomial equations in a single variable are said to be in **standard form** when one member is 0 and the other is a polynomial with its terms arranged in descending powers of the variable. The following are examples of polynomial equations in a single variable, in standard form:

$$x - 2 = 0 \quad \text{(linear)}$$

$$\left.\begin{aligned} x^2 - 3x + 5 &= 0 \\ 7x^2 - 4 &= 0 \end{aligned}\right\} \quad \text{(quadratic)}$$

$$x^3 - 4x - 1 = 0 \quad \text{(cubic)}$$

$$4x^4 - 6x^2 - 2x = 0 \quad \text{(quartic, or biquadratic)}$$

$$x^5 - 3x^4 + 2x^3 - 7x^2 + 4x + 3 = 0 \quad \text{(quintic)}$$

In the next several sections, we shall be concerned with solving second degree or quadratic equations in a single variable. Unless otherwise indicated, the replacement set, or domain of the variable, is assumed to be the set of complex numbers. The *general quadratic equation in standard form* is given as

$$ax^2 + bx + c = 0$$

where x is the variable and a, b, c represent real numbers, $a \neq 0$.

2. SOLVING QUADRATIC EQUATIONS BY FACTORING

In Section 5 of this chapter we shall develop a general formula for solving any quadratic equation. However, many such equations can also be solved more quickly by a factor method, the subject of this section. Underlying the factor method is the mathematical principle

$$\text{If } a \cdot b = 0, \text{ then } \begin{cases} a = 0, \text{ or} \\ b = 0, \text{ or} \\ \text{both } a = 0 \text{ and } b = 0 \end{cases}$$

Consequently, if the polynomial member of a quadratic equation in standard form can be factored, then each factor can be set equal to zero and the resulting equations solved.

To illustrate,

$$x^2 - 7x + 10 = 0$$

is a quadratic equation in standard form. The polynomial $x^2 - 7x + 10$ can be factored into $(x - 2)(x - 5)$. Thus

$$x^2 - 7x + 10 = 0$$

$$(x - 2)(x - 5) = 0$$

Clearly, when $x = 2$ or $x = 5$, the product of the two factors will be zero.

$$(x - 2)(x - 5)$$

$$x = 2: \quad (2 - 2)(2 - 5) = (0)(-3) = 0$$

$$x = 5: \quad (5 - 2)(5 - 5) = (3)(0) = 0$$

Because 2 and 5 satisfy the equation, they are its solutions.

$$x^2 - 7x + 10 = 0$$

$$x = 2: \quad (2)^2 - 7(2) + 10 \overset{?}{=} 0$$

$$4 - 14 + 10 = 0$$

$$x = 5: \quad (5)^2 - 7(5) + 10 \overset{?}{=} 0$$

$$25 - 35 + 10 = 0$$

PROGRAM 7.1

To solve a quadratic equation by factoring:

Step 1: Express the quadratic equation in standard form.

Step 2: Factor the polynomial member.

Step 3: Set each factor of Step 2 equal to zero.

Step 4: Solve the linear equations of Step 3 separately.

Step 5: Check by replacing the variable in the given equation with the solutions found in Step 4; those that satisfy the given equation are its solutions.

EXAMPLES

1. Solve $x^2 = 5x - 6$.

SOLUTION *Step 1:* Express the equation in standard form:

$$x^2 = 5x - 6$$

$$x^2 - 5x + 6 = 0$$

Step 2: Factor:

$$(x - 3)(x - 2) = 0$$

Step 3: Set each factor equal to 0:

$$(x - 3) = 0 \qquad (x - 2) = 0$$

Step 4: Solve the two equations developed in Step 3:

$$x - 3 = 0 \qquad x - 2 = 0$$
$$x = 3 \qquad x = 2$$

Step 5: Check the solutions obtained in Step 4 in the given equation:

$$x^2 = 5x - 6 \qquad\qquad x^2 = 5x - 6$$

$$x = 3: \quad (3)^2 \overset{?}{=} 5(3) - 6 \qquad x = 2: \quad (2)^2 \overset{?}{=} 5(2) - 6$$

$$9 = 15 - 6 \qquad\qquad 4 = 10 - 6$$

Both 3 and 2 satisfy the given equation, $x^2 = 5x - 6$; they are the solutions of that equation.

2. Solve $x^2 + x = 20$.

SOLUTION *Step 1:* Write the equation in standard form:

$$x^2 + x = 20$$
$$x^2 + x - 20 = 0$$

Step 2: Factor:

$$(x - 4)(x + 5) = 0$$

Step 3: Set the factors equal to 0:

$$(x - 4) = 0 \qquad (x + 5) = 0$$

Step 4: Solve the equations:

$$x - 4 = 0 \qquad x + 5 = 0$$
$$x = 4 \qquad x = -5$$

Step 5: Check:

$$x^2 + x = 20 \qquad\qquad x^2 + x = 20$$

$$x = 4: \quad (4)^2 + (4) \overset{?}{=} 20 \qquad x = -5: \quad (-5)^2 + (-5) \overset{?}{=} 20$$

$$16 + (4) = 20 \qquad\qquad 25 - 5 = 20$$

The solution set for $x^2 + x = 20$ is $\{4, -5\}$.

3. Solve $5x^2 = 3x$.

SOLUTION *Step 1:* $5x^2 = 3x$
$5x^2 - 3x = 0$

Step 2: $(x)(5x - 3) = 0$

Step 3: $(x) = 0 \qquad (5x - 3) = 0$

Step 4: $x = 0 \qquad 5x - 3 = 0$
$$5x = 3$$
$$x = \frac{3}{5}$$

Step 5:

$$5x^2 = 3x \qquad\qquad 5x^2 = 3x$$

$$x = 0: \quad 5(0)^2 \overset{?}{=} 3(0) \qquad\quad x = \frac{3}{5}: \quad 5\left(\frac{3}{5}\right)^2 \overset{?}{=} 3\left(\frac{3}{5}\right)$$

$$0 = 0 \qquad\qquad\qquad 5\left(\frac{9}{25}\right) \overset{?}{=} 3\left(\frac{3}{5}\right)$$

$$\frac{9}{5} = \frac{9}{5}$$

The solution set for $5x^2 = 3x$ is $\left\{0, \dfrac{3}{5}\right\}$.

Note: Resist the temptation in Step 1 to "simplify" the equation by dividing both members by the variable x. The result will be loss of one of the solutions, 0.

4. Solve $x^2 = 16$.

SOLUTION

Step 1: $x^2 = 16$

$\qquad\quad x^2 - 16 = 0$

Step 2: $(x - 4)(x + 4) = 0$

Step 3: $(x - 4) = 0 \qquad (x + 4) = 0$

Step 4: $\quad x - 4 = 0 \qquad\quad x + 4 = 0$

$\qquad\qquad\quad x = 4 \qquad\qquad\quad x = -4$

Step 5:

$$x^2 = 16 \qquad\qquad\qquad x^2 = 16$$

$$x = 4: \quad (4)^2 = 16 \qquad x = -4: \quad (-4)^2 = 16$$

Thus $\{4, -4\}$ is the solution set for $x^2 = 16$.

5. Show that $-\dfrac{1}{2}$ is the only solution of the quadratic equation, $4x^2 + 4x + 1 = 0$.

SOLUTION

$4x^2 + 4x + 1 = 0$

$(2x + 1)(2x + 1) = 0$

$2x + 1 = 0 \qquad\qquad 2x + 1 = 0$

$\qquad 2x = -1 \qquad\qquad\quad 2x = -1$

$\qquad\quad x = -\dfrac{1}{2} \qquad\qquad\quad x = -\dfrac{1}{2}$

Note: When the polynomial is a square trinomial, the quadratic equation has two equal solutions.

Solve for x and check.

1. $x^2 - 6x + 8 = 0$ 2. $x^2 + 2x - 15 = 0$
3. $x^2 + 10x + 24 = 0$ 4. $x^2 - x = 6$
5. $x^2 - 11x = -24$ 6. $x^2 + 11x + 18 = 0$
7. $x^2 + 13x + 36 = 0$ 8. $x^2 - 18x + 32 = 0$
9. $x^2 - 14x + 49 = 0$ 10. $x^2 + 8x + 16 = 0$
11. $x^2 = 25$ 12. $x^2 = 36$
13. $x^2 - 81 = 0$ 14. $49 - x^2 = 0$
15. $x^2 = x$ 16. $2x^2 = x$
17. $3x^2 = 2x$ 18. $5x^2 = 7x$
19. $x^2 = 6x + 7$ 20. $6x^2 - 5x + 1 = 0$
21. $3x^2 + 22x + 35 = 0$ 22. $9x^2 + 1 = 6x$
23. $15x^2 = 6 - x$ 24. $x^2 - 4ax + 4a^2 = 0$
25. $x^2 - 2ax - 15a^2 = 0$ 26. $25x^2 + 4 = 20x$
27. $6x^2 - 5ax + a^2 = 0$ 28. $a^2x^2 - ax - 6 = 0$
29. $3a^2x^2 + 7ax = 6$ 30. $4x^2 - 9x = 0$
31. $2x^2 - 8ax = 0$ 32. $3b^2x^2 = 5bx$

3. PURE QUADRATIC EQUATIONS

A quadratic equation of the form

$$ax^2 + c = 0 \qquad (a \neq 0)$$

is called a **pure quadratic equation**. We saw a few in the previous section and solved them by using Program **7.1**. An alternative method is available. It is based on the fact that every positive number has two square roots, one the negative of the other.

For instance, if

$$x^2 = 9$$

then $x = +\sqrt{9}$, or 3, and $x = -\sqrt{9}$, or -3. Thus the two solutions for $x^2 = 9$ are 3 and -3. Often such a pair is written ± 3 and read "plus or minus 3."

When the $\pm$ symbol appears in an equation, such as

$$x = \pm 3$$

a *pair* of equations is implied:

$$x = +3 \qquad x = -3$$

Program **7.2**, following, outlines this alternative method for solving a pure quadratic equation. Note in Step 3 that it is necessary to identify *all* roots, not just the principal square root.

PROGRAM 7.2

To solve a pure quadratic equation:

Step 1:   Transform the given equation so that all terms containing the variable are in one member and all other terms are in the other member; then simplify.

Step 2:   Divide both members of the equation of Step 1 by the coefficient of the variable term.

Step 3:   Reduce the variable to first-degree by taking the square root of both members; the square roots of the non-variable member are possible solutions.

Step 4:   Check by replacing the variable in the given equation with the solutions obtained in Step 3; those that satisfy the given equation are its solutions.

EXAMPLES

1.   Solve $4x^2 - 12 = x^2$.

SOLUTION   *Step 1:* Transform the equation so that all variable terms are in one member, and the constant is in the other member.

$$4x^2 - 12 = x^2$$

$$4x^2 - x^2 = 12$$

$$3x^2 = 12$$

Step 2: Divide both members so that the variable has a coefficient of 1:

$$\frac{3x^2}{3} = \frac{12}{3}$$

$$x^2 = 4$$

Step 3: Take the square root of each member:

$$\sqrt{x^2} = \sqrt{4}$$

$$x = 2, -2 \quad \text{(possible solutions)}$$

Step 4: Check the possible solutions of Step 3:

$4x^2 - 12 = x^2$	$4x^2 - 12 = x^2$
$x = 2:$ $4(2)^2 - 12 \overset{?}{=} (2)^2$	$x = -2:$ $4(-2)^2 - 12 \overset{?}{=} (-2)^2$
$4(4) - 12 \overset{?}{=} 4$	$4(4) - 12 \overset{?}{=} 4$
$16 - 12 = 4$	$16 - 12 = 4$

The solution set for $4x^2 - 12 = x^2$ is $\{2, -2\}$.

2. Solve $12x^2 - 5 = 3x^2 + 2$.

Step 1: Collect the variable terms into one member, constants in the other:

$$12x^2 - 5 = 3x^2 + 2$$
$$12x^2 - 3x^2 = 2 + 5$$
$$9x^2 = 7$$

Step 2: Divide to make the coefficient of the x^2 term 1:

$$\frac{9x^2}{9} = \frac{7}{9}$$

$$x^2 = \frac{7}{9}$$

Step 3: Solve by taking the square root:

$$x = \pm\sqrt{\frac{7}{9}} = \frac{\sqrt{7}}{3}, -\frac{\sqrt{7}}{3} \quad \text{(possible solutions)}$$

Step 4: Check:

$$12x^2 - 5 = 3x^2 + 2$$

$x = \dfrac{\sqrt{7}}{3}:$

$$12\left(\frac{\sqrt{7}}{3}\right)^2 - 5 \stackrel{?}{=} 3\left(\frac{\sqrt{7}}{3}\right)^2 + 2$$

$$12\left(\frac{7}{9}\right) - 5 \stackrel{?}{=} 3\left(\frac{7}{9}\right) + 2$$

$$\frac{28}{3} - 5 \stackrel{?}{=} \frac{7}{3} + 2$$

$$9\frac{1}{3} - 5 = 2\frac{1}{3} + 2$$

$x = -\dfrac{\sqrt{7}}{3}:$

$$12\left(-\frac{\sqrt{7}}{3}\right)^2 - 5 \stackrel{?}{=} 3\left(-\frac{\sqrt{7}}{3}\right)^2 + 2$$

$$12\left(+\frac{7}{9}\right) - 5 \stackrel{?}{=} 3\left(+\frac{7}{9}\right) + 2$$

$$\frac{28}{3} - 5 \stackrel{?}{=} \frac{7}{3} + 2$$

$$9\frac{1}{3} - 5 = 2\frac{1}{3} + 2$$

The solutions of $12x^2 - 5 = 3x^2 + 2$ are

$$\pm\frac{\sqrt{7}}{3} \quad \text{or} \quad \left\{\frac{\sqrt{7}}{3}, -\frac{\sqrt{7}}{3}\right\}$$

3. Solve $3x^2 + 21 = 0$.

$3x^2 + 21 = 0$

$$3x^2 = -21$$
$$x^2 = -7$$
$$\sqrt{x^2} = \sqrt{-7}$$
$$x = i\sqrt{7}, -i\sqrt{7} \quad \text{(possible solutions)}$$

$$3x^2 + 21 = 0$$

$x = i\sqrt{7}$:	$3(i\sqrt{7})^2 + 21 \overset{?}{=} 0$	$x = -i\sqrt{7}$:	$3(-i\sqrt{7})^2 + 21 \overset{?}{=} 0$
	$3(i^2)(\sqrt{7})^2 + 21 \overset{?}{=} 0$		$3(-i)^2(\sqrt{7})^2 + 21 \overset{?}{=} 0$
	$3(-1)(7) + 21 \overset{?}{=} 0$		$3(i^2)(7) + 21 \overset{?}{=} 0$
	$-21 + 21 = 0$		$3(-1)(7) + 21 \overset{?}{=} 0$
			$-21 + 21 = 0$

Thus the solution set for $3x^2 + 21 = 0$ is

$$\{i\sqrt{7},\ -i\sqrt{7}\}.$$

We may extend Program **7.2** to certain quadratic equations of the form

$$(x + p)^2 = c$$

where p and c are constants. The following examples are illustrative.

EXAMPLES

1. Solve for x: $(x + 1)^2 = 4$.

SOLUTION   Reduce the degree of the variable member by taking the square root of each member:

$$\sqrt{(x + 1)^2} = \sqrt{4}$$
$$(x + 1) = \pm 2$$

Then solve for x:

$$x + 1 = \pm 2$$
$$x = \pm 2 - 1$$

or

$$x = +2 - 1 = 1$$
$$x = -2 - 1 = -3$$

The solution set is $\{1, -3\}$.
Check:
$$(x + 1)^2 = 4$$

$x = 1$: $(1 + 1)^2 \overset{?}{=} 4$	$x = -3$: $(-3 + 1)^2 \overset{?}{=} 4$
$2^2 = 4$	$(-2)^2 = 4$

2. Solve $4x^2 - 12x + 9 = 16$.

SOLUTION   The left member of

$$4x^2 - 12x + 9 = 16$$

is recognized as a square trinomial, which can be expressed as the square of a binomial (Program **2.3**)

$$(2x - 3)^2 = 16$$

By taking the square root of each member, we obtain

$$2x - 3 = \pm 4$$

which may then be solved for x:

$$2x = 3 \pm 4$$

$$x = \frac{1}{2}(3 \pm 4)$$

$$x = \frac{7}{2}, \ -\frac{1}{2}$$

The solution set is $\left\{\dfrac{7}{2}, \ -\dfrac{1}{2}\right\}$.

Note: By Program **7.1**,

$$4x^2 - 12x + 9 = 16$$
$$4x^2 - 12x - 7 = 0$$
$$(2x + 1)(2x - 7) = 0$$
$$2x + 1 = 0 \qquad 2x - 7 = 0$$
$$x = -\frac{1}{2} \qquad x = \frac{7}{2}$$

3.   Solve for the variable: $y^2 - 2y + 1 = 3$.

SOLUTION   The left member is a square trinomial, that is, the square of a binomial:

$$(y - 1)^2 = 3$$

Take the square roots, and solve for the variable:

$$y - 1 = \pm\sqrt{3}$$
$$y = 1 \pm \sqrt{3}$$

The solution set is $\{1 + \sqrt{3}, 1 - \sqrt{3}\}$.

EXERCISE 7-2

Solve for x and check.

1. $9x^2 = 81$ **2.** $4x^2 = 36$

3. $4x^2 = 1$ **4.** $25x^2 = 4$

5. $3x^2 - 9 = 0$ **6.** $2x^2 - 12 = 0$

7. $x^2 + 1 = 0$

8. $x^2 + 4 = 0$

9. $5x^2 - 2 = 2x^2 + 1$

10. $6 - 2x^2 = 3x^2 - 14$

11. $3x^2 + 8 = 32 - x^2$

12. $6x^2 - 5 = 2x^2 + 20$

13. $x^2 + 3 = 2$

14. $3x^2 + 4 = x^2 - 2$

15. $5x^2 + 16 = x^2$

16. $4x^2 + 49 = 0$

17. $(x - 1)^2 = 9$

18. $(x - 2)^2 = 16$

19. $(x + 3)^2 = 25$

20. $(x + 2)^2 = 16$

21. $(x - 2)^2 = 5$

22. $(x - 3)^2 = 7$

23. $(x - 4)^2 = -4$

24. $(x + 2)^2 = -1$

25. $(2x - 1)^2 = 9$

26. $(3x + 1)^2 = 4$

27. $x^2 - 6x + 9 = 4$

28. $x^2 + 4x + 4 = 1$

29. $x^2 + 8x + 16 = 8$

30. $x^2 - 10x + 25 = 18$

31. $4x^2 - 4x + 1 = 3$

32. $4x^2 - 8x + 4 = -8$

33. $4x^2 - 12x + 9 = -4$

34. $9x^2 - 30x + 25 = 6$

4. COMPLETING THE SQUARE

It is possible to convert *any* quadratic equation into the squared-binomial form discussed in the latter part of the previous section by a technique known as **completing the square**. In effect, the method reverses Program **2.2** (To square a binomial), the second step of which calls for *doubling* the product of the two terms of the binomial.

For instance:

$$(x + 5)^2 = x^2 + 2\,(x)(5) + 5^2$$
$$= x^2 + 10x + 25$$

PROGRAM 7.3

To complete the square of an algebraic expression of the form $x^2 + bx$:

Step 1: Halve the coefficient (*b*) of the linear or first-degree term and square it.

Step 2: Add the result of Step 1 to the original expression to form a square trinomial.

EXAMPLES

1. Complete the square for $x^2 + 10x$.

SOLUTION *Step 1:* Halve the coefficient of the linear term, $10x$, then square it:

$$\frac{1}{2} \cdot 10 = 5; 5^2 = 25$$

Step 2: Add the result of Step 1 to the given expression to form a square trinomial:

$$x^2 + 10x + 25 = (x + 5)^2$$

2.   Complete the square for $x^2 - 6x$.

SOLUTION   *Step 1:* The coefficient of the linear term is -6:

$$\frac{1}{2} \cdot (-6) = -3; (-3)^2 = 9$$

Step 2: $x^2 - 6x + 9 = (x - 3)^2$

3.   Complete the square for $x^2 + 3x$.

SOLUTION   *Step 1:* $\frac{1}{2} \cdot 3 = \frac{3}{2}; \left(\frac{3}{2}\right)^2 = \frac{9}{4}$

Step 2: $x^2 + 3x + \frac{9}{4} = \left(x + \frac{3}{2}\right)^2$

4.   Complete the square for $x^2 - \frac{3}{5}x$.

SOLUTION   *Step 1:* $\left[\frac{1}{2}\left(-\frac{3}{5}\right)\right]^2 = \left(-\frac{3}{10}\right)^2 = \frac{9}{100}$

Step 2: $x^2 - \frac{3}{5}x + \frac{9}{100} = \left(x - \frac{3}{10}\right)^2$

Consider now the quadratic equation

$$x^2 - 6x - 1 = 0$$

The factors of $x^2 - 6x - 1$ are not readily apparent. So we reorganize the equation by putting the terms containing the variable in one member and the constant term in the other.

Then we use Program **7.3** to complete the square trinomial:

$$x^2 \; \boxed{-6} \; x = 1$$

By Program **7.3**,

$$\left[\frac{1}{2}(\,\boxed{-6}\,)\right]^2 = [-3]^2 = 9$$

$$x^2 - 6x + 9 = 1 + 9$$

Factor and solve.

$$(x - 3)^2 = 10$$
$$x - 3 = \pm\sqrt{10}$$
$$x = 3 \pm \sqrt{10}$$

Thus the solution set for $x^2 - 6x - 1 = 0$ is

$$\{3 + \sqrt{10}, 3 - \sqrt{10}\}$$

PROGRAM 7.4

To solve a quadratic equation by completing the square:

Step 1: Transform the equation so that the variable terms are in one member and the constant term is in the other member.

Step 2: Divide the members of the equation by the coefficient of the second-degree term.

Step 3: Add to both members whatever number is necessary to complete the square of the member containing the variables.

Step 4: Reduce the degree of the equation by taking the square root of each member and solve for the variable.

Step 5: Check the solutions obtained in Step 4 in the given quadratic equation.

EXAMPLES

1. Solve for x by completing the square.

$$2x^2 - 4x - 30 = 0$$

SOLUTION *Step 1:* Transform the equation:

$$2x^2 - 4x \qquad = 30$$

Step 2: Divide both members by the coefficient of the x^2 term, 2, so that the coefficient of the x^2 term in the resulting equation becomes 1:

$$\frac{2x^2 - 4x}{2} = \frac{30}{2}$$

$$x^2 - 2x = 15$$

Step 3: Complete the square using Program **7.3** and add to both members:

$$x^2 - 2x + 1 = 15 + 1$$

$$\left[\frac{1}{2}(-2)\right]^2$$

Step 4: Factor, take the square root of each member, and solve:

$$(x - 1)^2 = 16$$

$$\sqrt{(x - 1)^2} = \sqrt{16}$$

$$x - 1 = \pm 4$$

$$x = 1 \pm 4$$

$$x = 5, -3$$

Step 5: Check:

$$2x^2 - 4x - 30 = 0$$

$$x = 5: \qquad 2(5)^2 - 4(5) - 30 \overset{?}{=} 0$$

$$50 - 20 - 30 = 0$$

$$x = -3: \quad 2(-3)^2 - 4(-3) - 30 \overset{?}{=} 0$$

$$18 + 12 - 30 = 0$$

2. Solve for x

$$3x^2 + 8x - 3 = 0$$

by completing the square.

SOLUTION | *Step 1:* Variables in one member; constant in the other:

$$3x^2 + 8x = 3$$

Step 2: Divide by 3 to make 1 the coefficient of the x^2 term (i.e., to express the left member in the form $x^2 + bx$):

$$\frac{3x^2 + 8x}{3} = \frac{3}{3}$$

$$x^2 + \frac{8}{3}x = 1$$

Step 3: Complete the square of the left member, and add equivalently to the right member:

$$\left[\frac{1}{2}\left(\frac{8}{3}\right)\right]^2 = \left[\frac{4}{3}\right]^2 = \frac{16}{9}$$

$$x^2 + \frac{8}{3}x + \frac{16}{9} = 1 + \frac{16}{9}$$

Step 4: Solve for x:

$$\left(x + \frac{4}{3}\right)^2 = \frac{25}{9}$$

$$x + \frac{4}{3} = \pm\frac{5}{3}$$

$$x = -\frac{4}{3} \pm \frac{5}{3}$$

$$x = \frac{1}{3}, -3$$

Step 5: Check:

$$3x^2 + 8x - 3 = 0$$

$$x = \frac{1}{3}: \quad 3\left(\frac{1}{3}\right)^2 + 8\left(\frac{1}{3}\right) - 3 \overset{?}{=} 0$$

$$\frac{1}{3} + \frac{8}{3} - 3 = 0$$

$$x = -3: \quad 3(-3)^2 + 8(-3) - 3 \overset{?}{=} 0$$

$$27 - 24 - 3 = 0$$

3. Solve $x^2 + 2x - 1 = 0$ by completing the square.

SOLUTION

$$x^2 + 2x - 1 = 0$$

$$x^2 + 2x = 1$$

Add $\left[\frac{1}{2}(2)\right]^2$, or 1, to complete the square:

$$x^2 + 2x + 1 = 1 + 1$$

$$(x + 1)^2 = 2$$

$$x + 1 = \pm\sqrt{2}$$

$$x = -1 \pm \sqrt{2}$$

4. Solve $2x^2 + 2x + 1 = 0$.

SOLUTION

$$2x^2 + 2x + 1 = 0$$

$$2x^2 + 2x = -1$$

$$\frac{2x^2 + 2x}{2} = \frac{-1}{2}$$

$$x^2 + x = -\frac{1}{2}$$

Complete the square and solve.

$$x^2 + x + \frac{1}{4} = -\frac{1}{2} + \frac{1}{4}$$

$$\left(x + \frac{1}{2}\right)^2 = -\frac{1}{4}$$

$$x + \frac{1}{2} = \pm\frac{1}{2}i$$

$$x = -\frac{1}{2} \pm \frac{1}{2}i$$

or

$$x = \frac{-1 \pm i}{2}$$

EXERCISE 7-3

Complete the square (i.e., add a third term to make a square trinomial).

1. $x^2 - 6x + ?$ 2. $x^2 + 6x + ?$
3. $y^2 + 8x + ?$ 4. $m^2 - 10x + ?$
5. $x^2 + 7x + ?$ 6. $x^2 - 3x + ?$
7. $a^2 - 11a + ?$ 8. $p^2 - 9p + ?$
9. $x^2 - \frac{1}{2}x + ?$ 10. $x^2 + \frac{1}{3}x + ?$
11. $a^2 + \frac{3}{4}x + ?$ 12. $m^2 - \frac{2}{3}m + ?$

Solve by completing-the-square method.

13. $x^2 - 6x + 8 = 0$ 14. $x^2 - 8x + 12 = 0$
15. $x^2 - 2x - 15 = 0$ 16. $x^2 - 2x - 3 = 0$
17. $x^2 + 5x - 6 = 0$ 18. $x^2 + 7x - 8 = 0$
19. $k^2 - 3k - 18 = 0$ 20. $y^2 + 9y + 14 = 0$
21. $2x^2 - 7x + 3 = 0$ 22. $3x^2 - 10x + 3 = 0$
23. $3x^2 - 4x - 4 = 0$ 24. $2x^2 - x - 3 = 0$
25. $x^2 - 6x + 6 = 0$ 26. $x^2 - 4x + 2 = 0$
27. $b^2 - 6b + 2 = 0$ 28. $m^2 - 10m + 7 = 0$
29. $x^2 + 8x + 8 = 0$ 30. $x^2 + 6x - 3 = 0$
31. $a^2 + 2a + 5 = 0$ 32. $b^2 + 8b + 20 = 0$
33. $x^2 - 4x + 8 = 0$ 34. $x^2 + 6x + 10 = 0$
35. $9x^2 + 6x - 2 = 0$ 36. $2x^2 - 4x - 2 = 0$
37. $4x^2 = 8x - 12$ 38. $5a^2 + 5 = 6a$
39. $9x^2 + 5 = 12x$ 40. $3 - 4x + 3x^2 = 0$

5. QUADRATIC FORMULA

In the previous section we developed a method, Program **7.4**, for solving quadratic equations. Here we reduce that method to a formula. We start with the general quadratic equation

$$ax^2 + bx + c = 0 \qquad (a \neq 0)$$

and use Program **7.4** to solve for x.

Step 1:   Transform the equation:

$$ax^2 + bx = -c$$

Step 2:   Divide by the coefficient of the x^2 term, a:

$$x^2 + \frac{b}{a}x = -\frac{c}{a}$$

Step 3:   Complete the square of the left member with $\left[\frac{1}{2}\left(\frac{b}{a}\right)\right]^2$, or $\left(\frac{b}{2a}\right)^2$, and balance the equation.

$$x^2 + \frac{b}{a}x + \left(\frac{b}{2a}\right)^2 = -\frac{c}{a} + \left(\frac{b}{2a}\right)^2$$

Step 4:   Simplify, factor, and solve for the variable.

$$x^2 + \frac{b}{a}x + \frac{b^2}{4a^2} = -\frac{4ac}{4a^2} + \frac{b^2}{4a^2}$$

$$\left(x + \frac{b}{2a}\right)^2 = \frac{b^2 - 4ac}{4a^2}$$

$$x + \frac{b}{2a} = \pm\sqrt{\frac{b^2 - 4ac}{4a^2}}$$

$$x = -\frac{b}{2a} \pm \frac{\sqrt{b^2 - 4ac}}{2a}$$

Thus

$$x = \frac{-b + \sqrt{b^2 - 4ac}}{2a}, \quad x = \frac{-b - \sqrt{b^2 - 4ac}}{2a}$$

These two solutions of the general quadratic equation are frequently combined into a single expression

$$x = \frac{-b \pm \sqrt{b^2 - 4ac}}{2a}$$

called the **quadratic formula**. Upon replacement of a, b, and c by the specific numerical coefficients of a quadratic equation, the right member of the formula yields two numbers, directly, as possible solutions of the equation.

PROGRAM 7.5

To solve a quadratic equation by the quadratic formula:

Step 1: Express the quadratic equation equivalently in the form

$$ax^2 + bx + c = 0.$$

Step 2: Identify the numerical coefficients that correspond to a, b, and c in the equation of Step 1.

Step 3: Substitute the corresponding coefficients for a, b, and c in the quadratic formula

$$x = \frac{-b \pm \sqrt{b^2 - 4ac}}{2a}$$

to obtain possible solutions.

Step 4: Check by replacing the variable in the given equation with the solutions obtained in Step 3; those that satisfy the given equation are its solutions.

EXAMPLES

1. Solve by the quadratic formula: $3x^2 - 10x + 8 = 0$.

SOLUTION *Step 1:* The model and the given equation:

$$a\,x^2 + b\,x + c = 0$$

$$3\,x^2 - 10\,x + 8 = 0$$

Step 2: Identify the a, b, c values:

$$a = 3, \quad b = -10, \quad c = 8$$

Step 3: Substitute the values of Step 2 into the quadratic formula; solve for the variable:

$$x = \frac{-b \pm \sqrt{b^2 - 4ac}}{2a} = \frac{-(-10) \pm \sqrt{(-10)^2 - 4(3)(8)}}{2(3)}$$

$$= \frac{10 \pm \sqrt{100 - 96}}{6} = \frac{10 \pm \sqrt{4}}{6} = \frac{10 \pm 2}{6}$$

$$= \frac{12}{6}, \frac{8}{6}$$

$$= 2, \frac{4}{3} \quad \text{(possible solutions)}$$

Step 4: Check the possible solutions:

$$3x^2 - 10x + 8 = 0$$

$x = 2:$ $3(2)^2 - 10(2) + 8 \stackrel{?}{=} 0$ $\Bigg|$ $x = \dfrac{4}{3}:$ $3\left(\dfrac{4}{3}\right)^2 - 10\left(\dfrac{4}{3}\right) + 8 \stackrel{?}{=} 0$

$12 - 20 + 8 = 0$ $\Bigg|$ $\dfrac{16}{3} - \dfrac{40}{3} + \dfrac{24}{3} = 0$

The solution set for $3x^2 - 10x + 8 = 0$ is $\left\{2, \dfrac{4}{3}\right\}$.

2. Solve $2x^2 + x = 1$.

SOLUTION *Step 1:* Transform the given equation to standard form:

$$2x^2 + x = 1$$
$$2x^2 + x - 1 = 0$$

Step 2: Identify a, b, c values:

$$a = 2, \quad b = 1, \quad c = -1$$

Step 3: Substitute into the formula, and solve:

$$x = \frac{-b \pm \sqrt{b^2 - 4ac}}{2a} = \frac{-1 \pm \sqrt{(1)^2 - 4(2)(-1)}}{2(2)}$$

$$= \frac{-1 \pm \sqrt{1 + 8}}{4} = \frac{-1 \pm \sqrt{9}}{4} = \frac{-1 \pm 3}{4}$$

$$= \frac{2}{4}, \frac{-4}{4}$$

$$= \frac{1}{2}, -1 \quad \text{(possible solutions)}$$

Step 4: Check:

$$2x^2 + x = 1$$

$x = \dfrac{1}{2}:$ $2\left(\dfrac{1}{2}\right)^2 + \left(\dfrac{1}{2}\right) \stackrel{?}{=} 1$ $\Bigg|$ $x = -1:$ $2(-1)^2 + (-1) \stackrel{?}{=} 1$

$2\left(\dfrac{1}{4}\right) + \dfrac{1}{2} \stackrel{?}{=} 1$ $\Bigg|$ $\qquad 2 - 1 = 1$

$\dfrac{1}{2} + \dfrac{1}{2} = 1$ $\Bigg|$

The solution set is $\left\{\dfrac{1}{2}, -1\right\}$.

3. Solve: $x^2 - 3 = -4x$.

Step 1: $x^2 + 4x - 3 = 0$

Step 2: $a = 1, b = 4, c = -3$

Step 3: $x = \dfrac{-b \pm \sqrt{b^2 - 4ac}}{2a} = \dfrac{-4 \pm \sqrt{16 - 4(1)(-3)}}{2(1)}$

$$= \dfrac{-4 \pm \sqrt{28}}{2} = \dfrac{-4 \pm 2\sqrt{7}}{2} = -2 \pm \sqrt{7}$$

Step 4:
$$x^2 - 3 = -4x$$

$x = -2 + \sqrt{7}:$ $(-2 + \sqrt{7})^2 - 3 \overset{?}{=} -4(-2 + \sqrt{7})$

$$4 - 4\sqrt{7} + 7 - 3 \overset{?}{=} 8 - 4\sqrt{7}$$

$$8 - 4\sqrt{7} = 8 - 4\sqrt{7}$$

$x = -2 - \sqrt{7}:$ $(-2 - \sqrt{7})^2 - 3 \overset{?}{=} -4(-2 - \sqrt{7})$

$$4 + 4\sqrt{7} + 7 - 3 \overset{?}{=} 8 + 4\sqrt{7}$$

$$8 + 4\sqrt{7} = 8 + 4\sqrt{7}$$

The solution set is $\{-2 + \sqrt{7}, -2 - \sqrt{7}\}$.

4. Solve $x^2 + 3 = 2x$.

Step 1: $x^2 + 3 = 2x$
$$x^2 - 2x + 3 = 0$$

Step 2: $a = 1, b = -2, c = 3$

Step 3: $x = \dfrac{-b \pm \sqrt{b^2 - 4ac}}{2a}$

$$= \dfrac{-(-2) \pm \sqrt{(-2)^2 - 4(1)(3)}}{2(1)}$$

$$= \dfrac{2 \pm \sqrt{4 - 12}}{2}$$

$$= \dfrac{2 \pm \sqrt{-8}}{2} = \dfrac{2 \pm 2i\sqrt{2}}{2} = 1 \pm i\sqrt{2}$$

Step 4:
$$x^2 + 3 = 2x$$

$x = 1 \pm i\sqrt{2}:$ $(1 \pm i\sqrt{2})^2 + 3 \overset{?}{=} 2(1 \pm i\sqrt{2})$

$$(1 \pm 2i\sqrt{2} + 2i^2) + 3 \overset{?}{=} 2 \pm 2i\sqrt{2}$$

$$1 \pm 2i\sqrt{2} - 2 + 3 \overset{?}{=} 2 \pm 2i\sqrt{2}$$

$$2 \pm 2i\sqrt{2} = 2 \pm 2i\sqrt{2}$$

The solutions of $x^2 + 3 = 2x$ are $1 + i\sqrt{2}$ and its conjugate, $1 - i\sqrt{2}$.

Note: When the trinomial left member of $ax^2 + bx + c = 0$ is factorable with integral co-efficients, "$b^2 - 4ac$" of the quadratic formula is invariably the square of an integer (as in Examples 1 and 2) and is the basis for the "test" suggested on page 63.

| EXERCISE 7-4

Solve for x by using the quadratic formula. Express solutions in simplest radical form.

1. $x^2 - 5x + 6 = 0$
2. $x^2 + 8x + 15 = 0$
3. $x^2 - 6x + 9 = 0$
4. $6x^2 - x - 15 = 0$
5. $6x^2 - 17x + 12 = 0$
6. $3x^2 - 19x = 14$
7. $4x^2 + 6x + 1 = 0$
8. $5x^2 - 12x - 12 = 0$
9. $x^2 - 2x + 5 = 0$
10. $x^2 + 13 = 4x$
11. $3x^2 + 2x = \dfrac{2}{3}$
12. $2x^2 - 3x + 5 = 0$
13. $9x^2 = 3x + 1$
14. $x^2 - 16a = 0$
15. $5 + 8x^2 = 16x$
16. $(x - 2)^2 + (x + 2)^2 = 8c$
17. $2x^2 + 3x - k = 0$
18. $x^2 - 2cx + 3 = 0$
19. $4x^2 - 9a^2 = 0$
20. $x^2 - 6ax + 8a^2 = 0$
21. $x^2 = 2ax + c$
22. $2x^2 = 6x + 4$
23. $3x^2 - 16x + 22 = 0$
24. $ax^2 - bx = 7$
25. $2 + 4x^2 = x(x + 4)$
26. $3x^2 + 4x = 2(2x - 1) - 3x^2$

6. FRACTIONAL EQUATIONS

A fractional equation is one in which the variable appears in the denominator of at least one term of the equation. (Recall Section 4, Chapter 4.) When both members of the equation are multiplied by a common denominator of the fraction terms, a new equation results, called the derived equation.

The derived equation, however, is not always an equivalent of the original equation. There may be solutions of the derived equation that do not satisfy the given equation (e.g., involve zero denominators). So it is especially important to check the solutions obtained from the derived equation by substituting them for the variable in the original equation. Those solutions that do not satisfy the original equation are to be discarded.

In the following examples, the equations are fractional. When the two members are multiplied by the least common denominator of the fraction terms, the derived equation is quadratic, and it can be solved by one of the foregoing methods.

EXAMPLES

1. Solve for x: $x + \dfrac{2}{x} = 3$.

SOLUTION Multiply both members by the common denominator, x,

$$x\left(x + \frac{2}{x}\right) = x(3)$$

and simplify:

$$x^2 + 2 = 3x$$

Solve the derived quadratic equation by Program **7.1**:

$$x^2 - 3x + 2 = 0$$
$$(x - 2)(x - 1) = 0$$
$$x - 2 = 0 \qquad x - 1 = 0$$
$$x = 2 \qquad\quad x = 1$$

Check the solutions of the derived equation in the original equation.

$$x + \frac{2}{x} = 3$$

$$x = 2: \quad (2) + \frac{2}{(2)} \overset{?}{=} 3 \qquad\bigg|\qquad x = 1: \quad (1) + \frac{2}{(1)} \overset{?}{=} 3$$

$$2 + 1 = 3 \qquad\qquad\qquad\qquad 1 + 2 = 3$$

Thus the solution set is confirmed to be $\{2, 1\}$.

2. Solve for x: $\dfrac{x - 2}{x^2 - 4} = \dfrac{x}{x + 2}$.

SOLUTION LCD is $x^2 - 4$, or $(x - 2)(x + 2)$.

$$(x^2 - 4) \cdot \frac{x - 2}{x^2 - 4} = (x - 2)(x + 2) \cdot \frac{x}{x + 2}$$

$$x - 2 = (x - 2)(x)$$
$$x - 2 = x^2 - 2x$$
$$x^2 - 3x + 2 = 0 \quad \text{(derived equation)}$$

Using Program **7.1**,

$$(x - 1)(x - 2) = 0$$
$$x - 1 = 0 \qquad x - 2 = 0$$
$$x = 1 \qquad\quad x = 2$$

Check:

$$\frac{x-2}{x^2-4} = \frac{x}{x+2}$$

$x = 1$: $\dfrac{(1)-2}{(1)^2-4} \overset{?}{=} \dfrac{1}{(1)+2}$ $\qquad$ $x = 2$: $\dfrac{(2)-2}{(2)^2-4} \overset{?}{=} \dfrac{2}{(2)+2}$

$$\frac{-1}{-3} = \frac{1}{3} \qquad\qquad\qquad\qquad \frac{0}{0} \neq \frac{2}{4}$$

Since $\dfrac{0}{0}$ is meaningless, we reject the solution, 2, of the derived equation. The solution set has a single member: $\{1\}$.

3. Solve $\dfrac{x}{x+1} + \dfrac{3x}{x^2-1} = \dfrac{2}{x-1}$.

SOLUTION LCD: $x^2 - 1$, or $(x-1)(x+1)$
Derived equation:

$$\frac{x(x^2-1)}{x+1} + \frac{3x(x^2-1)}{x^2-1} = \frac{2(x^2-1)}{x-1}$$

$$x(x-1) + 3x = 2(x+1)$$

$$x^2 + 2x = 2x + 2$$

$$x^2 = 2$$

$$x = \pm\sqrt{2}$$

Both solutions, $+\sqrt{2}$ and $-\sqrt{2}$, of the derived equation lead to nonzero denominators when substituted for the variable in the given equation. Thus the solution set for

$$\frac{x}{x+1} + \frac{3x}{x^2-1} = \frac{2}{x-1}$$

is $\{\sqrt{2}, -\sqrt{2}\}$.

EXERCISE 7-5

Solve for the variable.

1. $x - 5 + \dfrac{6}{x} = 0$ $\qquad\qquad$ 2. $x - 1 - \dfrac{12}{x} = 0$

3. $2x + 3 = \dfrac{2}{x}$ $\qquad\qquad$ 4. $7 + \dfrac{3}{x} = 6x$

5. $6 = y + \dfrac{9}{y}$ $\qquad\qquad$ 6. $m = 8 - \dfrac{16}{m}$

7. $x = \dfrac{4x + 3}{x + 2}$

8. $a = \dfrac{3a + 1}{2a + 1}$

9. $\dfrac{x - 9}{x + 1} = x$

10. $x = \dfrac{2x - 5}{x + 2}$

11. $\dfrac{x}{x + 2} = 2x - 1$

12. $\dfrac{t}{t - 3} = 3t + 4$

13. $\dfrac{4x}{2x - 1} = 1 - 2x$

14. $\dfrac{3x}{x - 2} = x + 2$

15. $\dfrac{a + 3}{2a - 1} = 3a + 1$

16. $x + 1 = \dfrac{x - 2}{2x + 1}$

17. $\dfrac{2x - 1}{3x + 4} = \dfrac{2x - 1}{x + 2}$

18. $\dfrac{x + 2}{x - 1} = \dfrac{x + 3}{3x - 1}$

19. $\dfrac{3x - 1}{2x - 3} = \dfrac{x + 2}{x + 1}$

20. $\dfrac{x - 3}{3x - 1} = \dfrac{2x - 1}{2x - 3}$

21. $\dfrac{3}{x - 1} - \dfrac{2}{x - 4} = 3$

22. $\dfrac{2}{x - 3} - \dfrac{4}{x - 1} = 2$

23. $\dfrac{6}{x^2 - 9} = \dfrac{1}{x - 3} - \dfrac{1}{5}$

24. $\dfrac{4}{x} = \dfrac{6}{x + 4} - \dfrac{5}{x + 3}$

25. $\dfrac{1}{x - 1} + \dfrac{3x}{x^2 + x - 2} + \dfrac{x}{x + 2} = 0$

26. $\dfrac{x}{x + 2} + 2 + \dfrac{x - 6}{x^2 - 4} = 0$

7. APPLICATIONS

An equation that represents a problem situation is, in effect, a mathematical model of that situation. Sometimes the model is more general than the circumstances of the problem warrant—all the more reason to check the solutions of the equation in the problem. For instance,

> A rectangle is 4 in. longer than it is wide, and has an area of 60 sq in. What are the dimensions of its sides?

Applying Program **5.1,**

Step 1:   Let x = length of shorter side (width); then
$x + 4$ = length of longer side (length).

Step 2:   The equivalence relationship:

$$\text{length} \cdot \text{width} = \text{area}$$

Step 3:   The corresponding equation (mathematical model):

$$(x + 4)(x) = 60$$

Step 4:   Solving the equation,

$$x^2 + 4x = 60$$

$$x^2 + 4x - 60 = 0$$

$$(x + 10)(x - 6) = 0$$

$$x + 10 = 0 \qquad x - 6 = 0$$

$$x = -10 \qquad x = 6$$

Step 5:   Check: Note that both solutions, -10 and 6, satisfy the equation:

$$x^2 + 4x = 60$$

$$x = -10: \quad (-10)^2 + 4(-10) \overset{?}{=} 60$$

$$100 - 40 = 60$$

$$x = 6: \qquad (6)^2 + 4(6) \overset{?}{=} 60$$

$$36 + 24 = 60$$

But both solutions do not satisfy the problem: a dimension of "-10 in." is meaningless. So we reject it as a valid solution of the problem even though it is a valid solution of the equation. The other solution of the equation, 6, is acceptable in terms of the problem. Thus the dimensions of the rectangle are 6 in. and $(x + 4)$, or 10 in., for which the area is indeed 60 sq in.

Whatever the number of solutions generated by the mathematical model for the problem, each should be checked against the information given in the problem. In some cases all solutions of the equation will be applicable (as in Example 1 below), but not always, as we have just seen.

EXAMPLES

1.   One number is 6 larger than another number, and their product is 55. What are the numbers?

SOLUTION   Let x = the smaller number.
Then $x + 6$ = the larger number.
Their product: $(x)(x + 6)$.

The equation or mathematical model:

$$(x)(x + 6) = 55$$

$$x^2 + 6x = 55$$

$$x^2 + 6x - 55 = 0$$

$$(x - 5)(x + 11) = 0$$

$$x - 5 = 0 \qquad x + 11 = 0$$

$$x = 5 \qquad x = -11$$

Two solutions: $\{5, -11\}$.
When $x = 5$, $x + 6 = 11$.
When $x = -11$, $x + 6 = -5$.

Both solutions of the equation apply. There are two pairs of numbers that satisfy the statement of the problem:

$$5 \text{ and } 11, \text{ product } 55$$

$$-5 \text{ and } -11, \text{ product } 55$$

2. By doubling the sides of a square, the area it covers is increased by 48 sq in. How long is each side of the original square?

SOLUTION Let $x =$ length of side of original square; then
$\quad 2x =$ length of extended side.
Thus $(x)^2 =$ area of original square, and
$\quad (2x)^2 =$ area of extended square.
Increase in area: 48 sq in.

The equation:

$$(2x)^2 - (x)^2 = 48$$

$$4x^2 - x^2 = 48$$

$$3x^2 = 48$$

$$x^2 = 16$$

$$x = 4, -4$$

The -4 solution is meaningless in terms of the problem. The original square has 4-in. sides.

3. A tray is to be formed by cutting square corners from a rectangular sheet of metal, 12 in. by 16 in., and folding up the edges. If the area of the base is to be 60 sq in., how deep should each corner cut be?

SOLUTION Let $x =$ length of edge of the square corner cuts. (See Figure 7.1.) The area of the base is to be 60 sq in.

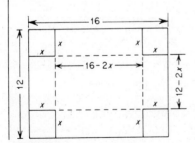

Figure 7.1

The equation:

$$(16 - 2x)(12 - 2x) = 60$$
$$192 - 56x + 4x^2 = 60$$
$$48 - 14x + x^2 = 15$$
$$x^2 - 14x + 33 = 0$$
$$(x - 11)(x - 3) = 0$$
$$x - 11 = 0 \qquad x - 3 = 0$$
$$x = 11 \qquad x = 3$$

Corner cuts of 11 in. would be impossible from a sheet 12 in. by 16 in. So we reject 11 in. as a solution. Corner cuts of 3 in. are possible, as shown in Figure 7.2. When the sides are folded up, a base of 60 sq in. is created.

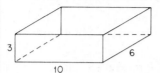

Figure 7.2

EXERCISE 7-6

Set up an equation for each problem and then solve.

1. Find two consecutive odd integers whose product is 63.
2. Two consecutive even integers have a product of 48. Which are they?
3. Two numbers differ by 5. Their product is 14. What are the numbers?
4. The product of two numbers, which differ by 8, is -15. What are the numbers?
5. The sum of two numbers is 5, and their product is -14. What are the numbers?
6. Find two numbers that have a sum of -4 and a product of -21.
7. One leg of a right triangle is 2 in. less than the other leg. If the area of the triangle is 24 sq in., what must be the length of the two legs?
8. The area of a rectangle is 40 square meters (m^2). If its length is 2 m less than three times its width, what are the side dimensions of the rectangle?
9. When the opposite sides of a square are extended 2 ft in the same direction and the other pair of sides is doubled, the area increases by 32 sq ft. What are the edge dimensions of the square?
10. If one pair of opposite sides of a square is increased by 3 m and the other pair is decreased by 2 m, the result is a rectangle with an area of 24 m^2. What was the area of the original square?

11. A framed picture covers 120 sq in. of wall. What is the width of the frame if the enclosed picture is 8 in. by 10 in.?

12. Plans call for 126 sq ft of carpet to be laid in a room that is 15 ft by 10 ft. If there is to be a uniform border of flooring exposed around the edge of the carpet, what should the dimensions of the carpet be?

13. The sum of the first n counting numbers, 1, 2, 3, 4, ..., n, is given by the formula $S = \frac{1}{2} n(n + 1)$. Find n for $S = 28$.

14. Use the formula of Exercise 13 to determine the consecutive counting numbers that have a sum of 15.

15. Some boy scouts hiked up a hill and back, a total distance of 8 mi, in 3 hr hiking time. The pace coming down was twice that going up. What was their speed going up?

16. A pilot flew 600 mi. Had he reduced his average speed by 40 mph, it would have taken him 45 min longer to make the trip. What must his average speed have been?

17. A man drove to a point 90 mi away at a certain rate of speed. On his return trip he increased his speed by 20 mph and cut 1 hr 12 min off his travel time. What were the two rates of speed?

18. A woman spends 54 min a day on a bus riding to and from her job, a distance of 10 mi one way. In the morning the bus averages 5 mph more than in the evening. What is the average speed of the bus in the morning?

19. A group of students decide that they can complete a job in 20 labor-hours. If they can get two more helpers, they figure that each would need to work $\frac{1}{2}$ hr less. How many students are in the group?

20. The membership of a club contracts to paint a barn for $200. It is figured that 72 labor-hours are needed to complete the job. However, if four of their members fail to show up, it will take each one who does come 3 hr more to complete the job. How many members are in the club?

21. It takes a helper 3 hr longer to assemble a machine than it does his boss. By working together, they can assemble the machine in 3 hr 36 min. How long will it take each to do the job alone?

22. One vent can empty a tank in 20 min less than another vent. If both vents are opened, a full tank can be emptied in $13\frac{1}{3}$ min. How long would it take each vent, alone, to empty the tank?

(Use the information following for problems 23–26.) The compound interest formula is

$$A = P(1 + i)^t$$

where A is the amount of money accumulated for an investment of P dollars, for t years, at $i\%$ interest per year, compounded (i.e., added to the account) annually.

Note: At 4%, $(1 + i) = (1.04)$.

23. How much will an investment of $1000 amount to after 2 years at a compound interest rate of 7% per year?

24. How much must you deposit into a savings account, where the interest rate is 5%, compounded annually, in order to have $22,050 saved by the end of 2 years?

25. What is the annual compound interest rate if, at the end of 2 years, an investment of $500 has grown to $720?

26. A woman bought a painting for $1000 and sold it 2 years later for $2250. Had she put the $1000 into a bank, at compound interest, instead of buying the painting, what annual interest rate would she have had to receive in order to do as well as she did in buying the painting?

8. RADICAL EQUATIONS

An equation in which the variable occurs as a radicand is called a **radical equation**. For instance,

$$\sqrt[3]{x} = 2$$

By cubing both members,

$$(\sqrt[3]{x})^3 = (2)^3$$

$$x = 8$$

we derive an equation for which the solution is obvious.

However, raising both members of an equation to a higher power does not always produce an equivalent equation. The derived equation may have more solutions than the original equation.

Solutions of the derived equation that are not solutions of the original equation are called **extraneous solutions**. So it is important that all solutions of the derived equation be checked as solutions of the original equation.

PROGRAM 7.6

To solve a radical equation:

Step 1: Transform the given equation so that one radical expression is alone in one member and the rest of the terms are in the other member.

Step 2: Raise both members to a power equal to the index of the isolated radical of Step 1 and simplify the equation.

Step 3: Repeat Steps 1 and 2, if necessary, to free the equation of all radicals.

Step 4: Solve the resulting equation.

Step 5: Check the solutions obtained in Step 4 by substituting each for the variable in the given equation.

1. Solve $x - 2 - \sqrt{x + 4} = 0$.

SOLUTION *Step 1:* Transform the equation so that a single radical expression appears in one member:

$$x - 2 - \sqrt{x + 4} = 0$$

$$x - 2 = \sqrt{x + 4}$$

Step 2: The index of the radical expression is 2; square both members and simplify.

$$(x - 2)^2 = (\sqrt{x + 4})^2$$

$$x^2 - 4x + 4 = x + 4$$

$$x^2 - 5x = 0$$

Step 3: There are no other radical expressions; go to the next step.

Step 4: Solve, using Program **7.1**:

$$(x)(x - 5) = 0$$

$$(x) = 0 \qquad (x - 5) = 0$$

$$x = 0 \qquad\qquad x = 5$$

Step 5: Possible solutions are 0 and 5. Check by replacing the variable in the given equation by each to see if a true statement results.

$$x - 2 - \sqrt{x + 4} = 0$$

$$x = 0: \quad (0) - 2 - \sqrt{0 + 4} \overset{?}{=} 0$$

$$-2 - \sqrt{4} \overset{?}{=} 0$$

$$-2 - 2 \overset{?}{=} 0$$

$$-4 \neq 0$$

So 0 is not a solution of the original equation.

$$x - 2 - \sqrt{x + 4} = 0$$

$$x = 5: \quad (5) - 2 - \sqrt{5 + 4} \overset{?}{=} 0$$

$$5 - 2 - \sqrt{9} \overset{?}{=} 0$$

$$5 - 2 - 3 = 0$$

Hence the solution set for $x - 2 - \sqrt{x + 4} = 0$ contains but a single number: 5.

2. Solve $\sqrt{2x + 3} + \sqrt{4x - 3} = 6$.

SOLUTION | To solve this equation, it will be necessary to square the members twice (Step 3).

Step 1: Transform the given equation so that a single radical expression appears in the left member:

$$\sqrt{2x + 3} + \sqrt{4x - 3} = 6$$

$$\sqrt{2x + 3} = 6 - \sqrt{4x - 3}$$

Step 2: Square both members to eliminate the radical in the left member; simplify:

$$(\sqrt{2x + 3})^2 = (6 - \sqrt{4x - 3})^2$$

$$2x + 3 = 36 - 12\sqrt{4x - 3} + (4x - 3)$$

$$12\sqrt{4x - 3} = 2x + 30$$

$$6\sqrt{4x - 3} = x + 15$$

Step 3: Square both members to eliminate all radicals:

$$(6\sqrt{4x - 3})^2 = (x + 15)^2$$

$$36(4x - 3) = x^2 + 30x + 225$$

$$144x - 108 = x^2 + 30x + 225$$

$$x^2 - 114x + 333 = 0$$

Step 4: Solve the resulting equation of Step 3:

$$(x - 3)(x - 111) = 0$$

$$(x - 3) = 0 \qquad (x - 111) = 0$$

$$x - 3 = 0 \qquad x - 111 = 0$$

$$x = 3 \qquad\qquad x = 111 \quad \text{(possible solutions)}$$

Step 5: Check:

$$\sqrt{2x + 3} + \sqrt{4x - 3} = 6$$

$x = 3$: | $x = 111$:

$$\sqrt{2(3) + 3} + \sqrt{4(3) - 3} \stackrel{?}{=} 6 \qquad \sqrt{2(111) + 3} + \sqrt{4(111) - 3} \stackrel{?}{=} 6$$

$$\sqrt{9} + \sqrt{9} \stackrel{?}{=} 6 \qquad\qquad \sqrt{225} + \sqrt{441} \stackrel{?}{=} 6$$

$$3 + 3 = 6 \qquad\qquad 15 + 21 \stackrel{?}{=} 6$$

$$36 \neq 6$$

Hence 3 is a solution of the equation, but 111 is not.

3. Solve $\sqrt{x-2} - \sqrt{2x+3} = 2$.

SOLUTION

$$\sqrt{x-2} - \sqrt{2x+3} = 2$$
$$\sqrt{x-2} = 2 + \sqrt{2x+3}$$
$$(\sqrt{x-2})^2 = (2 + \sqrt{2x+3})^2$$
$$x - 2 = 4 + 4\sqrt{2x+3} + (2x+3)$$
$$-4\sqrt{2x+3} = 4 + 2x + 3 - x + 2$$
$$-4\sqrt{2x+3} = x + 9$$
$$(-4\sqrt{2x+3})^2 = (x+9)^2$$
$$16(2x+3) = x^2 + 18x + 81$$
$$32x + 48 = x^2 + 18x + 81$$
$$0 = x^2 - 14x + 33$$
$$0 = (x-11)(x-3)$$
$$x - 11 = 0 \qquad x - 3 = 0$$
$$x = 11 \qquad x = 3$$

Check:

$$\sqrt{x-2} - \sqrt{2x+3} = 2$$

$x = 11$:

$$\sqrt{11-2} - \sqrt{2(11)+3} \overset{?}{=} 2$$
$$\sqrt{9} - \sqrt{25} \overset{?}{=} 2$$
$$3 - 5 \overset{?}{=} 2$$
$$-2 \neq 2$$

$x = 3$:

$$\sqrt{3-2} - \sqrt{2(3)+3} \overset{?}{=} 2$$
$$\sqrt{1} - \sqrt{9} \overset{?}{=} 2$$
$$1 - 3 \overset{?}{=} 2$$
$$-2 \neq 2$$

Neither 11 nor 3 is a solution of the given equation. In other words, the equation has *no solutions*; its solution set is the *empty set* (usually symbolized by $\varnothing$ or { }).

EXERCISE 7-7

Solve and verify all solutions.

1. $\sqrt{x+3} = 6$

2. $\sqrt{2x-4} + 2 = 0$

3. $2 + \sqrt{x-2} = 0$

4. $\sqrt{x^2+2} + x = 2$

5. $1 - 2x + \sqrt{4x+1} = 0$

6. $\sqrt{4x+3} - \sqrt{6x+2} = 0$

7. $3\sqrt{x-1} = 2\sqrt{2x-1}$

8. $3\sqrt{x^2-4} = \sqrt{3x^2+4x+6}$

9. $1 + \sqrt{2x} = \sqrt{2x+5}$

10. $\sqrt{x-2} - 2 = \sqrt{2x+3}$

11. $\sqrt{2x-5} = 1 + \sqrt{x-3}$

12. $\sqrt{x+2} + \sqrt{x-1} = \sqrt{2x+5}$

13. $\sqrt{x+8} = \sqrt{2-x} + \sqrt{5-x}$

14. $\sqrt{2x + \sqrt{2x-4}} = 2$

15. $\sqrt{\dfrac{3x-2}{9}} = \dfrac{x+2}{6}$

16. $\sqrt[3]{5x^2 + 2x + 3} - 3 = 0$

17. $\dfrac{\sqrt{x+4}}{x-1} = 2$

18. $\dfrac{1}{\sqrt{x}} = \dfrac{2}{\sqrt{3x+4}}$

19. $\dfrac{3}{\sqrt{2x-1}} = \dfrac{2}{\sqrt{x-3}}$

20. $\dfrac{3}{\sqrt{x-2}} = \dfrac{2}{\sqrt{x+3}}$

21. $\sqrt[3]{x^2+2} = 3$

22. $\sqrt[4]{x^2+x+4} = 2$

9. QUADRATIC INEQUALITIES★

A **quadratic inequality** in standard form is an inequality that can be expressed as

$$ax^2 + bx + c > 0 \quad \text{or} \quad ax^2 + bx + c < 0$$

where x represents the variable and a, b, and c represent real numbers, $a > 0$. Furthermore, every quadratic polynomial $ax^2 + bx + c$ can be expressed as a pair of factors.

For a simple instance, the quadratic inequality

$$x^2 - 5x + 6 > 0$$

can be expressed as

$$(x - 2)(x - 3) > 0$$

We are aware that when two factors are positive or two factors are negative, their product is positive (i.e., >0). So, in terms of our example, the $x - 2$ factor will be positive for all x replacements greater than 2 and negative for all x replacements less than 2 (see Figure 7.3).

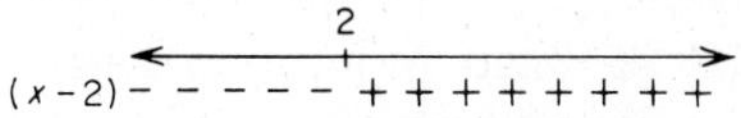

Figure 7.3

The other factor, $x - 3$, will be positive for all x replacements greater than 3 and negative for all x replacements less than 3 (see Figure 7.4).

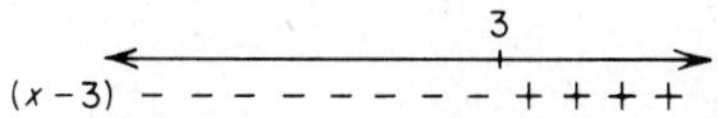

Figure 7.4

★ The content of this section, in more basic courses of study, may be postponed without significant effect upon the remaining parts of the book.

When we combine the two sets of facts illustrated in Figures 7.3 and 7.4, the result is as shown in Figure 7.5.

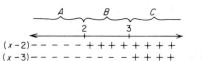

Figure 7.5

From Figure 7.5 we can see that both factors are negative in segment A (i.e., $x < 2$) and that both factors are positive in segment C (i.e., $x > 3$). In each instance, when x replacements are chosen from these segments, the *product* of the two factors, $(x - 2)(x - 3)$, will be positive (i.e., >0).

So the solution set for

$$x^2 - 5x + 6 > 0$$

must be

$$\{x < 2 \text{ or } x > 3\}$$

On the other hand, suppose that the original inequality had been

$$x^2 - 5x + 6 < 0$$

or in factor form

$$(x - 2)(x - 3) < 0$$

In order for the product of the two factors, $x - 2$ and $x - 3$, to be negative (i.e., <0), one factor must be positive and the other negative for the same x replacement. According to the scheme of Figure 7.5, this situation will occur only when the x replacements are taken from segment B.

So the solution set for

$$x^2 - 5x + 6 < 0$$

must be

$$\{2 < x < 3\}$$

For every quadratic inequality (in standard form) in which the factors are different real numbers, the scheme of Figure 7.5 can be simplified to that of Figure 7.6.

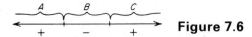

Figure 7.6

In general, if p and q are real number solutions of $ax^2 + bx + c = 0$, then the solution set for

$$ax^2 + bx + c < 0 \qquad (a > 0)$$

is

$$\{p < x < q\}$$

And the solution set for

$$ax^2 + bx + c > 0 \qquad (a > 0)$$

is

$$\{x < p \text{ or } x > q\}$$

PROGRAM 7.7

To solve a quadratic inequality expressed in standard form:

Step 1:   Set the quadratic polynomial of the inequality equal to zero and solve for the variable.

Step 2:   Record the solutions of Step 1 on a number line.

Step 3:   Select the middle segment of the number line of Step 2 as the graph of the solution set if the sign of the inequality is $<$; select the two outer segments as the graph of the solution set if the sign of the inequality is $>$.

EXAMPLES

1.   Solve $x^2 - 5x + 4 < 0$.

SOLUTION   *Step 1:* Set the quadratic polynomial of the inequality equal to zero and solve for the variable.

$$x^2 - 5x + 4 = 0$$
$$(x - 1)(x - 4) = 0$$
$$x - 1 = 0 \qquad x - 4 = 0$$
$$x = 1 \qquad\qquad x = 4$$

Step 2: Record the solutions of Step 1 on a number line.

Step 3: The sign of the given inequality is <(negative). So the graph of the solution set of the inequality is the middle segment

or

$$\{1 < x < 4\}$$

2. Solve $x^2 + 3x - 10 > 0$.

SOLUTION *Step 1:*
$$x^2 + 3x - 10 = 0$$
$$(x + 5)(x - 2) = 0$$
$$x + 5 = 0 \qquad x - 2 = 0$$
$$x = -5 \qquad x = 2$$

Step 2:

Step 3: Sign of the inequality is > (positive); so

or $\{x < -5$ or $x > 2\}$ is the solution set.

3. Solve $x^2 + 4x - 3 \leq 0$.

SOLUTION *Step 1:* $x^2 + 4x - 3 = 0$

$$x = \frac{-4 \pm \sqrt{(4)^2 - (4)(1)(-3)}}{2}$$

$$x = \frac{-4 \pm \sqrt{28}}{2}$$

$$x = -2 \pm \sqrt{7}$$

Step 2:

Step 3: Sign of the inequality is ≤ (negative or equal); so the solution set is

or

$$\{-2 - \sqrt{7} \leq x \leq -2 + \sqrt{7}\}.$$

When an inequality contains fractions, special care must be taken. If we multiply both members of an inequality by a common factor in order to eliminate denominators, a reversal in the sense of the inequality is necessary when the multiplying factor is negative.

Consider the inequality

$$\frac{x + 1}{x - 1} < 0$$

We cannot eliminate the denominator by multiplying both members by $x - 1$ (as we could in an equation) because we do not know whether $x - 1$ is positive or negative. It could be both, for different numbers in the solution set.

To avoid the difficulty, we multiply both members by the *square* of the denominator (which is certain to be positive) and keep the sense of the inequality as it is given.

Thus

$$(x - 1)^2 \cdot \frac{x + 1}{x - 1} < (x - 1)^2 (0)$$

$$(x - 1)(x + 1) < 0$$

By Program **7.7**,

$$x - 1 = 0 \qquad x + 1 = 0$$

$$x = 1 \qquad x = -1$$

The solution set for $\dfrac{x + 1}{x - 1} < 0$ is $\{-1 < x < 1\}$.

As with fractional equations, the solution set obtained from the derived inequality must be inspected for extraneous solutions. Any number that does not satisfy the given inequality (be careful of zero denominators) must be discarded.

EXAMPLES

1. Solve for x: $\dfrac{x + 2}{x - 3} \geq 0$.

SOLUTION | Multiply both members by the square of the denominator:

$$(x - 3)^2 \cdot \frac{x + 2}{x - 3} \geq (0)(x - 3)^2$$

$$(x - 3)(x + 2) \geq 0$$

Locate the interval points:

$$(x - 3)(x + 2) = 0$$

$$x - 3 = 0 \qquad x + 2 = 0$$

$$x = 3 \qquad x = -2$$

The solution set is $\{x \leq -2 \text{ or } x > 3\}$.

Note: $x = 3$ is discarded from the solution set because it produces a zero denominator when substituted for the variable in the original inequality.

2. Solve $\dfrac{2x - 1}{x} < 1$.

SOLUTION Multiply both members by the square of the denominator to obtain the derived inequality:

$$x^2 \cdot \frac{2x - 1}{x} < 1(x^2)$$

$$x(2x - 1) < x^2$$

$$2x^2 - x < x^2$$

$$x^2 - x < 0$$

$$x(x - 1) < 0$$

$$x = 0 \qquad x - 1 = 0$$

$$x = 1$$

The solution set is $\{0 < x < 1\}$.

Solve.

1. $x^2 - 4 < 0$ 2. $x^2 - 9 > 0$
3. $x^2 - 1 > 0$ 4. $x^2 - 16 < 0$
5. $m^2 < 3$ 6. $n^2 > 3$
7. $x^2 - 6x + 8 \leq 0$ 8. $p^2 - 7p + 12 \leq 0$
9. $n^2 - 8n + 15 > 0$ 10. $t^2 - 10t + 21 > 0$

11. $x^2 + x - 6 < 0$

12. $p^2 + p - 2 < 0$

13. $r^2 + 3r \geq 4$

14. $y^2 - 2y \geq 3$

15. $m^2 + 5m + 6 > 0$

16. $n^2 + 6n + 5 > 0$

17. $s^2 + 8s + 12 < 0$

18. $x^2 + 8x + 15 < 0$

19. $2x^2 - 7x - 4 \leq 0$

20. $3x^2 + 5x - 2 \geq 0$

21. $2x^2 + 15 > 13x$

22. $3x^2 + 10 < 11x$

23. $6y^2 + y - 2 < 0$

24. $8m^2 + 6m - 9 < 0$

25. $6n^2 + 13n + 5 > 0$

26. $8n^2 - 30n + 7 > 0$

27. $x^2 + 5 \leq 5x$

28. $y^2 - y < 1$

29. $2m^2 + 6m + 3 \geq 0$

30. $2p^2 - 6p + 1 > 0$

31. $4t^2 + 6t + 1 > 0$

32. $9x^2 + 6x - 2 > 0$

33. $4x^2 - 4x > -1$

34. $9p^2 + 12p \leq -4$

35. $12p + 9 < -4p^2$

36. $25n^2 - 20n + 4 > 0$

37. $2x^2 - 5x < 0$

38. $4t^2 + 2t > 0$

39. $6k^2 + 2k > 0$

40. $60y^2 < 140y$

41. $\dfrac{x + 2}{x - 2} < 0$

42. $\dfrac{x - 2}{x + 2} > 0$

43. $\dfrac{x - 7}{x + 7} \geq 0$

44. $\dfrac{x + 3}{x - 3} \leq 0$

45. $\dfrac{x + 3}{x} < 0$

46. $\dfrac{x - 4}{x} > 0$

47. $\dfrac{m - 2}{m + 3} \geq 0$

48. $\dfrac{y - 8}{y + 3} \leq 0$

49. $\dfrac{2x - 1}{x} \leq 1$

50. $\dfrac{3x - 2}{x} \geq 2$

51. $\dfrac{3x + 1}{x + 3} \geq 1$

52. $\dfrac{2x - 1}{x + 2} \leq 1$

53. $\dfrac{x - 1}{x + 2} \geq 2$

54. $\dfrac{2x + 1}{2x - 1} \leq 3$

REVIEW

PART A

Answer True or False.

1. The quadratic equation, $ax^2 + bx + c = 0$, becomes a linear equation when $a = 0$.

2. Every quadratic equation in a single variable has two solutions, even though both solutions may be the same.

3. If $A \cdot B = 0$, then it follows that A, or B, or both, is zero.

4. A useful first step in solving the quadratic equation, $3x^2 = 6x$, would be to divide both members by x, thereby producing the simpler equivalent equation, $3x = 6$.

5. The pure quadratic equation, $x^2 = -25$, has no solution among the real numbers, but it does among the complex numbers.

6. The quadratic formula may be used to solve most, but not all, quadratic equations.

7. A fractional equation (variable in the denominator) and its nonfractional derived equation will always have identical solution sets.

8. An extraneous solution is a false solution.

9. When $A \cdot B < 0$, one of the two factors is negative and the other is positive.

10. $-3 < x < +4$ means that the domain of x contains all numbers between (but not including) -3 and $+4$.

PART B

Solve for all values of x by factoring.

1. $x^2 - 13x + 30 = 0$

2. $x^2 + 10x + 16 = 0$

3. $x^2 = 3x + 18$

4. $x^2 + 10x + 25 = 0$

5. $12x^2 = 3 + 5x$

6. $x^2 + bx = 6b^2$

7. $15x^2 - 13ax + 2a^2 = 0$

8. $6 + ax = 15a^2x^2$

9. $2ax = 3x^2$

10. $2ax = 5a^2b^2x^2$

Solve for x.

11. $3x^2 = 27$

12. $5x^2 - 25 = 0$

13. $2x^2 + 8 = 0$

14. $4x^2 + 9 = x^2$

15. $5x^2 + 4 = x^2 + 13$

16. $3x^2 = 1$

17. $(x + 2)^2 = 16$

18. $(x - 1)^2 = -9$

19. $(x - 3)^2 = 3$

20. $(2x + 1)^2 = 5$

21. $9x^2 + 6x + 1 = 4$

22. $4x^2 + 4x + 1 = -4$

Insert a term in the empty parentheses to make the complete expression a square trinomial.

23. $x^2 - 12ax + (\ )$

24. $x^2 - 16x + (\ )$

25. $x^2 + \dfrac{x}{5} + (\ )$

26. $x^2 - 7x + (\ )$

27. $x^2 - abx + (\ )$

28. $x^2 - \dfrac{3}{8}x + (\ )$

Solve by completing the square.

29. $x^2 - 4x - 1 = 0$

30. $y^2 + 2y + 2 = 0$

31. $4a^2 - 12a = 0$

32. $4x^2 + 4x = 3$

Solve for x by means of the quadratic formula.

33. $4x^2 + 11x - 3 = 0$

34. $12x^2 - 11x + 2 = 0$

35. $x^2 - 7x + 2 = 0$

36. $2x^2 - 4x + 5 = 0$

37. $4x^2 + x - 14 = 0$

38. $5x^2 + 22x + 8 = 0$

39. $x^2 + 6x - 13 = 0$

40. $3x^2 + 10x + 2 = 0$

41. $x^2 - 6x + 10 = 0$

42. $3x^2 - 4x + 2 = 0$

43. $4x^2 = 4x + 5a$

44. $x^2 - 32c = 3x^2$

45. $(x - 2a)(x + 4a) = 9$

46. $(x - y)(x + 3y) = 2$

Solve for x.

47. $\dfrac{x}{2} + \dfrac{2}{x} = 2x$

48. $\dfrac{3}{x^2 - 1} = \dfrac{4}{x^2 + 2}$

49. $\dfrac{1}{x^2} = 1 - \dfrac{2}{x}$

50. $\dfrac{3}{x} - 2 = \dfrac{1}{x^2}$

51. $\dfrac{x + 3}{x - 2} = x$

52. $\dfrac{x}{x - 4} = 2x + 1$

53. $\dfrac{2}{x - 2} + \dfrac{3}{x + 1} = 2$

54. $\dfrac{3}{x - 1} = 3 + \dfrac{2}{x - 4}$

55. $\dfrac{x}{x - 1} + \dfrac{x + 2}{2x^2 - 3x + 1} = \dfrac{10}{2x - 1}$

56. $\dfrac{2x - 1}{x - 5} + \dfrac{2x + 1}{x - 1} = 2$

57. The hypotenuse of a right triangle is 15 ft and one leg is 3 ft longer than the other. What is the length of the longest leg?

58. A department store buyer paid $96 for a group of similar articles. If she could have gotten the unit price down by $4, she could have purchased four more articles for the same money. How many articles did she purchase?

59. Twenty-one is the product of two factors. One factor is a negative number increased by 2, and the other is that negative number decreased by 2. What is the negative number?

60. The base of a triangle is 4 in. longer than its altitude. If the area of the triangle is 160 sq in., what must be the length of its base?

61. A boat with a small outboard motor can travel at 8 mph in still water. A trip to a point 6 mi upstream and back takes 2 hr. How fast is the river flowing?

62. A landlord has 80 bedrooms which are now fully rented for $120 per month each. For every $2 increase in rent, he expects to suffer one vacancy. If he wants to increase his total rental income to $9782 a month, by how much should he increase the rents?

63. A man hiked 6 mi to the top of a hill and 6 mi back down again, in a total of 5 hr hiking time. His rate coming down was 1 mph faster than his rate going up. What was his rate each way?

64. There is a set of three consecutive integers such that twice the square of the middle integer is 26 more than the product of the other two integers. Name them.

65. An open box is formed by cutting a square from each corner of a 5 in. by 7 in. rectangular piece of metal and bending up the edges. If the box is to have a base area of 19 sq in., how long should the corner cuts be?

66. The edges of two cubes differ by 2 in. and their volumes differ by 296 cu in. What are the dimensions of each?

Solve and verify solutions.

67. $2\sqrt{x+1} - 4\sqrt{x-5} = 0$ **68.** $\sqrt{x^2+3} - 3x = 5$

69. $2\sqrt{x^2+3} = \sqrt{3x^2+5x+8}$ **70.** $2 + \sqrt{2x+4} = \sqrt{x}$

71. $\dfrac{3}{\sqrt{x-1}} = \dfrac{2}{\sqrt{x-6}}$ **72.** $\dfrac{1}{\sqrt{x-1}} = \dfrac{\sqrt{3}}{\sqrt{2x+1}}$

73. $\sqrt{x+1} - \sqrt{2x+3} + \sqrt{x+2} = 0$

74. $\sqrt{2x} - \sqrt{5x+1} = \sqrt{2}$ **75.** $x^2 + 2x - 15 > 0$

76. $p^2 - 16 > 0$ **77.** $2x^2 - 11x + 12 < 0$

78. $n^2 - 4n - 21 \leq 0$ **79.** $x^2 + 2x \geq 5$ **80.** $6t^2 > 15t$

81. $4e^2 + 1 < 4e$ **82.** $2c^2 + 1 < 5c$ **83.** $x^2 + 4 \leq 0$

84. $9 + k^2 \geq 6k$ **85.** $\dfrac{x+5}{x-5} > 0$ **86.** $\dfrac{x-3}{x+3} < 0$

87. $\dfrac{2a+1}{a-3} \geq 0$ **88.** $\dfrac{3x-2}{x} > 2$

89. $\dfrac{3y-1}{2y+3} \leq 1$ **90.** $\dfrac{2x+1}{x-1} \geq -2$

91. The product of two consecutive positive even integers is to be less than 35. Describe the set of possible pairs.

92. The product of two consecutive odd integers is to be less than 15. Describe the set of possible pairs.

93. Name the pairs of integers that satisfy this requirement: The integers are to differ by 3, and their product must be between 18 and 40 (i.e., not including 18 and 40).

94. Two consecutive integers have a product between, but not including, 42 and 72. Name them.

95. Strips of equal width are to be cut from the four sides of a rectangular piece of cardboard that is 8 in. by 10 in. If the resulting area is to be greater than 35 sq in. but less than 63 sq in., how wide might the strips be?

96. A grass plot 5 m by 8 m is to be extended equally in length and width. How long should the extensions be if the area is to fall between 88 m² and 130 m²?

8 EQUATIONS AND INEQUALITIES IN TWO VARIABLES

1. ORDERED PAIRS

In this chapter we expand the concept of mathematical equations and inequalities to include those in two variables. The replacement set for both variables, unless otherwise specified, will be the set of real numbers. Whereas solutions of equations in a single variable are single numbers, solutions of equations in two variables are pairs of numbers. For example,

$$x + 2y = 7$$

is an equation in two variables, x and y. Among the pairs of real numbers that satisfy the equation are:

(x, y)	$x + 2y = 7$
$(5, 1)$	$5 + 2(1) = 7$
$(3, 2)$	$3 + 2(2) = 7$
$(-1, 4)$	$-1 + 2(4) = 7$

The equation, $x + 2y = 7$, is also satisfied by the pair (1, 3), but not by the reverse of that pair, (3, 1):

$$(1, 3): \quad 1 + 2(3) = 7 \quad \text{(true)}$$
$$(3, 1): \quad 3 + 2(1) = 7 \quad \text{(false)}$$

That is, when 1 replaces x and 3 replaces y, the equation is satisfied; but not when 3 replaces x and 1 replaces y. Consequently, the pair (1, 3) is a solution of the equation, $x + 2y = 7$, but the pair (3, 1) is not a solution.

Use of the notation (1, 3) and (3, 1) involves an arbitrary assumption. The first, or left, component of the pair is assumed to be a replacement for x. The second, or right, component is assumed to be a replacement for y. Pairs of numbers for which their order is important are called **ordered pairs**.

EXAMPLES

1. Decide which of the following ordered pairs are solutions of the equation $2x + y = 12$: (3, 6), (6, 3), (4, −4), (−5, −2), (7, −2), (0, 12).

SOLUTION

$$2x + y = 12$$

(3 , 6):	$2(3) + 6 \overset{?}{=} 12$	Yes; (3, 6) is a solution
(6, 3):	$2(6) + 3 \overset{?}{=} 12$	No; (6, 3) is not a solution
(4, −4):	$2(4) + (−4) \overset{?}{=} 12$	No; not a solution
(−5, −2):	$2(−5) + (−2) \overset{?}{=} 12$	No; not a solution
(7, −2):	$2(7) + (−2) \overset{?}{=} 12$	Yes; solution
(0, 12):	$2(0) + (12) \overset{?}{=} 12$	Yes; solution

2. Find the ordered pairs that are solutions of the equation $2x − y = 8$ in which the first component of the pair (the x replacement) is (a) − 2; (b) 0; (c) 3.

SOLUTION (a) Substitute −2 for x; solve for y.

$$2x − y = 8$$
$$(−2, ?) \quad \text{or} \quad \left.\begin{array}{l} x = −2 \\ y = ? \end{array}\right\} \qquad 2(−2) − y = 8$$
$$−y = 8 + 4$$
$$y = −12$$

(b) Substitute 0 for x; solve for y.

$$(0, ?) \quad \text{or} \quad \left.\begin{array}{l} x = 0 \\ y = ? \end{array}\right\} \qquad 2(0) − y = 8$$
$$y = −8$$

(c) $\qquad (3, ?) \quad \text{or} \quad \left.\begin{array}{l} x = 3 \\ y = ? \end{array}\right\} \qquad 2(3) − y = 8$$
$$−y = 8 − 6$$
$$y = −2$$

Thus among the ordered pairs that are solutions of $2x - y = 8$ are $(-2, -12)$, $(0, -8)$, and $(3, -2)$.

EXERCISE 8-1

Select from among the ordered pairs given for each of the equations those that are solutions.

1. $2x + y = 4$; $(1, 2), (2, 3), (2, 0), (0, 2)$
2. $x - 3y = -2$; $(1, 1), (3, 2), (2, 3), (4, 2)$
3. $2x - y = 5$; $(1, -3), (2, -1), (3, 2), (4, 3)$
4. $2x - 3y = 4$; $(1, 1), (2, 0), (5, 2), (4, 4)$
5. $3x - 5y + 7 = 0$; $(1, 3), (-2, 4), (-4, -1), (-1, -3)$
6. $3x - 2y = 4$; $\left(0, 2\frac{1}{2}\right), \left(-1, -3\frac{1}{2}\right), (4, 8), \left(1, -\frac{1}{2}\right)$
7. $2x^2 - y = 4$; $(1, -2), (2, 4), (3, 11), (-1, -2)$
8. $2x^2 - 3y^2 = 4$; $(0, 2), (3, 4), (1, 2), (-1, -2)$

Replace the ? so as to make the ordered pair a solution of the given equation.

9. $x - 3y = 5$; $(?, 1), (-1, ?), (?, -3)$
10. $3x - 2y = 7$; $(3, ?), (-1, ?), (?, 4)$
11. $5x - 7y = 6$; $(?, -1), (4, ?), (?, 0)$
12. $x^2 - 3y = 4$; $(?, -1), (4, ?), (?, 0)$

2. CARTESIAN COORDINATE SYSTEM

The solution set of an equation or inequality in two variables is often expressed more easily by a graph in the **Cartesian (or rectangular) coordinate system**. Basic to the system are two perpendicular number lines, called **axes**, which divide the coordinate plane into four **quadrants**. Ordinarily, the quadrants are numbered as shown in Figure 8.1.

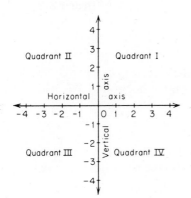

Figure 8.1

Every point in the coordinate plane will lie horizontally or vertically opposite some numbered point on each axis. The first number is taken from the horizontal axis. The second number is taken from the vertical axis. The result is an ordered pair of numbers, called the **coordinates** of the point. That point at which the two axes intersect is called the **origin**. Its ordered pair is (0, 0), as shown in Figure 8.2 below.

The first component of the ordered pair is also called the **abscissa** of the point. The second component is called the **ordinate** of the point.

When the variables are x and y, the usual convention is to associate the horizontal axis with x (thus the x *axis*) and the vertical axis with y (thus the y *axis*).

EXAMPLES

1.   In Figure 8.2, consider point (1, 3) in Quadrant I.

Its abscissa is 1; its ordinate is 3.
The point is located above 1 on the horizontal axis and opposite 3 on the vertical axis.

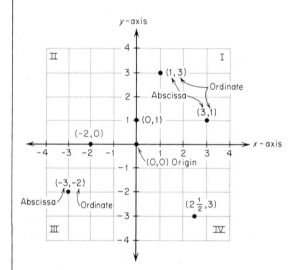

Figure 8.2

Consider point (3, 1) in Quadrant I:

Its abscissa and ordinate are the reverse of (1, 3).
The point is located above 3 on the horizontal axis and opposite 1 on the vertical axis.

Consider point $(-3, -2)$ in Quadrant III:

Its abscissa, or x value, is -3; its ordinate, or y value, is -2.
The point is located below -3 on the horizontal or x axis and opposite -2 on the vertical or y axis.

Consider point (0, 1) on the axis between Quadrants I and II:

> Its abscissa is 0; its ordinate is 1.
> All points on the vertical axis have 0 for an x value.

Consider point $(-2, 0)$ on the axis between Quadrants II and III:

> Its abscissa is -2; its ordinate is 0.
> All points on the horizontal axis have 0 for a y value.

Consider $\left(2\frac{1}{2}, -3\right)$ in Quadrant IV:

> Its positive x value, $2\frac{1}{2}$, means the point is located $2\frac{1}{2}$ units to the right of the y axis, and its negative y value, -3, means that the point is located 3 units below the x axis.

2. Estimate the coordinates of the lettered points on the grid of Figure 8.3.

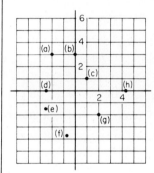

Figure 8.3

SOLUTION

(a) $(-2, 3)$

(b) $(0, 3)$

(c) $(1, 1)$

(d) $\left(-2\frac{1}{2}, 0\right)$

(e) $\left(-2\frac{1}{2}, -1\frac{1}{2}\right)$

(f) $(-0.8, -3.7)$

(g) $(2, -2)$

(h) $\left(4\frac{1}{4}, 0\right)$

EXERCISE 8-2

Plot the following points.

1. Abscissa 3, ordinate 2
2. Abscissa 4, ordinate -3
3. Abscissa -5, ordinate -3
4. Abscissa -2, ordinate -1
5. Abscissa -2, ordinate 0
6. $(-3, 2)$

7. $\left(2, 1\frac{1}{2}\right)$ **8.** $\left(-3, -2\frac{1}{2}\right)$

9. $\left(4, -3\frac{1}{2}\right)$ **10.** $\left(-\frac{1}{2}, \frac{1}{2}\right)$

11. Estimate the coordinates for the points labeled A, B, C, D, and E.

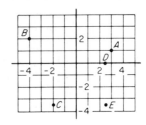

12. Draw the triangle ABC whose vertices are $A: (2, -2)$, $B: (0, 4)$, $C: (-4, -1)$.

13. Connect these points in order: $A: (2, -3)$, $B: (4, 1)$, $C: (0, 3)$, $D: (-2, -1)$. What kind of figure is $ABCD$?

14. Draw the two diagonals in the figure of Exercise 13 and estimate the co-ordinates of the point of intersection.

15. The lower left corner of a square five units on edge is at $(-3, -1)$. If the sides are parallel to the axes, what are the coordinates of the other three corners?

16. Draw the diagonals of the square of Exercise 15 and estimate the coordinates of their point of intersection.

3. DISTANCE BETWEEN TWO POINTS

The **distance** between any two points on a number line is the absolute value of their difference. For example, the distance between P_1 and P_2, shown on the number line of Figure 8.4, is 3:

$$\left.\begin{array}{l} P_1 = 1 \\ P_2 = 4 \end{array}\right\} \quad \begin{array}{l} d = |P_2 - P_1| \\ = |4 - 1| = |3| \\ = 3 \end{array}$$

Figure 8.4

Because direction has no bearing on distance, we can reverse P_1 and P_2 and get the same result.

$$d = |P_1 - P_2|$$
$$= |1 - 4| = |-3| = 3$$

EXAMPLE

Compute the distance between P_a and P_b.

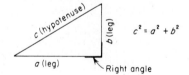

SOLUTION

$$d = |P_a - P_b| \quad \text{or} \quad d = |P_b - P_a|$$
$$= |(-2) - (3)| \qquad\qquad = |(3) - (-2)|$$
$$= |-5| \qquad\qquad\qquad = |3 + 2|$$
$$= 5 \qquad\qquad\qquad\qquad = 5$$

Important to computing the distance between *any* two points in a plane is the **Pythagorean Theorem**:

> *If the legs of a right triangle are of lengths a and b, then the hypotenuse is of length c, where $c^2 = a^2 + b^2$.*

Figure 8.5 illustrates the Theorem.

c (hypotenuse)   *b* (leg) $c^2 = a^2 + b^2$

a (leg)   Right angle   **Figure 8.5**

EXAMPLES

1. Compute the length of the hypotenuse, *c*, of a right triangle in which the legs are three and four units long.

SOLUTION
$$c^2 = a^2 + b^2$$
$$= (4)^2 + (3)^2$$
$$= 16 + 9$$
$$= 25$$
$$c = \sqrt{25}$$
$$= 5, -5$$

a = 4 units
b = 3 units
c = ? units

Note: We take only the principal square root of 25, namely 5. A length of -5 units is meaningless.

2. Compute the length of the hypotenuse, c.

SOLUTION
$$c^2 = a^2 + b^2$$
$$= 9 + 4$$
$$= 13$$
$$c = \sqrt{13} \quad \text{(approx. 3.6; see Table I)}$$

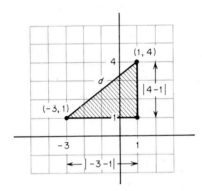

$a = 3$
$b = 2$
$c = ?$

Suppose, now, that we wish to compute the distance between two points, say $(-3, 1)$ and $(1, 4)$, as shown in Figure 8.6. Other data are included in the figure to illustrate our approach to the solution.

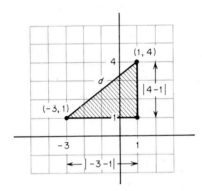

Figure 8.6

Think of the distance between the two given points $(-3, 1)$ and $(1, 4)$ as the hypotenuse, d, of a right triangle (shaded).

The length of the horizontal leg of the triangle is

$$|-3 - 1| = |-4| = 4$$

The length of the vertical leg of the triangle is

$$|4 - 1| = |3| = 3$$

By the Pythagorean Theorem,

$$d^2 = (3)^2 + (4)^2$$

$$= 25$$

and

$$d = 5$$

The distance between $(-3, 1)$ and $(1, 4)$, therefore, is 5 units.

To put this in more general terms, we choose any two points in the Cartesian plane, and call them (x_1, y_1) and (x_2, y_2). A line segment drawn between the two points represents the distance between them. See Figure 8.7(a).

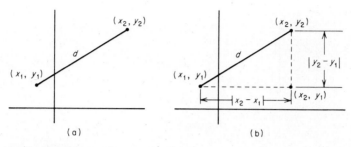

(a)   (b)

Figure 8.7

Now imagine the line segment of Figure 8.7(a) as the hypotenuse of a right triangle in which the legs run parallel to the axes, as shown in Figure 8.7(b). The legs meet to form a right angle at a point that must be (x_2, y_1).

The length of the horizontal leg is $|x_2 - x_1|$. The length of the vertical leg is $|y_2 - y_1|$. By the Pythagorean Theorem, it follows that

$$d^2 = (|x_2 - x_1|)^2 + (|y_2 - y_1|)^2$$

or more simply

$$d = \sqrt{(x_2 - x_1)^2 + (y_2 - y_1)^2}$$

Program **8.1** below is based on the logic of the foregoing discussion. It outlines a procedure for computing the distance between *any* two points in a plane for which the coordinates are known.

PROGRAM 8.1

To determine the distance between any two points in a plane:

Step 1:   Designate one (either) of the two points as (x_1, y_1).

Step 2:   Designate the other of the two points as (x_2, y_2).

Step 3:   Substitute the appropriate coordinate numbers for the x's and y's in the formula

$$d = \sqrt{(x_2 - x_1)^2 + (y_2 - y_1)^2}$$

and compute the desired distance, d.

1.   Determine the distance between $(-3, 1)$ and $(1, 4)$. See Figure 8.8.

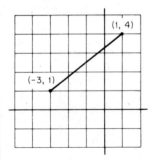

(1, 4)

(-3, 1)

Figure 8.8

SOLUTION   *Step 1:* We elect to designate $(-3, 1)$ as (x_1, y_1). That is, $x_1 = -3$ and $y_1 = 1$.

Step 2: The other point, $(1, 4)$, we designate as (x_2, y_2). That is, $x_2 = 1$, $y_2 = 4$.

Step 3: Substitute in the formula:

$$d = \sqrt{(x_2 - x_1)^2 + (y_2 - y_1)^2}$$
$$= \sqrt{[1 - (-3)]^2 + (4 - 1)^2}$$
$$= \sqrt{[4]^2 + (3)^2}$$
$$= \sqrt{16 + 9}$$
$$= \sqrt{25}$$
$$= 5$$

The distance between $(-3, 1)$ and $(1, 4)$ is 5 (units).

2.   Compute the distance between $(2, 2)$ and $(-4, -2)$. See Figure 8.9.

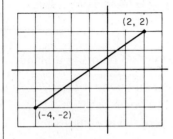

(2, 2)

(-4, -2)

Figure 8.9

SOLUTION   *Step 1:* We designate $(2, 2)$ as (x_1, y_1).

Step 2: We designate the other point, $(-4, -2)$, as (x_2, y_2).

Step 3:

$$d = \sqrt{(x_2 - x_1)^2 + (y_2 - y_1)^2}$$

$$\left.\begin{array}{l} x_1 = 2 \\ y_1 = 2 \\ x_2 = -4 \\ y_2 = -2 \end{array}\right\}$$
$$= \sqrt{(-4 - 2)^2 + (-2 - 2)^2}$$
$$= \sqrt{(-6)^2 + (-4)^2}$$
$$= \sqrt{36 + 16}$$
$$= \sqrt{52}$$
$$= 2\sqrt{13} \quad \text{(or 7.2 approx.)}$$

3.   Compute the distance between $(-4, -1)$ and $(3, -1)$. These two points are on the same horizontal line. See Figure 8.10.

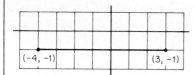

Figure 8.10

SOLUTION   Program **8.1** still applies.
Let

$$(x_1, y_1) = (-4, -1)$$
$$(x_2, y_2) = (3, -1)$$

Then

$$d = \sqrt{(x_2 - x_1)^2 + (y_2 - y_1)^2}$$
$$= \sqrt{[(3) - (-4)]^2 + [(-1) - (-1)]^2}$$
$$= \sqrt{[7]^2 + [0]^2}$$
$$= \sqrt{49}$$
$$= 7$$

EXERCISE 8-3

Plot each of the following pairs of points; then use the formula to find the distance between them.

1. $(3, -2), (3, 7)$ **2.** $(4, 1), (-3, 1)$ **3.** $(-3, -3), (6, -3)$
4. $(4, 0), (4, -9)$ **5.** $(0, 5), (0, 2)$ **6.** $(-7, 0), (1, 0)$
7. $(0, 0), (7, 0)$ **8.** $(0, -10, (0, 0)$ **9.** $(3, 7), (4, 2)$
10. $(4, 6), (-1, -3)$ **11.** $(-3, 7), (5, 7)$ **12.** $(4, 7), (4, -4)$
13. $(-1, -6), (3, -2)$ **14.** $(-3, -8), (-1, 4)$

15. $(6, 3), (-4, 7)$ **16.** $(5, 12), (0, 0)$ **17.** $(0, 0), (-3, 4)$

18. $(8, -7), (-4, -2)$ **19.** $(7, 7), (-7, -7)$

20. $(-1, -1), (0, 0)$ **21.** $(3, 7), (0, 0)$ **22.** $(3, -4), (-3, 4)$

23. Show that the triangle with vertices at $A(3, 8)$, $B(-11, 3)$, and $C(-8, -2)$ is isosceles (i.e., two sides equal). Plot the triangle.

24. Find the lengths of the sides of the triangle having vertices at $A(-4, 3)$, $B(2, -3)$, and $C(-1, 2)$. Is the triangle isosceles?

25. Find the lengths of the sides of the triangle having vertices at $(3, 3)$, $(1, 6)$, and $(7, 10)$. Is it a right triangle? [*Hint:* See if the lengths of the sides satisfy the Pythagorean Theorem.]

26. Show that the points $A(7, 5)$, $B(2, 3)$, and $C(6, -7)$ are the vertices of a right triangle.

4. GRAPHING AN EQUATION IN TWO VARIABLES

When the real numbers are the replacement set for an equation in two variables, an infinite number of ordered pairs satisfy the equation. We can plot such a solution set on a Cartesian grid. Collectively, this set of points is known as the **graph of the equation**. The abscissa and ordinate of each point of the graph correspond to a pair of variable replacements that satisfy the equation.

Because the solution set of the equation contains infinitely many members, it is impossible to tabulate or to plot all of them. However, by plotting occasional solution pairs and then connecting these points with a smooth, continuous line, we can develop a reasonably accurate picture of the solution set of the equation.

PROGRAM 8.2

To graph an equation in two variables:

Step 1:   Solve the equation for one of its variables.

Step 2:   Replace the variable in the "solution" member of the equation of Step 1 with numbers selected from the replacement set, then determine the corresponding replacement for the other variable necessary to make the equation true.

Step 3:   Plot the solution pairs of Step 2 on a Cartesian coordinate grid.

Step 4:   Connect the plotted points of Step 3, left to right, with a smooth, continuous line.

1. Graph $3x - y = 2$.

Step 1: Solve the equation for one of its variables:

$$3x - y = 2$$

$$y = 3x - 2$$

Step 2: Select numbers to replace the variable x: 0, 1, 2, say. Determine the corresponding replacements for the other variable, y:

$$y = 3x - 2$$

$$x = 0: \quad y = 3(0) - 2 = -2$$

$$x = 1: \quad y = .3(1) - 2 = 1$$

$$x = 2: \quad y = 3(2) - 2 = 4$$

Step 3: The ordered pair solutions resulting from Step 2 are $(0, -2)$, $(1, 1)$, and $(2, 4)$. They are plotted in Figure 8.11(a).

Step 4: Connect the points of Step 3, left to right, with a smooth, continuous line, as in Figure 8.11(b). The graph is that of $3x - y = 2$.

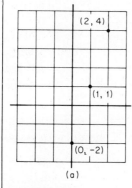
(a)

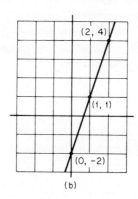
(b)

Figure 8.11

2. Graph $x^2 - y = -2$.

Step 1: Solve the equation for one of the variables:

$$x^2 - y = -2$$

$$y = x^2 + 2$$

Steps 2 and 3: These two steps can be combined by constructing a table as shown below. First the x replacements are decided on; then the corresponding y replacements are computed. If the x replacements are recorded along the top row (even though we may sometimes substitute for y first) and the corresponding y replacements along the bottom row, then the vertical top–bottom pairs in the table are the ordered pairs that we wish to plot.

x	-2	-1	0	1	2
y	6	3	2	3	6

Step 4: Plot the points of Steps 2 and 3, and connect them with a smooth continuous line for the graph of $x^2 - y = -2$ (Figure 8.12).

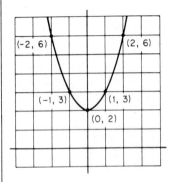

Figure 8.12

3.   Graph $x + y^2 + 2 = 0$.

SOLUTION | *Step 1:* In this case it will be simpler to choose the y replacements first, then solve for the corresponding x replacements. Consequently, we solve the given equation for x:

$$x + y^2 + 2 = 0$$
$$x = -y^2 - 2$$

Steps 2 and 3:

y	$-y^2 - 2$	x		x	y
0	$-(0)^2 - 2$	-2		-2	0
1	$-(1)^2 - 2$	-3		-3	1
-1	$-(-1)^2 - 2$	-3		-3	-1
2	$-(2)^2 - 2$	-6		-6	2
-2	$-(-2)^2 - 2$	-6		-6	-2

Step 4: See Figure 8.13.

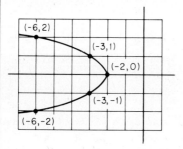

Figure 8.13

4. Graph (a) $y = 6$; (b) $x = -2$.

SOLUTION   (a) The solution set for $y = 6$ is a set of ordered pairs in which *every* y component is 6, no matter what the x component is. Thus $(0, 6), (-5, 6)$, and $(125, 6)$ are among the members of the solution set. Graphically, points corresponding to such a set of ordered pairs occur along a line that is parallel to the horizontal or x axis and 6 units above it. The line contains all points for which the ordinate is 6. See Figure 8.14.

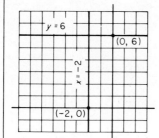

Figure 8.14

(b) The solution set for $x = -2$ is a set of ordered pairs in which every x component is -2, no matter what the y component is. So $(-2, -6), (-2, 6), \left(-2, 3\frac{1}{2}\right)$, and $(-2, 635)$ are among the members of the solution set. Graphically, points corresponding to such a set of ordered pairs occur along a line parallel to the vertical or y axis, and 2 units to the left of it. The line contains all points for which the abscissa is -2. See Figure 8.14.

EXERCISE 8-4

The graph of each of the following is a straight line; so only two points are necessary for its determination. However,

(a) Graph each equation with a minimum of four points.

(b) *Estimate the coordinates of two other points on the graph and see if they satisfy the equation.*

1. $y = x$	2. $y = x + 1$	3. $x + y = 1$
4. $x = -y$	5. $x - 4 = 0$	6. $y + 2 = 0$
7. $2x + y = 3$	8. $x - 3y = 1$	9. $4x - 2y = 5$
10. $2x + 3y = 6$	11. $x = \dfrac{3}{4} y$	12. $y = -\dfrac{1}{2} x$

13. Plot (a) $y = -x$, (b) $x + y = 3$, and (c) $2x + 2y = 4$ on the same grid. What geometric property do these lines seem to exhibit?

14. Write the equation of the straight line that is parallel to the x axis and four units above it.

15. Write the equation of the straight line that is parallel to the y axis and eight units to its left.

16. Write the equation of the line that coincides with the x axis.

Graph each of the following with no less than eight points; keep the variable replacements between $+10$ and -10 (i.e., $-10 < x < +10$, $-10 < y < +10$).

17. $y = 2x^2$	18. $y = x^2 - x$	19. $y = 2x^2 - 3$
20. $y = x^3$	21. $x = y^2 - 4y + 3$	22. $y^2 + 2x = x + 3$
23. $x^2 + y^2 = 16$	24. $2x^2 + y^2 = 16$	

5. INTERCEPTS

A point at which the graph of an equation intersects an axis of the Cartesian coordinate system is called an **intercept**. An intersection of the x axis is called an **x intercept**. An intersection of the y axis is called a **y intercept**.

Because the abscissa (x value) of every point along the y axis is 0, the y intercept for an equation is readily found by replacing x by 0 and solving for y. For instance, consider the equation

$$2x + 3y = 6$$

$$x = 0: \quad 2(0) + 3y = 6$$

$$3y = 6$$

$$y = 2$$

Thus the y intercept for $2x + 3y = 6$ is 2. See Figure 8.15.

Similarly, the x intercept becomes known when the y variable is replaced by 0 (since the y value of every point along the x axis is 0).

$$2x + 3y = 6$$
$$y = 0: \quad 2x + 3(0) = 6$$
$$2x = 6$$
$$x = 3$$

The x intercept for $2x + 3y = 6$ is 3. See Figure 8.15.

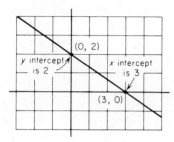

Figure 8.15

EXAMPLES

1. Determine the x and y intercepts for $2x - 3y = 6$.

SOLUTION To find the x intercept, substitute 0 for y:

$$2x - 3y = 6$$
$$2x - 3(0) = 6$$
$$2x = 6$$
$$x = 3$$

To find the y intercept, substitute 0 for x:

$$2x - 3y = 6$$
$$2(0) - 3y = 6$$
$$-3y = 6$$
$$y = -2$$

The x intercept is 3, and the y intercept is -2. See Figure 8.16.

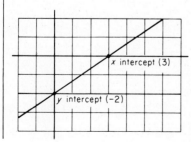

Figure 8.16

2. Compute the x and y intercepts for $y + x^2 = x + 2$.

x intercept ($y = 0$):

$$y + x^2 = x + 2$$
$$y = 0: \quad 0 + x^2 = x + 2$$
$$x^2 - x - 2 = 0$$
$$(x - 2)(x + 1) = 0$$
$$x - 2 = 2 \qquad x + 1 = 0$$
$$x = 2 \qquad\qquad x = -1$$

There are two x intercepts: $(2, 0)$ and $(-1, 0)$.

y intercept ($x = 0$):

$$y + x^2 = x + 2$$
$$x = 0: \quad y + 0 = 0 + 2$$
$$y = 2$$

The graph of the equation is shown in Figure 8.17. Note that it has one y intercept, at $(0, 2)$, and two x intercepts.

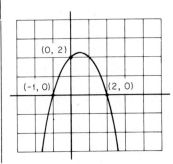

Figure 8.17

EXERCISE 8-5

Find the x intercept and the y intercept for each of these linear equations.

1. $3x + 4y = 12$
2. $2x - 4y = 4$
3. $5x - y = 5$
4. $2x - 3y = 3$
5. $2x - 3y + 7 = 0$
6. $3x - 10 = 4y$
7. $x = -4y$
8. $3y = 2x$
9. $2x + 3y + 3 = 0$
10. $2x + y = -4$
11. $3x = 5$
12. $2y = 4$

The following nonlinear equations have graphs similar in shape (but not necessarily in direction) to that shown in Figure 8.17. Determine their x and y intercepts.

13. $y = 4 - x^2$
14. $y = 9 - x^2$
15. $y = x^2 - 1$

16. $y = x^2 - \dfrac{1}{4}$ **17.** $y = 8 + 2x - x^2$ **18.** $y = 6 - x - x^2$

19. $y^2 - y = x + 2$ **20.** $x - 3 = y^2 + 4y$

6. SLOPE

An important property of a line is its inclination, or "tilt," with respect to the horizontal axis. This property is called the **slope** of the line. Slope is usually denoted by the letter m. For any two points on a line, (x_1, y_1) and (x_2, y_2),

$$m \text{ (slope)} = \frac{y_2 - y_1}{x_2 - x_1}$$

Thus the slope number, m, expresses a ratio comparing the change in y values to the change in x values between the two points. (The order in which the points are taken is not important.)

The slope ratio may also be regarded as a comparison of "rise" to "run":

$$m = \frac{\text{rise}}{\text{run}}$$

where rise is the vertical change in y values ($y_2 - y_1$) and run is the horizontal change in x values ($x_2 - x_1$). See Figure 8.18.

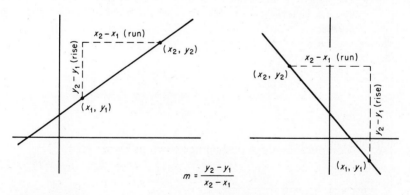

Figure 8.18

Lines and line segments that incline to the *right* of vertical have *positive* numbers for slopes ("the run is to the right; positive"). Those that incline to the *left* of vertical have *negative* numbers for slopes ("the run is to the left; negative"). Both are illustrated in the following examples.

1. Compute the slope of a line containing the points (1, 3) and (5, 6).

SOLUTION | We consider (1, 3) to be (x_1, y_1) and (5, 6) to be (x_2, y_2), then use the slope formula.

$$m = \frac{y_2 - y_1}{x_2 - x_1}$$

$$= \frac{6 - 3}{5 - 1}$$

$$= \frac{3}{4}$$

The slope of the line containing the points (1, 3) and (5, 6) is $\frac{3}{4}$. See Figure 8.19.

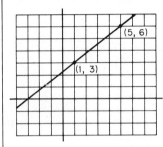

(5, 6)

(1, 3)

Figure 8.19

Note: The slope of a line is not affected by the choice of points designated as (x_1, y_1) and (x_2, y_2). Here we reverse them from the order above, and yet obtain the same slope number:

$$(x_1, y_1) = (5, 6)$$

$$(x_2, y_2) = (1, 3)$$

$$m = \frac{y_2 - y_1}{x_2 - x_1} = \frac{3 - 6}{1 - 5} = \frac{-3}{-4} = \frac{3}{4}$$

2. Compute the slope of the line shown in Figure 8.20.

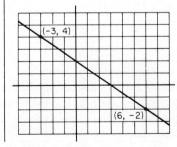

(-3, 4)

(6, -2)

Figure 8.20

Arbitrarily,

$$(x_1, y_1) = (6, -2)$$
$$(x_2, y_2) = (-3, 4)$$

Then

$$m = \frac{y_2 - y_1}{x_2 - x_1} = \frac{4 - (-2)}{(-3) - 6} = \frac{6}{-9} = -\frac{2}{3}$$

3. For the line in Figure 8.21,

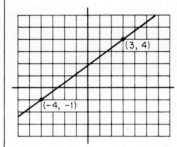

(3, 4)

(-4, -1)

Figure 8.21

$$m = \frac{y_2 - y_1}{x_2 - x_1} = \frac{4 - (-1)}{3 - (-4)} = \frac{5}{7}$$

4. For the line in Figure 8.22,

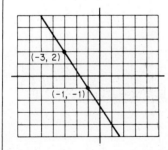

(-3, 2)

(-1, -1)

Figure 8.22

$$m = \frac{y_2 - y_1}{x_2 - x_1} = \frac{2 - (-1)}{-3 - (-1)} = \frac{3}{-2} = -1\frac{1}{2}$$

What about the slope of lines parallel to the axes?

The ordinate, or y value, of every point of a line parallel to the horizontal axis is the same number. This means that the numerator of the slope ratio,

$\dfrac{y_2 - y_1}{x_2 - x_1}$, will equal 0. Thus the slope of all lines parallel to the horizontal axis is 0. (See Example 1 below.)

On the other hand, the abscissa, or x value, of every point of a line parallel to the vertical axis is the same number. This means that the denominator of the slope ratio, $\dfrac{y_2 - y_1}{x_2 - x_1}$, will equal 0. Since division by 0 is undefined, the slope of all lines parallel to the vertical axis is undefined. (See Example 2 below.)

In brief, horizontal lines have a slope of 0. Vertical lines have no defined slope.

EXAMPLES

1.   For the line in Figure 8.23,

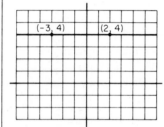

Figure 8.23

$$m = \frac{y_2 - y_1}{x_2 - x_1} = \frac{4 - 4}{2 - (-3)} = \frac{0}{5} = 0$$

2.   For the line in Figure 8.24,

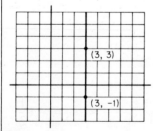

Figure 8.24

$$m = \frac{y_2 - y_1}{x_2 - x_1} = \frac{3 - (-1)}{3 - 3} = \frac{4}{0} \quad \text{(undefined)}$$

274 Chapter 8 EQUATIONS AND INEQUALITIES IN TWO VARIABLES

Plot the line determined by each of the following pairs of points and compute its slope.

1. (3, 2), (1, 4)   **2.** (1, 1), (3, 6)   **3.** (1, 0), (−2, 5)

4. (−3, −2), (0, 4)   **5.** (−2, −3), (0, 4)   **6.** (−3, −1), (0, 0)

7. (−4, −2), (3, −2)   **8.** (4, 6), (4, −3)   **9.** (2, 3), (4, −6)

10. (−3, −3), (0, −3) **11.** (−5, −2), (−5, 2) **12.** (−3, −5), (−2, −8)

Find two ordered pairs that satisfy the equation; then use them to find the slope of the line defined by the equation.

13. $y = 2x + 1$ **14.** $y = 3x$ **15.** $2y = x + 4$

16. $3y = 2x − 1$ **17.** $2x + 5y = 8$ **18.** $3y + x − 7 = 0$

19. $2x − 3y = 4$ **20.** $y − \dfrac{1}{2}x = 6$

21. Solve each equation in Exercises 13 through 20 for y.

(a) For Exercise 13 through 20, compare the slope number you found with the coefficient of the x term in the transformed equation. Are they the same?

(b) Compare the constant term in the transformed equation with the ordinate (y value) of the point at which the graph of the equation crosses the y axis (the y intercept). Are they the same?

7. WRITING EQUATIONS FOR LINES

Every linear equation has a corresponding line for a graph. Every line has a corresponding linear equation (of which many equivalents are possible). When certain facts are known about the line, we may write an equation for it directly.

For instance, if we know the slope and the coordinates of one point of a line, we have enough data to define the line and to write its equation. Say that the known slope is m and that the known point is (x_1, y_1). We choose any other point on the line and call it (x, y), as shown in Figure 8.25. Then we use our formula for slope to write the equation.

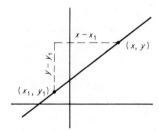

Figure 8.25

$$\frac{y - y_1}{x - x_1} = m$$

This fractional equation can be transformed to

(A) $$y - y_1 = m(x - x_1)$$

Equation (A) is called the **point-slope form** of a linear equation.

1. Write an equation for the line that has a slope of $\frac{3}{4}$ and that contains the point (2, 1).

SOLUTION Use $m = \frac{3}{4}$ and $(x_1, y_1) = (2, 1)$ in the point-slope form.

$$\left. \begin{array}{l} m = \dfrac{3}{4} \\[2mm] x_1 = 2 \\[2mm] y_1 = 1 \end{array} \right\} \qquad \begin{array}{l} y - y_1 = m(x - x_1) \\[2mm] y - 1 = \dfrac{3}{4}(x - 2) \\[2mm] 4y - 4 = 3x - 6 \end{array}$$

or

$$3x - 4y - 2 = 0$$

The graph of the equation having the given characteristics is shown in Figure 8.26.

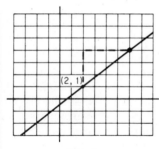

$m = \dfrac{3 \,(\text{rise})}{4 \,(\text{run})}$ **Figure 8.26**

Note: To plot a line when given a point and the slope, first locate the point. Here: (2, 1). Then interpret the slope as rise/run. Here, from (2, 1), "rise" 3 and "run" 4 (positive, right) to a second point on the line.

2. Write an equation for the line that has a slope of -3 and that contains the point $(-2, 0)$.

SOLUTION

$$m = -3, \qquad (x_1, y_1) = (-2, 0)$$
$$y - y_1 = m(x - x_1)$$
$$y - 0 = -3[x - (-2)]$$

or

$$3x + y + 6 = 0$$

See Figure 8.27 for its graph.

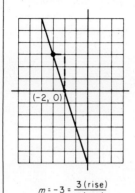

(-2, 0)

$m = -3 = \dfrac{3 \text{ (rise)}}{-1 \text{ (run)}}$ **Figure 8.27**

Note: A negative slope means the "run" is to the left in locating a second point.

A special case of the point-slope form occurs when the known point lies on the vertical axis, say at $(0, b)$, as shown in Figure 8.28.

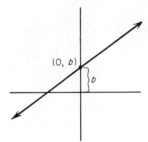

(0, b)

b

Figure 8.28

The number b is the y intercept. Substituting $(0, b)$ for (x_1, y_1), we obtain

$$y - y_1 = m(x - x_1)$$
$$y - b = m(x - 0)$$
$$y - b = mx$$

(B) $$y = mx + b.$$

Equation (B) is called the **slope-intercept form** of a linear equation.

The important correspondence between the slope and the coefficient of the x term and between the y intercept and the constant term, when a linear equation is in slope-intercept form, was previously noted in Problem 21 of Exercise 8-6. Further use of this relationship will be made in the next section.

1. Write an equation for the line that has a slope of 2 and contains the point (0, 4).

SOLUTION Substitute 2 for m (slope) and 4 for b (y intercept) in the slope-intercept form.

$$y = mx + b$$

$$\left.\begin{array}{l} m = 2 \\ b = 4 \end{array}\right\} \quad y = 2x + 4$$

The graph of the equation having the given characteristics is shown in Figure 8.29.

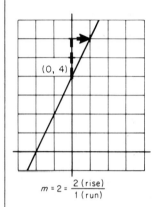

(0, 4)

$m = 2 = \dfrac{2 \text{ (rise)}}{1 \text{ (run)}}$ 　　　　**Figure 8.29**

2. Write an equation for the line that has a slope of $-\dfrac{2}{3}$ and an intercept on the vertical axis of -1.

SOLUTION

$$y = mx + b$$

$$\left.\begin{array}{l} m = -\dfrac{2}{3} \\ b = -1 \end{array}\right\} \quad y = -\dfrac{2}{3}x + (-1)$$

$$3y = -2x - 3$$

$$2x + 3y + 3 = 0$$

See Figure 8.30 for its graph.

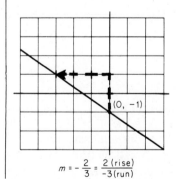

(0, −1)

$m = -\dfrac{2}{3} = \dfrac{2 \text{ (rise)}}{-3 \text{ (run)}}$ 　　　**Figure 8.30**

3. Parallel lines have equal slopes. Tell if these two lines are parallel.

$$A: \quad 6x - 2y = 0$$

$$B: \quad 3x - y = 2$$

SOLUTION Transform each equation to slope-intercept form.

$$A: \quad 6x - 2y = 0 \qquad B: \quad 3x - y = 2$$

$$-2y = -6x \qquad\qquad -y = 2 - 3x$$

$$y = 3x \qquad\qquad\quad y = 3x - 2$$

Both equations in slope-intercept form have the same coefficient for x (namely, 3). So both equations have the same slope, 3; their graphs are parallel lines. (See Figure 8.31.)

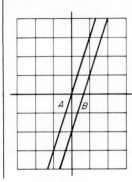

Figure 8.31

EXERCISE 8-7

Write an equation in the form $ax + by + c = 0$, *where a, b, c are integers, for the line for which a point and the slope are given.*

1. $(2, 3)$; 3(slope)   **2.** $(4, -1)$; 2 **3.** $(3, -2)$; -3

4. $(-2, 1)$; -1 **5.** $(5, 0)$; $\dfrac{1}{2}$ **6.** $(6, -2)$; $-1\dfrac{1}{3}$

7. $(-2, 0)$; $-\dfrac{1}{3}$ **8.** $(-3, -4)$; $-1\dfrac{1}{2}$ **9.** $(0, -2)$; $-\dfrac{3}{4}$

10. $(0, 6)$; $1\dfrac{1}{2}$ **11.** $(0, 5)$; -1 **12.** $\left(0, 9\dfrac{1}{3}\right)$; $-\dfrac{5}{3}$

13. $(0, -4)$; 1 **14.** $(0, -6)$; -2

Write an equation of the form $ax + by + c = 0$, *where a, b, c are integers, for the line that has the given slope and y intercept.*

15. $m = 3$; $(0, 2)$ **16.** $m = 1$; $(0, 3)$

17. $m = -2$; $(0, -3)$ **18.** $m = -4$; $(0, -1)$

19. $m = 0; (0, -2)$ **20.** $m = 0; (0, 4)$

21. $m = \dfrac{3}{5}; (0, 6)$ **22.** $m = \dfrac{2}{3}; (0, -3)$

23. $m = -1\dfrac{1}{2}; (0, 1)$ **24.** $m = 1\dfrac{1}{4}; (0, -2)$

Parallel lines have equal slopes. Write an equation for the line that contains the given point and is parallel to the line specified by the given equation.

25. $(3, 1); y = 3x - 1$ **26.** $(-2, 0); 2x - y = 4$
27. $(-1, 4); y = 2$ **28.** $(0, 0); 2y = x + 3$
29. $(-3, -1); 3x - 2y = 1$ **30.** $(-3, 0); x - 5 = 0$

8. GRAPHING BY THE SLOPE-INTERCEPT METHOD

The slope-intercept form of the linear equation

$$y = mx + b$$

has many applications, both in mathematics and in other disciplines. Moreover, the form is especially useful for sketching the graph of *any* linear equation.

PROGRAM 8.3

To use the slope-intercept form to graph a linear equation:

Step 1:   Express the linear equation in slope-intercept form

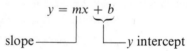

Step 2:   Locate the *y* intercept (*b*) as a point on the vertical axis.

Step 3:   Consider the slope number (*m*) as a fraction

$$\frac{\text{rise}}{\text{run}}$$

If the slope number is negative, associate the minus sign ($-$) with the denominator.

Step 4:   Locate a second point of the graph as follows: From the *y* intercept point of Step 2, move up the *y* axis a number of units equal to the numerator of Step 3 (rise), and then horizontally a number of units equal to the denominator of Step 3 (run)—left if negative, right if positive.

Step 5:   Draw a line segment through the points of Step 2 and Step 4.

EXAMPLES

1. Graph $3x + 2y = 8$.

SOLUTION | *Step 1:* Express the equation in slope-intercept form:

$$3x + 2y = 8$$

$$2y = -3x + 8$$

$$y = -\frac{3}{2}x + 4$$

$$\underbrace{\phantom{y = -\frac{3}{2}}}_{\text{slope}} \qquad \underbrace{}_{y \text{ intercept}}$$

Step 2: The y intercept is $+4$. Locate it on the vertical axis; see Figure 8.32(a).

Step 3: The slope number is $-\dfrac{3}{2}$; associate the minus sign with the denominator

$$\frac{3}{-2}\left(\frac{\text{rise}}{\text{run}}\right)$$

Step 4: From the point of Step 2, "rise" 3 units and "run (left)" 2 units, to locate a second point; see Figure 8.32(b).

Step 5: Draw a line segment through the points of Steps 2 and 4; see Figure 8.32(c).

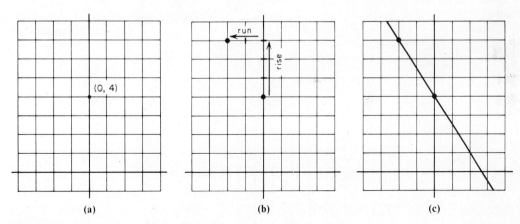

(a)   (b)   (c)

Figure 8.32

2. Sketch the graph of $y = 3x - 2$.

SOLUTION | *Step 1:* Equation is already in slope-intercept form.

$$y = 3x - 2$$

$$\underbrace{}_{\text{slope}} \qquad \underbrace{}_{y \text{ intercept}}$$

The graph is shown in Figure 8.33.

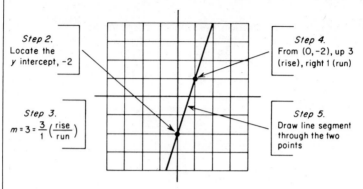

Step 2.
Locate the
y intercept, -2

Step 4.
From (0,-2), up 3
(rise), right 1 (run)

Step 3.
$m = 3 = \frac{3}{1}\left(\frac{\text{rise}}{\text{run}}\right)$

Step 5.
Draw line segment
through the two
points

Figure 8.33

3. Graph $2x - 3y = 5$.

SOLUTION Express the equation in slope-intercept form:

$$2x - 3y = 5$$
$$-3y = -2x + 5$$
$$y = \frac{2}{3}x - \frac{5}{3}$$

$$\text{Slope:} \frac{2}{3}\left(\frac{\text{rise}}{\text{run}}\right); \ y \text{ intercept:} -\frac{5}{3} = -1\frac{2}{3}$$

The graph is shown in Figure 8.34.

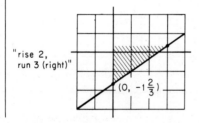

"rise 2,
run 3 (right)"

$(0, -1\frac{2}{3})$

Figure 8.34

| EXERCISE 8-8

Sketch the graph of these equations, using the slope-intercept method.

1. $y = \frac{1}{2}x + 3$

2. $y = \frac{2}{3}x + 1$

3. $y = \frac{2}{3}x - 1$

4. $y = \frac{1}{2}x - 3$

5. $y = -\frac{3}{4}x + 2$

6. $y = -\frac{4}{5}x + 1$

282 Chapter 8 EQUATIONS AND INEQUALITIES IN TWO VARIABLES

7. $y = 3x - 1$ 8. $y = 2x - 2$ 9. $y = -2x + 5$
10. $y = -3x - 1$ 11. $3x - 2y = 2$ 12. $4x - 3y = 6$
13. $4x + 3y = 3$ 14. $3x + 2y = -4$ 15. $6y - 4x + 3 = 0$
16. $x - 5y + 3 = 0$ 17. $3x - 8y = 0$ 18. $5x - 2y = 0$
19. $21x + 6y = -2$ 20. $10x + 6y = 27$

9. ABSOLUTE-VALUE EQUATIONS*

A simple example of an absolute-value equation in two variables is:

$$y = |x - 3|$$

Because "$|x - 3|$" will always be nonnegative, whatever the replacement for x, y will always be nonnegative. This means the graph of the equation will never go below the x axis, where all ordinates (y values) are negative.

We may use Program **8.2** (To graph an equation in two variables) to produce the graph. That is, we generate ordered pairs, (x, y), by substituting numbers for one variable, say x, and computing the corresponding value for the other variable.

The graph of

$$y = |x - 3|$$

is shown in Figure 8.35. Its V-shape is typical of such first-degree absolute-value equations.

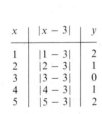

x	$|x - 3|$	y
1	$|1 - 3|$	2
2	$|2 - 3|$	1
3	$|3 - 3|$	0
4	$|4 - 3|$	1
5	$|5 - 3|$	2

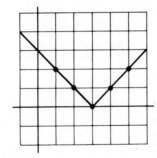

Figure 8.35

Of special importance is the place where a graph changes direction. Choosing values for the variables at random may or may not lead to that part of the graph. In the case of the absolute-value equation, $y = |x - 3|$, the change in direction takes place when the expression within the absolute-value symbol equals 0:

$$x - 3 = 0$$

$$x = 3$$

* The content of this section, in more basic courses of study, may be postponed without significant effect upon the remaining parts of the book.

A refinement on Program **8.2**, then, is to study the equation first, to determine the region of special activity (in this case, change of direction). Then select specific points to be graphed.

1. Graph $y = |15 - 3x|$.

SOLUTION The graph will be V-shaped. It will not go below the x axis (i.e., y will always be nonnegative because the member containing the x variable, $15 - 3x$, is totally within the absolute-value symbol).

The x coordinate of the point where the graph changes direction will be where $y = 0$, or where $x = 5$:

$$15 - 3x = 0$$
$$-3x = -15$$
$$x = 5$$

Select x coordinates of points to be graphed in the neighborhood of $x = 5$. Then use Program **8.2** to carry out the graphing. See Figure 8.36.

x	$\lvert 15 - 3x \rvert$	y
1	$\lvert 15 - 3(1) \rvert$	12
3	$\lvert 15 - 3(3) \rvert$	6
5	$\lvert 15 - 3(5) \rvert$	0
7	$\lvert 15 - 3(7) \rvert$	6
9	$\lvert 15 - 3(9) \rvert$	12

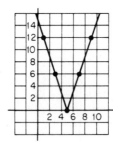

Figure 8.36

2. Graph $y = -|2x - 3|$.

SOLUTION Because the expression within the absolute value symbol is first-degree, the graph will be V-shaped. However, it is apparent from the right member of the equation that y can never be positive. This means that the graph will never go *above* the x axis.

The region of special interest will be near where $y = 0$; that is, where

$$2x - 3 = 0$$
$$2x = 3$$
$$x = \frac{3}{2}$$

The graph is given in Figure 8.37

x	$-\lvert 2x - 3 \rvert$	y
0	$-\lvert 2(0) - 3 \rvert = -\lvert -3 \rvert$	-3
1	$-\lvert 2(1) - 3 \rvert = -\lvert -1 \rvert$	-1
$\dfrac{3}{2}$	$-\left\lvert 2\!\left(\dfrac{3}{2}\right) - 3 \right\rvert = -\lvert 0 \rvert$	0
2	$-\lvert 2(2) - 3 \rvert = -\lvert 1 \rvert$	-1
3	$-\lvert 2(3) - 3 \rvert = -\lvert 3 \rvert$	-3

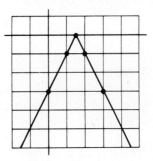

Figure 8.37

3.   Graph $3y = \lvert 2 - 3x \rvert$.

SOLUTION Determine the x coordinate of the vertex (point of the V) of the graph:

$$2 - 3x = 0$$

$$x = \frac{2}{3}$$

Solve the equation for y:

$$3y = \lvert 2 - 3x \rvert$$

$$y = \frac{1}{3}\lvert 2 - 3x \rvert$$

Then substitute x values in the neighborhood of $\dfrac{2}{3}$:

x	$\dfrac{1}{3}\lvert 2 - 3x \rvert$	y
-1	$\dfrac{1}{3}\lvert 2 - 3(-1) \rvert$	$\dfrac{5}{3}$
0	$\dfrac{1}{3}\lvert 2 - 3(0) \rvert$	$\dfrac{2}{3}$
$\dfrac{2}{3}$	$\dfrac{1}{3}\left\lvert 2 - 3\!\left(\dfrac{2}{3}\right) \right\rvert$	0
1	$\dfrac{1}{3}\lvert 2 - 3(1) \rvert$	$\dfrac{1}{3}$
2	$\dfrac{1}{3}\lvert 2 - 3(2) \rvert$	$\dfrac{4}{3}$

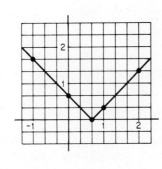

Figure 8.38

The graph appears in Figure 8.38.

4. Graph $y = 2 - |x|$.

SOLUTION

x	$2 - \|x\|$	y
-3	$2 - \|-3\|$	-1
-2	$2 - \|-2\|$	0
-1	$2 - \|-1\|$	1
0	$2 - \|0\|$	2
1	$2 - \|1\|$	1
2	$2 - \|2\|$	0
3	$2 - \|3\|$	-1

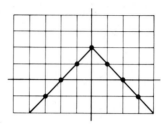

Figure 8.39

See Figure 8.39.

EXERCISE 8-9

Determine the point at which the graph changes direction.

1. $y = |3x - 3|$ 2. $y = |2x - 4|$ 3. $y = |1 - 2x|$
4. $y = |3 - 4x|$ 5. $y = -|2x + 6|$ 6. $y = -|5x - 10|$
7. $y = |x|$ 8. $y = -|x|$ 9. $y = 2|x - 3|$
10. $y = 3|2 - 5x|$

Graph.

11. $y = |3 - x|$ 12. $y = |2 - x|$ 13. $y = |2x + 5|$

14. $y = |3 + 4x|$ 15. $y = -|6 - 3x|$ 16. $y = -|4 + 4x|$

17. $2y = |2x - 1|$ 18. $\frac{1}{2}y = |3 - 2x|$ 19. $y = -|12 + 6x|$

20. $y = -|4 - 3x|$ 21. $y = 2|3 + 4x|$ 22. $y = 4\left|\frac{1}{2} - \frac{1}{4}x\right|$

23. $\frac{1}{3}y = -|x - 1|$ 24. $2y = -|2x - 3|$ 25. $y = |x| - 3$

26. $y = |x| + 1$ 27. $y = |x - 2| - 2$ 28. $y = -|x + 1| + 2$

10. INEQUALITIES IN TWO VARIABLES*

A straight line divides the Cartesian coordinate plane into two half-planes. A simple example is illustrated in Figure 8.40. The line, represented by the equation $y = x$, divides the whole plane into two separate half-planes.

★ The content of this section, in more basic courses of study, may be postponed without significant effect upon the remaining parts of the book.

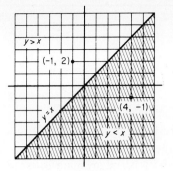

Figure 8.40

The half-plane above and to the left of $y = x$ contains all points for which the y value (ordinate) is greater than the x value (abscissa)—that is, $y > x$.

As an example, the point $(-1, 2)$ lies in that upper half-plane, and $2 > -1$.

The half-plane below and to the right of $y = x$ contains all points for which the y value (ordinate) is less than the x value (abscissa)—that is, $y < x$. The point $(4, -1)$, for example, lies in that lower half-plane, and $-1 < 4$.

Similarly, the graph of every linear equation in two variables separates the points of the Cartesian plane into three distinct sets:

1. The set of points that corresponds to the solution set of the equation and is the graph of the equation.
2. The set of points in the half-plane on one side of the graph of the equation.
3. The set of points in the half-plane on the other side of the graph of the equation.

Sets 2 and 3 correspond to the solution sets of two inequalities. The two inequalities are term-by-term duplicates of the equation whose graph is given by Set 1 except that the $=$ sign is replaced by $>$ or $<$.

EXAMPLES

1.

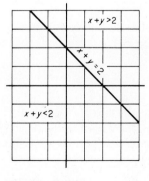

Figure 8.41

2.

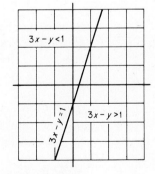

Figure 8.42

PROGRAM 8.4

To graph the solution set of an inequality in two variables:

Step 1: Rewrite the given inequality as an equation by replacing the inequality sign with an $=$ sign.

Step 2: Graph the equation of Step 1.

Step 3: Select any point in one of the half-planes determined by the graph of Step 2 and substitute its abscissa and ordinate for the appropriate variables of the given inequality. If the given inequality is satisfied, then the points of that half-plane correspond to the solution set of the inequality. If the inequality is not satisfied, the points of the other half-plane correspond to the solution set of the given inequality.

EXAMPLES

1. Graph $x + 2y > 1$.

SOLUTION   *Step 1:* Rewrite the inequality as an equation:

$$x + 2y = 1$$

Step 2: Graph the equation of Step 1, as in Figure 8.43(a).

Step 3: Select $(2, 3)$ as the check point in the half-plane above the graph of $x + 2y = 1$ and substitute for the appropriate variables in the given inequality.

$$x + 2y > 1$$

$$(2, 3): \quad 2 + 2(3) > 1 \quad \text{(true)}$$

Because the check point satisfies the inequality, the points of the half-plane above the graph of $x + 2y = 1$ correspond to the solution set of $x + 2y > 1$. See Figure 8.38(b).

Note: The graph of the solution set of an inequality in two variables is usually illustrated as a shaded half-plane, with the bounding line dashed, as shown in Figure 8.43(b). Thus the graph consists of all points in the shaded part but *not* the points of the bounding line. On the other hand, when the inequality is compound (i.e., $\geq$ or $\leq$), then the set of points of the bounding line is included and indicated by a solid line, as shown in Figure 8.43(c).

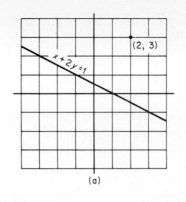

(a)

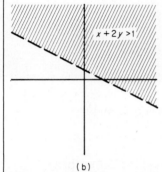

(b)

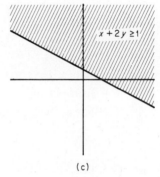

(c)

Figure 8.43

2.   Graph $2x - y < 4$.

SOLUTION   *Step 1:* Replace $<$ with $=$:

$$2x - y = 4$$

Step 2: Graph $2x - y = 4$. Use a dashed line, since the points that represent $2x - y = 4$ are not included in $2x - y < 4$. See Figure 8.44.

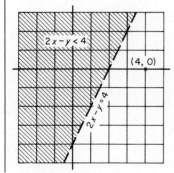

Figure 8.44

Step 3: Select (4, 0) as a convenient check point, substitute, and test the truth of the resulting statement:

$$2x - y < 4$$

$$(4, 0): \quad 2(4) - 0 < 4 \quad \text{(false)}$$

The fact that (4, 0) does not satisfy the inequality implies that the graph of the solution set is the half-plane *other* than the one that includes (4, 0). See Figure 8.44.

3. Graph $2x \le 9 + 3y$.

SOLUTION | *Step 1:* Replace $\le$ with $=$:

$$2x = 9 + 3y$$

Step 2: Graph $2x = 9 + 3y$. Use a solid line, since the points that represent $2x = 9 + 3y$ are included in $2x \le 9 + 3y$. See Figure 8.45.

Step 3: Select (0, 0) as a check point, substitute, and test.

$$2x \le 9 + 3y$$

$$(0, 0): \quad 2(0) \le 9 + 3(0) \quad \text{(true)}$$

The fact that (0, 0) satisfies the inequality implies that the graph of the solution set is the half-plane that includes (0, 0). See Figure 8.45.

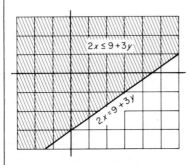

Figure 8.45

4. Graph the inequality $x^2 - y > -2$.

SOLUTION | Any graph that separates the coordinate plane into two distinct plane regions can be handled according to Program **8.4**.

Step 1: $x^2 - y = -2$

Step 2: This equation was plotted in Section 4, Example 2, page 266.

Step 3: (0, 0) is selected as the check point.

$$x^2 - y > -2$$

$$(0, 0): \quad 0^2 - 0 > -2 \quad \text{(true)}$$

The solution set is represented by the shaded part of Figure 8.46.

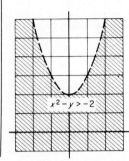

Figure 8.46

| EXERCISE 8-10

Graph the solution set for each inequality.

1. $y > 2x$	**2.** $y < 2x$	**3.** $y < x + 1$
4. $y > x + 1$	**5.** $y \geq 3x - 1$	**6.** $y \leq 3x - 1$
7. $x + y < 1$	**8.** $x + y > 1$	**9.** $2x + y \leq 3$
10. $2x + y \geq 3$	**11.** $y \geq 3$	**12.** $x \leq -2$
13. $4x < 5 + 2y$	**14.** $4x > 5 + 2y$	**15.** $y < 2x^2$
16. $y \geq 2x^2$	**17.** $y \leq x^2 - x$	**18.** $y > x^2 - x$
19. $y > x^3$	**20.** $y < x^3$	

| REVIEW

PART A

Answer True or False.

1. (2, 4) and (4, 2) are the same ordered pair.
2. The graph of $(-2, -3)$ occurs in Quadrant III of the Cartesian coordinate system.
3. Points along the horizontal axis have the same abscissa.
4. For every right triangle, the square of the length of the longest side is equal to the sum of the squares of the lengths of each of the two other sides.
5. The graph of an equation in two variables is actually the graph of its solution set.
6. The point at which the graph of an equation in x and y crosses the y axis is called the y intercept.

7. The slope of all lines parallel to the vertical axis is 0.
8. The graph of the equation $2y = 3x - 1$, has a slope of 3 and a y intercept of -1.
9. At no place will the graph of $y = |3 - x|$ come below the x axis.
10. The solution set for $y \leq 3x + 2$ includes the solutions of $y = 3x + 2$.

PART B

Select from among the ordered pairs given for each equation those that are solutions of the equation.

1. $3x - y = 5$; $(0, 5)$, $(3, 4)$, $(1, -2)$, $(2, -1)$
2. $2x - 4y + 3 = 0$; $\left(0, \dfrac{3}{4}\right)$, $\left(-1, \dfrac{1}{4}\right)$, $\left(3, 2\dfrac{1}{2}\right)$, $\left(-3, \dfrac{3}{4}\right)$
3. $3x - 5y = 6$; $(2, 0)$, $\left(\dfrac{4}{3}, 2\right)$, $(3, 4)$, $(7, 3)$
4. $2x^2 - 3y^2 = 8$; $(3, 2)$, $(-4, 0)$, $\left(5\dfrac{1}{2}, -1\right)$, $(\sqrt{10}, -2)$

Replace the "?" so as to make the ordered pair a solution of the given equation.

5. $4x - 5y = 11$; $(?, 3)$, $(-1, ?)$, $(?, -2)$
6. $x^2 - 4y = 4$; $(?, 3)$, $(-6, ?)$, $(?, 0)$
7. Connect the four points in order with line segments: A: $(0, 4)$; B: $(-2, 2)$; C: $(2, -1)$; D: $(4, 1)$.

   (a) What kind of figure is $ABCD$?
   (b) Draw the diagonals and estimate their point of intersection.
   (c) Using each of the sides of the figure as a hypotenuse, visualize a right triangle in the squares of the grid, and determine the length of each side.
8. The ends of the hypotenuse of two right triangles whose legs parallel the axes are at $(-2, -3)$ and $(4, 0)$. What are the coordinates of the right-angle vertices?

The mirror image of an object appears to be behind the surface of the mirror at a distance equal to that between the surface of the mirror and the actual object. Consider the horizontal and vertical axes to be mirror surfaces, and give the coordinates of the two points which are "mirror images" of each of the following:

9. $(3, 4)$ 10. $(-2, 3)$ 11. $(-4, -4)$ 12. $(3, -1)$

Compute the distance between these pairs of points.

13. $(4, 1)$, $(-5, 1)$ 14. $(6, 2)$, $(6, -5)$
15. $(4, 2)$, $(0, 0)$ 16. $(7, 10)$, $(1, 2)$
17. $(5, -2)$, $(-1, -2)$ 18. $(-3, -4)$, $(-1, -2)$

Compute the length of the hypotenuse of the right triangle defined by the given points.

19. $(-3, 4)$, $(0, 0)$, $(-3, 0)$ 20. $(-3, -2)$, $(3, 2)$, $(-5, 1)$

Graph the equation, using Program **8.2**.

21. $2x + y = 1$ **22.** $3x - y = 2$ **23.** $x = 2y + 2$

24. $2x = y - 1$ **25.** $3 - y = 0$ **26.** $x - 2 = 0$

27. $y = 2x^2 - 3x + 1$ **28.** $y = x^3 - x$

29. $y = \dfrac{12}{x}$ **30.** $x + 3y = x^2 - 7$

Determine the x and y intercepts.

31. $2x - 3y = 6$ **32.** $3x + 5y = 0$

33. $x = -3y$ **34.** $3x + 2y = 1$

35. $3x - 2y - 8 = 0$ **36.** $2y = 3x + 18$

37. $y = x^2 - 4$ **38.** $x^2 = y + 9$

39. $x + y^2 = 2y + 3$ **40.** $y^2 = x + 5$

Compute the slope of the line containing the pair of points.

41. $(5, 6), (3, 3)$ **42.** $(2, -1), (-3, 4)$ **43.** $(4, 0), (0, 3)$

44. $(6, 2), (0, 2)$ **45.** $(3, 2), (3, -4)$ **46.** $(1, -4), (-3, -4)$

Given a point and a slope. Write an equation for the line that satisfies both.

47. $(2, 4)$; 3 (slope)   **48.** $(1, -1)$; -2 **49.** $\left(0, -\dfrac{1}{2}\right); \dfrac{3}{2}$

50. $(-1, 0); \dfrac{2}{5}$ **51.** $(1, 0); -1\dfrac{1}{2}$ **52.** $(2, -2); -\dfrac{2}{5}$

Write an equation of the form, ax + by + c = 0, where a, b, c are integers, having the given slope and y intercept.

53. $m = 3$; y intercept, 2   **54.** $m = -\dfrac{1}{2}$; y intercept, -2

55. $m = -1$; y intercept, 0   **56.** $m = 0$; y intercept, 3

57. $m = -\dfrac{3}{4}$; y intercept, -5 **58.** $m = \dfrac{3}{8}$; y intercept, $-\dfrac{1}{2}$

Write an equation for the line containing the given point and parallel to the line specified by the given equation.

59. $(1, 2)$; $y = 2x + 1$ **60.** $(0, -3)$; $y = -3x - 2$

61. $(2, -1)$; $2x + 3y = 2$ **62.** $(-2, 0)$, $x + y = 0$

Sketch the graph, using slope-intercept data.

63. $y = \dfrac{1}{3}x - 2$ **64.** $y = -\dfrac{2}{3}x + 1$ **65.** $y = -2x + 1$

66. $y = 3x - 2$ **67.** $2y + 3 = x$ **68.** $2x - y = 1$

69. $3y - 2x = 0$ **70.** $2x = -5y$

Sketch the graph in the vicinity of the vertex.

71. $y = |3x - 2|$ **72.** $y = -|2x - 1|$ **73.** $y = -|4 + 2x|$

74. $y = |6 + 2x|$ **75.** $|x| = y + 1$ **76.** $y = -|x| - 4$

77. $y = -|x + 1| - 3$ **78.** $y = 2 - |3x + 1|$

Graph the solution set.

79. $x + y \geq 2$ **80.** $x - y \leq 1$ **81.** $2x - 3y < -1$

82. $y - 2x > 3$ **83.** $x + 3 < y^2$ **84.** $y^2 \leq 2x$

85. $2y + x > x^2$ **86.** $y \geq x^2 - 3x + 2$

SYSTEMS OF EQUATIONS AND INEQUALITIES

9

1. SOLUTION BY GRAPH

A **system of equations** is a set of equations. The simplest system consists of a pair of linear, first-degree equations; for instance:

$$\begin{cases} 2x - y = 7 \\ x + y = 5 \end{cases}$$

The solution of such a system is an ordered pair of numbers that satisfies *both* equations. For the preceding system, the solution is the ordered pair (4, 1).

$$\left. \begin{array}{l} x = 4 \\ y = 1 \end{array} \right\} \qquad \begin{array}{l} 2x - y = 7 \\ 2(4) - (1) = 7 \end{array} \qquad \begin{array}{l} x + y = 5 \\ (4) + (1) = 5 \end{array}$$

One direct way to identify the solution of a system is to graph the equations and then note the x and y coordinates of each point where the graphs intersect. Because we can never draw a perfect graph, what appears to be a solution should be checked in each equation of the system.

PROGRAM 9.1

To solve a system of two linear equations in two variables by graphing:

Step 1: Graph each equation of the system on the same grid.

Step 2: Estimate the ordered pair coordinates of the point at which the graphs of Step 1 intersect.

Step 3: Substitute the ordered pair of Step 2 for the appropriate variables in each equation of the system to check that it is a common solution of the equations of the system.

EXAMPLES

1. Solve the system of equations graphically: $\begin{cases} 2x - y = 7 \\ x + y = 5 \end{cases}$

SOLUTION   *Step 1:* Graph each equation of the system by Program **8.2**, as shown in Figure 9.1.

$2x - y = 7$

x	0	$3\frac{1}{2}$	2
y	-7	0	-3

$x + y = 5$

x	0	5		
y	5	0		

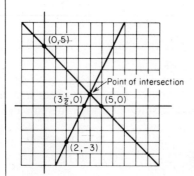

Figure 9.1

Step 2: Estimate the coordinates of the point where the graphs intersect: (4, 1).

Step 3: Check:

$$2x - y = 7 \qquad\qquad x + y = 5$$

$$\left.\begin{matrix} x = 4 \\ y = 1 \end{matrix}\right\} \quad 2(4) - 1 = 7 \qquad \left.\begin{matrix} x = 4 \\ y = 1 \end{matrix}\right\} \quad (4) + (1) = 5$$

Replacing x and y with 4 and 1, respectively, satisfies both equations in the system; thus, (4, 1) is the solution of the system.

2. Solve by graphing: $\begin{cases} x + 2y = -3 \\ 2x - y = 4 \end{cases}$

SOLUTION *Step 1:* Use the slope-intercept method (Program **8.3**) to graph the equations of the system; see Figure 9.2:

$$x + 2y = -3$$

$$2y = -x - 3$$

$$y = -\frac{1}{2}x - \frac{3}{2}$$

$$2x - y = 4$$

$$y = 2x - 4$$

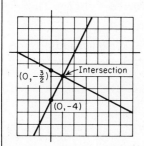

Figure 9.2

Step 2: Estimate the point of intersection: $(1, -2)$.

Step 3: Check:

$$x + 2y = -3 \qquad\qquad 2x - y = 4$$

$$\left.\begin{matrix} x = 1 \\ y = -2 \end{matrix}\right\} \quad (1) + 2(-2) = -3 \qquad 2(1) - (-2) = 4$$

3. Solve the system: $\begin{cases} x = 7 \\ x + 3y - 4 = 0 \end{cases}$

SOLUTION | *Step 1:* The graph of $x = 7$ is a line perpendicular to the x axis along which all points have an x coordinate of 7, as shown in Figure 9.3. For the other equation of the system, we identify two points and plot its graph:

$$x + 3y - 4 = 0$$

x	0	4
y	$\dfrac{4}{3}$	0

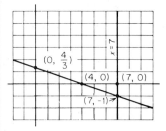

Figure 9.3

Step 2: $(7, -1)$ appears to be the point of intersection.

Step 3: Check:

$$x = 7$$
$$\left.\begin{array}{l} x = 7 \\ y = -1 \end{array}\right\} \quad (7) = 7$$

$$x + 3y - 4 = 0$$
$$\left.\begin{array}{l} x = 7 \\ y = -1 \end{array}\right\} \quad (7) + 3(-1) - 4 = 0$$

The solution of the system is the ordered pair $(7, -1)$.

A system of two linear equations may have one solution, no solution, or an infinite number of solutions. This situation becomes apparent when the equations are seen in slope-intercept form

$$y = mx + b$$

To illustrate, reconsider the equations of the system of Example 1 above

$$\begin{cases} 2x - y = 7 \\ x + y = 5 \end{cases}$$

rewritten in slope-intercept form

$$\begin{cases} y = 2x - 7 \\ y = -x + 5 \end{cases}$$

The slope of the top equation is 2. The slope of the bottom equation is -1. Lines in a plane that have different slope numbers are not parallel and will intersect at some point. So when the two linear equations of a system have different slopes, there is a single solution. Such systems are said to be **independent**.

Consider now the system

$$\begin{cases} 2x - y + 7 = 0 \\ 2x = y - 4 \end{cases}$$

in slope-intercept form

$$\begin{cases} y = 2x + 7 \\ y = 2x + 4 \end{cases}$$

Here the slopes are the same (2), but the y-intercepts (7 and 4) differ. The lines are parallel, as shown in Figure 9.4, and will never intersect.

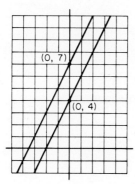

Figure 9.4

Therefore when the two linear equations of a system have the same slope but different y intercepts, the system has no solution. Such systems are said to be **inconsistent**.

Finally, consider the system

$$\begin{cases} 2x = y - 3 \\ 4y = 8x + 12 \end{cases}$$

rewritten in slope-intercept form

$$\begin{cases} y = 2x + 3 \\ y = \dfrac{8}{4}x + \dfrac{12}{4} \end{cases}$$

Here the slope numbers are equal $\left(2 \text{ and } \dfrac{8}{4}\right)$, and so are the y intercepts $\left(3 \text{ and } \dfrac{12}{4}\right)$. Clearly, the two equations are equivalent to one another. They both have the same solution sets and their line graphs "intersect" (really coincide) at every point.

Thus when two linear equations of a system have the same slope and y intercept, the system has an infinite number of solutions. Such systems are said to be **dependent**.

1. Solve graphically: $\begin{cases} x + 2y - 4 = 0 \\ 2x + 4y - 12 = 0 \end{cases}$

SOLUTION The system, with equations simplified and in slope-intercept form, is

$$\begin{cases} y = -\dfrac{1}{2}x + 2 \\ y = -\dfrac{1}{2}x + 3 \end{cases}$$

The slopes are equal $\left(-\dfrac{1}{2}\right)$, but the y intercepts differ (2 and 3). The system is inconsistent—that is, has no solution. See Figure 9.5.

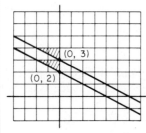

(0, 3)

(0, 2)

Figure 9.5

2. Solve graphically: $\begin{cases} 8x - 2y = 10 \\ 12x - 3y = 15 \end{cases}$

SOLUTION The system, with equations in slope-intercept form, is

$$\begin{cases} y = \dfrac{8}{2}x - \dfrac{10}{2} \\ y = \dfrac{12}{3}x - \dfrac{15}{3} \end{cases}$$

Slopes are equal $\left(\dfrac{8}{2} \text{ and } \dfrac{12}{3}\right)$ and y intercepts are the same $\left(-\dfrac{10}{2} \text{ and } -\dfrac{15}{3}\right)$. The system is dependent—that is, has an infinite number of solutions. See Figure 9.6.

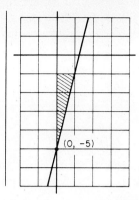

(0, −5)

Figure 9.6

Apply the slope-intercept analysis to decide which system will have a single, unique solution.

1. $\begin{cases} y = 2x - 5 \\ y = -2x - 5 \end{cases}$
2. $\begin{cases} y = 3x - 2 \\ y = 3x + 2 \end{cases}$
3. $\begin{cases} y = 5x - 2 \\ -y = -5x + 2 \end{cases}$

4. $\begin{cases} y = \dfrac{2}{3}x - 1 \\ y = \dfrac{1}{3}x - 1 \end{cases}$
5. $\begin{cases} 2x - 3y = 1 \\ 4x - 6y = -1 \end{cases}$
6. $\begin{cases} 2x - 3y = 2 \\ x - y = 0 \end{cases}$

7. $\begin{cases} y + 3x = 4 \\ 2x - y = 1 \end{cases}$
8. $\begin{cases} 2x - \dfrac{1}{2}y = 3 \\ x = 7 \end{cases}$

9. $\begin{cases} y = 4 \\ x = 4 \end{cases}$
10. $\begin{cases} \dfrac{x}{4} - \dfrac{y}{2} = \dfrac{1}{4} \\ 6y = 3(x - 1) \end{cases}$

Solve the system of equations by graphing.

11. $\begin{cases} 2x + y = 7 \\ x + y = 5 \end{cases}$
12. $\begin{cases} 3x + y = 7 \\ 2x + y = 6 \end{cases}$

13. $\begin{cases} x + 2y - 4 = 0 \\ x + y - 1 = 0 \end{cases}$
14. $\begin{cases} x - 2y - 1 = 0 \\ 2x + y + 8 = 0 \end{cases}$

15. $\begin{cases} x - 2y - 5 = 0 \\ x + 2y - 3 = 0 \end{cases}$
16. $\begin{cases} x - 3y = 2 \\ 2x - 6y = 4 \end{cases}$

17. $\begin{cases} 4x + 3y + 7 = 0 \\ 2x + 5y = 0 \end{cases}$
18. $\begin{cases} 3x - 2y = 8 \\ 6x - 4y = 10 \end{cases}$

19. $\begin{cases} x + y + 8 = 0 \\ x - y - 1 = 0 \end{cases}$
20. $\begin{cases} 3x - y - 6 = 0 \\ 3x + 2y + 12 = 0 \end{cases}$

Chapter 9 SYSTEMS OF EQUATIONS AND INEQUALITIES **301**

21. $\begin{cases} 4x - y = -8 \\ x + 2 = 0 \end{cases}$ **22.** $\begin{cases} 4x + 3y - 3 = 0 \\ 8x + 6y = 4 \end{cases}$

23. $\begin{cases} x = -3 \\ 2x - 4y = 0 \end{cases}$ **24.** $\begin{cases} 3x - y = 1 \\ y = 2 \end{cases}$

2. SOLUTION BY SUBSTITUTION

Several algebraic methods exist for solving systems of two linear equations in two variables. One method, known as **substitution**, is a natural next step to analyzing the equations of a system in slope-intercept form.

To illustrate, consider the system

$$\begin{cases} x - y = 3 \\ x + 2y = 6 \end{cases}$$

rewritten in slope-intercept form

$$\begin{cases} y = x - 3 \\ y = -\dfrac{1}{2}x + 3 \end{cases}$$

The slopes of the two equations are different, and so the system is independent—that is, has one solution. By equating the right members of the two equations (both are equal to y), we obtain an equation in a single variable, x:

$$x - 3 = -\frac{1}{2}x + 3$$

which can be readily solved:

$$x + \frac{1}{2}x = 3 + 3$$

$$\frac{3}{2}x = 6$$

$$x = 4$$

Thus 4 is the x component of the solution.

The y component can now be found by substituting 4 for x in *either* of the original equations:

$$x - y = 3$$

$$x = 4: \quad (4) - y = 3$$

$$y = 1$$

The y component of the solution is 1.

The solution of the system, then, is (4, 1). It can be checked by substituting for the variables in the *other* equation of the system:

$$x + 2y = 6$$

$$\left. \begin{array}{l} x = 4 \\ y = 1 \end{array} \right\} \quad (4) + 2(1) \stackrel{?}{=} 6$$

$$4 + 2 = 6$$

Had we rewritten the equations of the original system in terms of the other variable, x, then equated, solved, and substituted, we would have arrived at the same solution:

$$\begin{cases} x - y = 3 \\ x + 2y = 6 \end{cases} \Rightarrow \begin{cases} x = y + 3 \\ x = -2y + 6 \end{cases}$$

$$y + 3 = -2y + 6$$

$$3y = 3$$

$$y = 1$$

Solving for x,

$$x - y = 3$$

$$y = 1: \quad x - (1) = 3$$

$$x = 4$$

Solution: (4, 1).

When we solve one of the two linear equations of the system in terms of one of the variables and then replace that variable in the other equation with the solution, we always obtain an equation in a single variable. In terms of our illustration

$$\begin{cases} x - y = 3 \\ x + 2y = 6 \end{cases}$$

From the top equation:

$$y = x - 3 \qquad\qquad x = y + 3$$

Substituted in the bottom equation:

$$x + 2(\,x - 3\,) = 6 \qquad\qquad y + 3 + 2y = 6$$

$$x + 2x - 6 = 6 \qquad\qquad 3y = 3$$

$$3x = 12 \qquad\qquad y = 1 \quad \text{etc.}$$

$$x = 4 \quad \text{etc.}$$

Program **9.2** summarizes the substitution procedure.

PROGRAM 9.2

To solve a system of two equations in two variables by substitution:

Step 1:   Solve one of the equations for one of the variables in terms of the other.

Step 2:   Substitute the solution of Step 1 for the appropriate variable in the other equation.

Step 3:   Solve the derived equation of Step 2 for one component of the ordered pair solution of the system.

Step 4:   Substitute the solution of Step 3 for the appropriate variable in one of the equations of the system and solve for the remaining variable.

Step 5:   Check by replacing the variables of the equation of the system not used in Step 4 with the ordered pair determined in Steps 3 and 4.

EXAMPLES

1.   Solve by substitution: $\begin{cases} 2x - y = 7 \\ x + y = 5 \end{cases}$

SOLUTION   *Step 1:* Solve $2x - y = 7$ for y:

$$y = 2x - 7$$

Step 2: Substitute $2x - 7$ for y in the other equation.

$$x + \boxed{y} = 5$$

$$y = 2x - 7: \quad x + \boxed{2x - 7} = 5$$

Step 3: Solve the derived equation of Step 2 for x:

$$x + 2x - 7 = 5$$

$$3x = 12$$

$$x = 4$$

Step 4: Substitute 4 for x in one of the equations of the system, and solve for y:

$$2x - y = 7$$

$$x = 4: \quad 2(4) - y = 7$$

$$y = 1$$

Step 5: Check, using the other equation of the system:

$$x + y = 5$$

$$\left. \begin{array}{l} x = 4 \\ y = 1 \end{array} \right\} \quad (4) + (1) \overset{?}{=} 5$$

$$5 = 5$$

304 Chapter 9 SYSTEMS OF EQUATIONS AND INEQUALITIES

The ordered pair (4, 1) satisfies both equations and, therefore, is the solution of the system.

2.   Solve by substitution: $\begin{cases} 6x - 3y = 15 \\ x - 4y = 6 \end{cases}$

SOLUTION   *Step 1:* Solve the bottom equation for x:

$$x - 4y = 6$$

$$x = 4y + 6$$

Step 2: Substitute $4y + 6$ for x in the top equation:

$$6x - 3y = 15$$

$$x = 4y + 6: \quad 6(4y + 6) - 3y = 15$$

Step 3: Solve the derived equation of Step 2:

$$24y + 36 - 3y = 15$$

$$24y - 3y = 15 - 36$$

$$21y = -21$$

$$y = -1$$

Step 4: Replace y in one of the original equations by -1, and solve for x:

$$x - 4y = 6$$

$$y = -1: \quad x - 4(-1) = 6$$

$$x = 6 - 4$$

$$x = 2$$

Step 5: Check:

$$6x - 3y = 15$$

$$\left. \begin{array}{l} x = 2 \\ y = -1 \end{array} \right\} \quad 6(2) - 3(-1) \overset{?}{=} 15$$

$$12 + 3 = 15$$

The solution of the system is $(2, -1)$.

3.   Solve by substitution: $\begin{cases} 4x + 3y = 7 \\ 8x - 3y = 41 \end{cases}$

Step 1: Solve for one of the variables:

$$4x + 3y = 7$$

$$4x = 7 - 3y$$

$$x = \frac{7 - 3y}{4}$$

$$= \left(\frac{1}{4}\right)(7 - 3y)$$

Step 2: Use the solution of Step 1 in the other equation to derive an equation in a single variable:

$$8x - 3y = 41$$

$$x = \frac{1}{4}(7 - 3y): \quad 8\left(\frac{1}{4}\right)(7 - 3y) - 3y = 41$$

Step 3: Solve the derived equation of Step 2:

$$2(7 - 3y) - 3y = 41$$

$$14 - 6y - 3y = 41$$

$$-6y - 3y = 41 - 14$$

$$-9y = 27$$

$$y = -3$$

Step 4: Solve for the other variable:

$$4x + 3y = 7$$

$$y = -3: \quad 4x + 3(-3) = 7$$

$$4x - 9 = 7$$

$$4x = 16$$

$$x = 4$$

Step 5: Check:

$$8x - 3y = 41$$

$$\left.\begin{array}{l} x = 4 \\ y = -3 \end{array}\right\} \quad 8(4) - 3(-3) \overset{?}{=} 41$$

$$32 + 9 = 41$$

4. Solve by substitution:
$$\begin{cases} \dfrac{x + 3}{3} + \dfrac{y + 1}{2} = 2 \\ \\ \dfrac{x + 3y}{3} = 1 \end{cases}$$

SOLUTION | First, clear fractions and simplify the system:

$$\begin{cases} 2(x + 3) + 3(y + 1) = 6(2) \\ x + 3y = 3(1) \end{cases}$$

$$\begin{cases} 2x + 6 + 3y + 3 = 12 \\ x + 3y = 3 \end{cases}$$

$$\begin{cases} 2x + 3y = 3 \\ x + 3y = 3 \end{cases}$$

Apply Program **9.2**, by solving the bottom equation for x:

$$x = 3 - 3y$$

and substitute $3 - 3y$ for x in the top equation:

$$2x + 3y = 3$$

$$x = 3 - 3y: \quad 2(3 - 3y) + 3y = 3$$

$$6 - 6y + 3y = 3$$

$$-3y = -3$$

$$y = 1$$

Use $y = 1$ to solve for x:

$$x + 3y = 3$$

$$y = 1: \quad x + 3(1) = 3$$

$$x = 0$$

The solution of the original system is $(0, 1)$.

| **EXERCISE 9-2**

Solve the systems by substitution.

1. $\begin{cases} 2x + 3y = 13 \\ x + 2y = 8 \end{cases}$
2. $\begin{cases} 2x - y = 3 \\ 3x - 2y = 1 \end{cases}$
3. $\begin{cases} 4x + 2y = 5 \\ 3x + y = 2 \end{cases}$

4. $\begin{cases} x + 2y = -1 \\ 2x + 5y = 0 \end{cases}$
5. $\begin{cases} x + y = 6 \\ x + 2y = 7 \end{cases}$
6. $\begin{cases} x + y = 2 \\ 3x - y = 14 \end{cases}$

7. $\begin{cases} 4x - 3y = 1 \\ x - 2y = 4 \end{cases}$
8. $\begin{cases} 3x - 4y = -9 \\ 2x + y = -6 \end{cases}$
9. $\begin{cases} 2x - 5y = 9 \\ 3x - 8y = 13 \end{cases}$

10. $\begin{cases} 4x + 7y = 2 \\ 3x + 4y = -1 \end{cases}$
11. $\begin{cases} 3a - 2b = 6 \\ 5a + 3b = -9 \end{cases}$
12. $\begin{cases} 4m - 2n = 5 \\ 8m + 3n = -4 \end{cases}$

13. $\begin{cases} \dfrac{1}{2}x + y = 1 \\ \dfrac{1}{4}x + \dfrac{1}{3}y = \dfrac{5}{12} \end{cases}$
14. $\begin{cases} x + \dfrac{5}{2}y = 2 \\ y + \dfrac{1}{2}x = \dfrac{1}{2} \end{cases}$

15. $\begin{cases} \dfrac{3}{4}x + \dfrac{1}{2}y = -\dfrac{1}{4} \\ \dfrac{1}{3}x - \dfrac{1}{2}y = \dfrac{4}{3} \end{cases}$ **16.** $\begin{cases} \dfrac{2}{3}x - \dfrac{3}{4}y = -\dfrac{5}{8} \\ \dfrac{4}{3}x + \dfrac{3}{2}y = -2\dfrac{3}{4} \end{cases}$

17. $\begin{cases} \dfrac{x-2}{2} + \dfrac{y+3}{5} = -3 \\ \dfrac{x+3}{3} - \dfrac{y-2}{2} = -1 \end{cases}$ **18.** $\begin{cases} \dfrac{x+1}{7} + \dfrac{y+4}{3} = 1 \\ \dfrac{2x-1}{3} + \dfrac{y-3}{2} = -3 \end{cases}$

19. $\begin{cases} \dfrac{m+3}{2} - \dfrac{2n-1}{3} = 3\dfrac{1}{2} \\ \dfrac{4-m}{3} + \dfrac{n-5}{2} = -2\dfrac{1}{3} \end{cases}$ **20.** $\begin{cases} \dfrac{a+b}{2} - \dfrac{b-a}{3} = 1 \\ \dfrac{2a+b}{3} + \dfrac{b+2a}{5} = \dfrac{8}{5} \end{cases}$

3. SOLUTION BY ADDITION

The key to solving a system of two linear equations in two variables by substitution (Program **9.2**) is to reduce the system to a single equation in a single variable. There is another way to reduce such systems to a single equation in a single variable, called the **addition method**. We illustrate with the system

$$\begin{cases} 3x - 2y = 5 \\ x + 2y = 3 \end{cases}$$

If we add the two equations of this system together, term by term, the y terms "add out":

$$(+)\ \frac{\begin{array}{r} 3x - 2y = 5 \\ x + 2y = 3 \end{array}}{4x + 0y = 8}$$

The result is an equation in a single variable, x:

$$4x = 8$$

However, most systems consist of equations that, when added together term by term, will *not* produce an equation in a single variable. For those, a preliminary step is needed.

> *Multiply the equations by numbers that will make the coefficients of one of the variables negatives of one another.*

This step produces equations that are equivalent to the original equations (no change in solution sets) but makes possible the elimination of one of the variables upon addition. For example, reconsider the system of Example 7 previous section:

$$\begin{cases} 6x - 3y = 15 \\ x - 4y = 6 \end{cases}$$

Addition of the two equations, term by term, will not produce an equation in a single variable. But by multiplying the bottom equation through by -6 and then adding it with the top equation, we obtain the desired result—a single equation in a single variable, y:

$$6x - 3y = 15 \quad \xrightarrow{\text{same}} \quad 6x - 3y = 15$$

$$x - 4y = 6 \quad \xrightarrow{\text{times } (-6)} \quad (+) \; \frac{-6x + 24y = -36}{0x + 21y = -21}$$

Or we could have multiplied the top equation by 4 and the bottom equation by -3 and then added to produce a single equation in a single variable—this time, x:

$$6x - 3y = 15 \quad \xrightarrow{\text{times } 4} \quad 24x - 12y = 60$$

$$x - 4y = 6 \quad \xrightarrow{\text{times } (-3)} \quad (+) \; \frac{-3x + 12y = -18}{21x + 0y = 42}$$

In the first instance, we "eliminated" the x term; in the second instance, we "eliminated" the y term.

PROGRAM 9.3

To solve a system of two linear equations in two variables by addition:

Step 1: Multiply, if necessary, the terms of the equations by numbers that will make the coefficients of one of the variables negatives of one another.

Step 2: Add the two equations of Step 1, term by term, to produce a single equation in a single variable.

Step 3: Solve the resulting equation of Step 2 for one component of the solution of the system.

Step 4: Substitute the solution of Step 3 for the appropriate variable in one of the equations of the system and solve for the other component of the solution of the system.

Step 5: Check the solution by replacing the variables of the equation of the system not used in Step 4 with the components of the solution determined in Steps 3 and 4.

1. Solve by the addition method: $\begin{cases} 4x + 3y = 6 \\ 8x - 3y = 30 \end{cases}$

Step 1: No multiplication is necessary, since the coefficients of the y terms are negatives of each other.

Step 2: Add the two equations, term by term:

$$4x + 3y = 6$$
$$(+) \quad \underline{8x - 3y = 30}$$
$$12x \quad\quad = 36$$

Step 3: Solve for x: $\quad\quad x \quad\quad = 3$

Step 4: Replace the variable x in the first equation with 3; solve for y:

$$4x + 3y = 6$$
$$x = 3: \quad 4(3) + 3y = 6$$
$$3y = 6 - 12$$
$$y = -2$$

Step 5: Check by replacing x and y of the second equation with 3 and -2, respectively:

$$\left. \begin{array}{l} x = 3 \\ y = -2 \end{array} \right\} \quad \begin{array}{l} 8x - 3y = 30 \\ 8(3) - 3(-2) \overset{?}{=} 30 \\ 24 + 6 = 30 \end{array}$$

The solution of the given system is the ordered pair $(3, -2)$.

2. Solve by the addition method: $\begin{cases} 6x + 2y = 11 \\ 4x + 3y = 14 \end{cases}$

Step 1: Multiply all terms of the top equation by 2 and of the bottom equation by -3:

$$2 \cdot (6x + 2y = 11): \quad 12x + 4y = 22$$
$$-3 \cdot (4x + 3y = 14): \quad \underline{-12x - 9y = -42}$$

Step 2: Add: $\quad\quad\quad\quad\quad\quad\quad\quad\quad -5y = -20$

Step 3: Solve for y: $\quad\quad\quad\quad\quad\quad\quad\quad y = 4$

Step 4: Substitute 4 for y in one of the equations of the system and solve for x:

$$6x + 2y = 11$$
$$y = 4: \quad 6x + 2(4) = 11$$
$$6x + 8 = 11$$
$$6x = 3$$
$$x = \frac{1}{2}$$

Step 5: Check, using the equation of the system not used in Step 4:

$$4x + 3y = 14$$

$$\left. \begin{array}{l} x = \dfrac{1}{2} \\[2mm] y = 4 \end{array} \right\} \quad 4\left(\dfrac{1}{2}\right) + 3(4) \overset{?}{=} 14$$

$$2 + 12 = 14$$

Solution of the system: $\left(\dfrac{1}{2}, 4\right)$.

3. Solve: $\begin{cases} 5x - 2y = 5 \\ 2x + 3y = 21 \end{cases}$

SOLUTION | *Step 1:* Multiply the top equation by 3 and the bottom equation by 2:

$$3 \cdot (5x - 2y = 5): \quad 15x - 6y = 15$$

$$2 \cdot (2x + 3y = 21): \quad \underline{4x + 6y = 42}$$

Step 2: Add: $19x \qquad = 57$

Step 3: Solve for x: $x \qquad = 3$

Step 4: Substitute 3 for x in one of the equations of the system and solve for y:

$$5x - 2y = 5$$

$$x = 3: \quad 5(3) - 2y = 5$$

$$15 - 2y = 5$$

$$-2y = -10$$

$$y = 5$$

Step 5: Check, using the other of the two equations of the system:

$$2x + 3y = 21$$

$$\left. \begin{array}{l} x = 3 \\ y = 5 \end{array} \right\} \quad 2(3) + 3(5) \overset{?}{=} 21$$

$$6 + 15 = 21$$

Solution of the system: (3, 5).

The preferred way to determine whether a system of two linear equations is other than independent is to use the slope-intercept analysis as discussed in the first section of this chapter. It is instructive, however, to note what happens when we apply Program **9.3** to dependent or inconsistent systems.

When the system is dependent, both variables are eliminated and the constant terms add to zero, as demonstrated in Example 1 below.

When the system is inconsistent, both variables are also eliminated, but the constant terms add to a number other than zero, as in Example 2 below.

EXAMPLES

1. Solve: $\begin{cases} 8x - 2y = 10 \\ 12x - 3y = 15 \end{cases}$

SOLUTION *Step 1:* Multiply the top equation by 3 and the bottom equation by -2:

$$3 \cdot (8x - 2y = 10): \quad 24x - 6y = 30$$

$$-2 \cdot (12x - 3y = 15): \quad \underline{-24x + 6y = -30}$$

Step 2: Add: $\quad 0x + 0y = 0$

$$0 = 0$$

The system is dependent, and *any* ordered pair that satisfies one equation of the system satisfies the other.

2. Solve: $\begin{cases} 4x - 5y = 8 \\ 8x - 10y = 2 \end{cases}$

SOLUTION *Step 1:* Multiply the top equation by -2:

$$-2 \cdot (4x - 5y = 8): \quad -8x + 10y = -16$$

$$\underline{8x - 10y = 2}$$

Step 2: Add: $\quad 0x + 0y = -14$

$$0 = -14$$

The result of Step 2, $0 = -14$, is false. The system is inconsistent. There is *no* ordered pair that satisfies *both* equations of the system.

EXERCISE 9-3

Solve the systems by the addition method.

1. $\begin{cases} x + y = 7 \\ x - y = 5 \end{cases}$ **2.** $\begin{cases} x + y = 1 \\ x - y = 7 \end{cases}$ **3.** $\begin{cases} x - 2y = 7 \\ 2x + y = 9 \end{cases}$

4. $\begin{cases} 2x + y = 4 \\ x - 2y = 7 \end{cases}$ **5.** $\begin{cases} 3x - 2y = 4 \\ 3x + 2y = 8 \end{cases}$ **6.** $\begin{cases} 3x - 2y = 9 \\ 3x - 5y = 18 \end{cases}$

7. $\begin{cases} 3x - 4y = 6 \\ 4x - 3y = 1 \end{cases}$ **8.** $\begin{cases} 2x - 3y - 20 = 0 \\ 3x + 5y - 11 = 0 \end{cases}$ **9.** $\begin{cases} 2x - 3y = 12 \\ 3x - 5y = 19 \end{cases}$

10. $\begin{cases} 3x + 2y = -4 \\ 2x + 5y = 12 \end{cases}$ **11.** $\begin{cases} 4x - 2y = 7 \\ 3y + 6 = 0 \end{cases}$ **12.** $\begin{cases} 3x - 6y = 3 \\ -2x + 7y = -4 \end{cases}$

13. $\begin{cases} \dfrac{3}{4}x - \dfrac{2}{3}y = -7 \\ \dfrac{1}{2}x - 3y = -20 \end{cases}$ **14.** $\begin{cases} \dfrac{2}{3}a + \dfrac{1}{2}b = \dfrac{1}{2} \\ \dfrac{4}{5}a + \dfrac{3}{5}b = \dfrac{1}{3} \end{cases}$

15. $\begin{cases} \dfrac{4m}{3} - \dfrac{2n}{9} = \dfrac{2}{3} \\[2mm] \dfrac{3m}{2} - \dfrac{n}{4} = \dfrac{3}{4} \end{cases}$

16. $\begin{cases} \dfrac{2a}{3} + \dfrac{3b}{2} = -\dfrac{2}{3} \\[2mm] \dfrac{-4a}{3} + \dfrac{b}{2} = -1 \end{cases}$

17. $\begin{cases} \dfrac{y-2}{3} - \dfrac{2+x}{4} = -2 \\[2mm] \dfrac{x-3}{5} + \dfrac{2-y}{3} = 2 \end{cases}$

18. $\begin{cases} \dfrac{x-y}{2} - \dfrac{2y+x}{3} = \dfrac{2}{3} \\[2mm] \dfrac{2x+y}{4} + \dfrac{y-x}{6} = -2\dfrac{5}{6} \end{cases}$

Solve the systems by either addition or substitution.

19. $\begin{cases} 7x + 3y = 4 \\ 14x + 9y = 7 \end{cases}$
20. $\begin{cases} x - 2y = 4 \\ x + 4y = 1 \end{cases}$
21. $\begin{cases} 3x + 6y = 9 \\ 4x + 8y = 12 \end{cases}$

22. $\begin{cases} x - 4y = 1 \\ 3x - 8y = 7 \end{cases}$
23. $\begin{cases} 2x - 3y = 5 \\ -6y + 4x = 8 \end{cases}$
24. $\begin{cases} 6x + y - 4 = 0 \\ y + 8 = 0 \end{cases}$

25. $\begin{cases} x - 5y = -6\dfrac{1}{2} \\[2mm] 2x + 2y = 17 \end{cases}$
26. $\begin{cases} x + y = -2 \\[2mm] \dfrac{3x}{10} + \dfrac{7y}{10} = -5 \end{cases}$
27. $\begin{cases} 4x + 2y = 7 \\ y = -2x + 3 \end{cases}$

28. $\begin{cases} 6x - 5y + 4 = 0 \\ x - 10y = 1 \end{cases}$

4. DETERMINANTS*

Another method for solving systems of linear equations, discussed in the next section, is to make use of a square array of numbers, called a **determinant**. For instance,

$$\begin{vmatrix} 6 & -1 \\ 2 & 0 \end{vmatrix}$$

or, in general,

$$\begin{vmatrix} a_1 & b_1 \\ a_2 & b_2 \end{vmatrix}$$

As in every determinant, the one shown immediately above consists of:

> **elements**—the numbers a_1, a_2, b_1, b_2
> **rows**—the elements in a horizontal line (same subscripts—e.g., a_2, b_2)
> **columns**—the elements in a vertical line (e.g., b_1, b_2), and

---

* The content of this section, in more basic courses of study, may be postponed without significant effect upon the remaining parts of the book.

principal diagonal—the elements occurring along a diagonal that starts at the upper left element and ends at the lower right element—for example,

$$\begin{vmatrix} a_1 & b_1 \\ a_2 & b_2 \end{vmatrix}$$

When there are two elements in each row and in each column, the determinant is called a **second-order determinant**.

The **expansion** of a determinant is a polynomial. By definition, the expansion of

$$\begin{vmatrix} a_1 & b_1 \\ a_2 & b_2 \end{vmatrix} = (a_1)(b_2) - (a_2)(b_1)$$

PROGRAM 9.4

To expand a second-order determinant:

Step 1: Multiply the elements along the principal diagonal.
Step 2: Multiply the elements along the other diagonal.
Step 3: Subtract the product of Step 2 from that of Step 1.

EXAMPLES

1. Expand $\begin{vmatrix} a_1 & b_1 \\ a_2 & b_2 \end{vmatrix}$.

SOLUTION *Step 1:* Multiply the elements along the principal diagonal:

$$\begin{vmatrix} a_1 & b_1 \\ a_2 & b_2 \end{vmatrix}$$

$$a_1 \cdot b_2 = a_1 b_2$$

Step 2: Multiply the elements along the other diagonal:

$$a_2 \cdot b_1 = a_2 b_1$$

$$\begin{vmatrix} a_1 & b_1 \\ a_2 & b_2 \end{vmatrix}$$

Step 3: Subtract:

$$\begin{vmatrix} a_1 & b_1 \\ a_2 & b_2 \end{vmatrix} = a_1 b_2 - a_2 b_1$$

2. Expand $\begin{vmatrix} 2 & 7 \\ 5 & 3 \end{vmatrix}$.

Step 1: Principal diagonal:

$$\begin{vmatrix} 2 & 7 \\ 5 & 3 \end{vmatrix}$$

(2)(3)

Step 3: Other diagonal:

(5)(7)

$$\begin{vmatrix} 2 & 7 \\ 5 & 3 \end{vmatrix}$$

Step 4: Subtract:

$$\begin{vmatrix} 2 & 7 \\ 5 & 3 \end{vmatrix} = (2)(3) - (5)(7) = 6 - 35 = -29$$

3. Evaluate the determinant $\begin{vmatrix} 4 & 3 \\ -2 & 1 \end{vmatrix}$.

$\begin{vmatrix} 4 & 3 \\ -2 & 1 \end{vmatrix} = (4)(1) - (-2)(3) = 4 - (-6) = 4 + 6 = 10$

4. Evaluate $\begin{vmatrix} 3 & 2 \\ 0 & -5 \end{vmatrix}$.

$\begin{vmatrix} 3 & 2 \\ 0 & -5 \end{vmatrix} = (3)(-5) - (0)(2) = -15$

EXERCISE 9-4

Complete the expansions of the following second-order determinants.

1. $\begin{vmatrix} 2 & 1 \\ 3 & 4 \end{vmatrix} = (2)(\quad) - (3)(\quad) = 5$ **2.** $\begin{vmatrix} 3 & 0 \\ 1 & 6 \end{vmatrix} = (3)(\quad) - (\quad)(\quad) = 18$

3. $\begin{vmatrix} 2 & 3 \\ -1 & -4 \end{vmatrix} = (\quad)(-4) - (\quad)(\quad) = -5$

4. $\begin{vmatrix} 1 & 4 \\ -3 & -1 \end{vmatrix} = (\quad)(\quad) - (\quad)(4) = 11$

Evaluate.

5. $\begin{vmatrix} -3 & 2 \\ 4 & 1 \end{vmatrix}$ **6.** $\begin{vmatrix} 2 & 4 \\ 1 & 3 \end{vmatrix}$ **7.** $\begin{vmatrix} 6 & 3 \\ 2 & -4 \end{vmatrix}$

8. $\begin{vmatrix} 7 & -1 \\ -3 & 2 \end{vmatrix}$ **9.** $\begin{vmatrix} 3 & 4 \\ 0 & 5 \end{vmatrix}$ **10.** $\begin{vmatrix} -2 & 4 \\ -3 & 6 \end{vmatrix}$

11. $\begin{vmatrix} 3 & -5 \\ 3 & -5 \end{vmatrix}$ **12.** $\begin{vmatrix} -7 & 8 \\ 4 & 2 \end{vmatrix}$ **13.** $\begin{vmatrix} 0 & 0 \\ -2 & 3 \end{vmatrix}$

14. $\begin{vmatrix} 1 & 0 \\ 0 & 1 \end{vmatrix}$ **15.** $\begin{vmatrix} -3 & 6 \\ -2 & -1 \end{vmatrix}$ **16.** $\begin{vmatrix} 3 & 0 \\ -4 & 0 \end{vmatrix}$

17. $\begin{vmatrix} 2 & x \\ 3 & x \end{vmatrix}$ **18.** $\begin{vmatrix} 1 & x \\ x & 4 \end{vmatrix}$ **19.** $\begin{vmatrix} 1 & b \\ a & 0 \end{vmatrix}$

20. $\begin{vmatrix} a & p \\ 0 & 0 \end{vmatrix}$

5. SOLUTION BY DETERMINANTS*

Consider now this system of two linear equations in two variables, x and y:

$$\begin{cases} a_1 x + b_1 y = c_1 \\ a_2 x + b_2 y = c_2 \end{cases}$$

If we multiply the top equation by b_2 and the bottom equation by $-b_1$ and then add, the y terms are eliminated (Program **9.3**). We can then solve for x.

$$a_1 b_2 x + b_1 b_2 y = b_2 c_1$$
$$(+) \frac{-a_2 b_1 x - b_1 b_2 y = -b_1 c_2}{a_1 b_2 x - a_2 b_1 x = b_2 c_1 - b_1 c_2}$$
$$(a_1 b_2 - a_2 b_1)x = b_2 c_1 - b_1 c_2$$
$$x = \frac{b_2 c_1 - b_1 c_2}{a_1 b_2 - a_2 b_1}$$

Similarly, we can multiply the top equation of the system by $-a_2$, the bottom equation by a_1, then add to eliminate the x terms, and arrive at the following expression for y.

$$y = \frac{a_1 c_2 - a_2 c_1}{a_1 b_2 - a_2 b_1}$$

Notice that for both x and y, just derived, the denominators are the same, $a_1 b_2 - a_2 b_1$, which is the expansion of the determinant

$$D = \begin{vmatrix} a_1 & b_1 \\ a_2 & b_2 \end{vmatrix} = a_1 b_2 - a_2 b_1$$

Moreover, the numerator of the x and y equivalents, respectively, are expansions of the determinants D_x and D_y:

$$D_x = \begin{vmatrix} c_1 & b_1 \\ c_2 & b_2 \end{vmatrix} = c_1 b_2 - c_2 b_1 \qquad D_y = \begin{vmatrix} a_1 & c_1 \\ a_2 & c_2 \end{vmatrix} = a_1 c_2 - a_2 c_1$$

---

* The content of this section, in more basic courses of study, may be postponed without significant effect upon the remaining parts of the book.

Thus we can express x and y as quotients of determinants.

$$x = \frac{b_2c_1 - b_1c_2}{a_1b_2 - a_2b_1} = \frac{D_x}{D} = \frac{\begin{vmatrix} c_1 & b_1 \\ c_2 & b_2 \end{vmatrix}}{\begin{vmatrix} a_1 & b_1 \\ a_2 & b_2 \end{vmatrix}}$$

$$y = \frac{a_1c_2 - a_2c_1}{a_1b_2 - a_2b_1} = \frac{D_y}{D} = \frac{\begin{vmatrix} a_1 & c_1 \\ a_2 & c_2 \end{vmatrix}}{\begin{vmatrix} a_1 & b_1 \\ a_2 & b_2 \end{vmatrix}}$$

In this we have the basis for another method for solving a system of two linear equations in two variables. It is known as **Cramer's Rule**.

PROGRAM 9.5

To solve a system of two linear equations in two variables, x and y, by determinants:

Step 1: Arrange the terms in each equation of the system so that the variables appear in the same order in the left member of each equation and the constant terms in the right member.

Step 2: Form a determinant, D, that has for elements the numerical coefficients to the variables as they occur in the equations.

Step 3: Form a determinant, D_x, similar to the one in Step 2 except replace the elements that are the coefficients of the x terms by the constant terms of the respective equations.

Step 4: Form a determinant, D_y, similar to the one in Step 2 except replace the elements that are the coefficients of the y terms by the constant terms of the respective equations.

Step 5: Compute the solution of the system, using $x = \dfrac{D_x}{D}$ and $y = \dfrac{D_y}{D}$.

EXAMPLES

1. Solve: $\begin{cases} 2x - y - 7 = 0 \\ x + y - 5 = 0 \end{cases}$

SOLUTION *Step 1:* Transform the equations of the system:

$$\begin{cases} 2x - y = 7 \\ x + y = 5 \end{cases}$$

Step 2: Form determinant D: $\begin{vmatrix} 2 & -1 \\ 1 & 1 \end{vmatrix}$

Step 3: Form determinant D_x: $\begin{vmatrix} 7 & -1 \\ 5 & 1 \end{vmatrix}$

Step 4: Form determinant D_y: $\begin{vmatrix} 2 & 7 \\ 1 & 5 \end{vmatrix}$

Step 5: Solve for x and y.

$$x = \frac{D_x}{D} = \frac{\begin{vmatrix} 7 & -1 \\ 5 & 1 \end{vmatrix}}{\begin{vmatrix} 2 & -1 \\ 1 & 1 \end{vmatrix}} = \frac{(7) - (-5)}{(2) - (-1)} = \frac{12}{3} = 4$$

$$y = \frac{D_y}{D} = \frac{\begin{vmatrix} 2 & 7 \\ 1 & 5 \end{vmatrix}}{\begin{vmatrix} 2 & -1 \\ 1 & 1 \end{vmatrix}} = \frac{(10) - (7)}{(2) - (-1)} = \frac{3}{3} = 1$$

The solution is (4, 1). Compare with Example 1, page 304.

2. Solve: $\begin{cases} 5x - 2y = 5 \\ 2x + 3y = 21 \end{cases}$

SOLUTION | *Step 1:* Equations are already properly arranged.

Step 2: $D = \begin{vmatrix} 5 & -2 \\ 2 & 3 \end{vmatrix} = (15) - (-4) = 19$

Step 3: $D_x = \begin{vmatrix} 5 & -2 \\ 21 & 3 \end{vmatrix} = (15) - (-42) = 57$

Step 4: $D_y = \begin{vmatrix} 5 & 5 \\ 2 & 21 \end{vmatrix} = (105) - (10) = 95$

Step 5: $x = \frac{D_x}{D} = \frac{57}{19} = 3$

$y = \frac{D_y}{D} = \frac{95}{19} = 5$

Solution is (3, 5). Compare with Example 3, page 311.

3. Solve: $\begin{cases} 3x + 2y = 4 \\ 3y = 6 \end{cases}$

SOLUTION | *Step 1:* Equations properly arranged.

Note: The bottom equation may be written, $0x + 3y = 6$.

$$\text{Step 2: } D = \begin{vmatrix} 3 & 2 \\ 0 & 3 \end{vmatrix} = 9 - 0 = 9$$

$$\text{Step 3: } D_x = \begin{vmatrix} 4 & 2 \\ 6 & 3 \end{vmatrix} = 12 - 12 = 0$$

$$\text{Step 4: } D_y = \begin{vmatrix} 3 & 4 \\ 0 & 6 \end{vmatrix} = 18 - 0 = 18$$

$$\text{Step 5: } x = \frac{D_x}{D} = \frac{0}{9} = 0$$

$$y = \frac{D_y}{D} = \frac{18}{9} = 2$$

Solution: (0, 2).

4. Solve: $\begin{cases} 8x - 2y = 10 \\ 12x - 3y = 15 \end{cases}$

SOLUTION On page 312 we found this system to be dependent. Notice what happens to D, D_x, and D_y in such systems.

$$D = \begin{vmatrix} 8 & -2 \\ 12 & -3 \end{vmatrix} = (-24) - (-24) = 0$$

$$D_x = \begin{vmatrix} 10 & -2 \\ 15 & -3 \end{vmatrix} = (-30) - (-30) = 0$$

$$D_y = \begin{vmatrix} 8 & 10 \\ 12 & 15 \end{vmatrix} = (120) - (120) = 0$$

Consequently, $x = \dfrac{D_x}{D}$ and $y = \dfrac{D_y}{D}$ are undefined fractions, $\dfrac{0}{0}$. When $D = D_x = D_y = 0$, the system is dependent.

5. Solve: $\begin{cases} 4x - 5y = 8 \\ 8x - 10y = 2 \end{cases}$

SOLUTION On page 312 we found this system of equations to be inconsistent. Notice what happens to D, D_x and D_y in such systems.

$$D = \begin{vmatrix} 4 & -5 \\ 8 & -10 \end{vmatrix} = (-40) - (-40) = 0$$

$$D_x = \begin{vmatrix} 8 & -5 \\ 2 & -10 \end{vmatrix} = (-80) - (-10) = -70$$

$$D_y = \begin{vmatrix} 4 & 8 \\ 8 & 2 \end{vmatrix} = (8) - (64) = -56$$

Consequently, $x = \dfrac{D_x}{D} = \dfrac{-70}{0}$ and $y = \dfrac{D_y}{D} = \dfrac{-56}{0}$, both undefined. When $D = 0$, $D_x \ne 0$, and $D_y \ne 0$, the system is inconsistent.

EXERCISE 9-5

Solve, using determinants.

1. $\begin{cases} 2x + 3y = 13 \\ x + 2y = 8 \end{cases}$
2. $\begin{cases} 2x - y = 3 \\ 3x - 2y = 1 \end{cases}$
3. $\begin{cases} 4x + 2y = 5 \\ 3x + y = 2 \end{cases}$

4. $\begin{cases} x + 2y = -1 \\ 2x + 5y = 0 \end{cases}$
5. $\begin{cases} x - y = 4 \\ x + 2y = 7 \end{cases}$
6. $\begin{cases} x + y = 2 \\ 3x - y = 14 \end{cases}$

7. $\begin{cases} 4x - 3y = 1 \\ x - 2y = 4 \end{cases}$
8. $\begin{cases} 3x - 4y = -9 \\ 2x + y = -6 \end{cases}$
9. $\begin{cases} 2x - 5y = 9 \\ 3x - 8y = 13 \end{cases}$

10. $\begin{cases} 4x + 7y = 2 \\ 3x + 4y = -1 \end{cases}$
11. $\begin{cases} 3a - 2b = 6 \\ 5a + 3b = -9 \end{cases}$
12. $\begin{cases} 4m - 2n = 5 \\ 8m + 3n = -4 \end{cases}$

13. $\begin{cases} 3b - 2c = 4 \\ 6b - 4c = 5 \end{cases}$
14. $\begin{cases} 4x - 2y = 6 \\ 6x - 3y = 9 \end{cases}$
15. $\begin{cases} 4x - y = -5 \\ \quad\quad y = -7 \end{cases}$

16. $\begin{cases} 3x - 2y = -4 \\ 6y - 9x = 0 \end{cases}$
17. $\begin{cases} 2x - 3y = 1 \\ \dfrac{1}{2}x - \dfrac{3}{4}y = \dfrac{1}{4} \end{cases}$
18. $\begin{cases} x - 3 = 1 \\ 2x - 5y = 18 \end{cases}$

19. $\begin{cases} 3x = 6 \\ 2y = 1 \end{cases}$
20. $\begin{cases} 4y = 0 \\ 2x = 5 \end{cases}$

6. SOLVING SYSTEMS OF THREE LINEAR EQUATIONS

In the same sense that a solution of an equation in two variables is an ordered pair, the solution of an equation in three variables is an **ordered triple**. For instance, the ordered triple, $(2, -2, 1)$, is a solution of the equation

$$x + y + 3z = 3$$

$$\left.\begin{array}{l} x = 2 \\ y = -2 \\ z = 1 \end{array}\right\} \quad (2) + (-2) + 3(1) = 3$$

There are several ways to solve systems of three linear equations in three variables. One is to reduce the larger system to an equivalent smaller system to be solved. The following program summarizes the method.

PROGRAM 9.6

To solve a system of three linear equations in three variables:

Step 1:   Select a pair of equations from the system and eliminate one of the variables.

Step 2:   Select another pair of equations from the system and eliminate the same variable as in Step 1.

Step 3:   Solve the system consisting of the equations derived in Steps 1 and 2.

Step 4:   Replace the variables in one of the original equations of the system with the solutions obtained in Step 3 and solve to complete the solution of the system.

EXAMPLES

1.   Solve: $\begin{cases} x + y - 3z = -2 \\ 2x - 2y + z = -2 \\ x - y + 2z = 2 \end{cases}$

SOLUTION   *Step 1:*   Other choices are possible, but here we select the top and bottom equations and eliminate y by addition:

$$x + y - 3z = -2$$

$$(+)\ \frac{x - y + 2z = 2}{2x \qquad -z = 0}$$

Step 2:   To eliminate y from another pair of equations, we elect to add twice the top equation to the middle equation:

$$\begin{array}{c} 2 \cdot (x + y - 3z = -2) \\ \underline{2x - 2y + z = -2} \end{array} \Rightarrow \begin{array}{c} 2x + 2y - 6z = -4 \\ (+)\ \dfrac{2x - 2y + z = -2}{4x \qquad -5z = -6} \end{array}$$

Step 3:   Combine the derived equations of Steps 1 and 2 into a system and solve:

$$\begin{cases} 2x - z = 0 \\ 4x - 5z = -6 \end{cases}$$

We may begin by solving the top equation for z, using Program **9.2**:

$$2x - z = 0$$

$$z = 2x$$

Then replace z by $2x$ in the bottom equation, and solve for x:

$$4x - 5z = -6$$

$$z = 2x: \quad 4x - 5(2x) = -6$$

$$4x - 10x = -6$$

$$-6x = -6$$

$$x = 1$$

Then solve for z:

$$2x - z = 0$$

$$x = 1: \quad 2(1) - z = 0$$

$$z = 2$$

Step 4: Finally, select one of the original equations, substitute the solutions found for x and z in Step 3, and solve for y:

$$x + y - 3z = -2$$

$$\left.\begin{array}{r} x = 1 \\ z = 2 \end{array}\right\} \quad (1) + y - 3(2) = -2$$

$$y - 5 = -2$$

$$y = 3$$

The ordered triple, (1, 3, 2), is the solution of the given system of three linear equations in three variables—and may be readily checked by substitution.

2. Solve: $\begin{cases} 3x + 6y + 2z = -4 \\ 11x + 3y + 2z = 37 \\ x - 3y - 2z = 11 \end{cases}$

SOLUTION *Step 1:* Eliminate z by adding the top and bottom equations.

$$3x + 6y + 2z = -4$$

$$(+)\, \underline{\quad x - 3y - 2z = 11\quad}$$

$$4x + 3y \qquad = 7$$

Step 2: Eliminate z by adding the middle equation to the negative of the top equation.

$$11x + 3y + 2z = 37$$

$$(+)\, \underline{\quad -3x - 6y - 2z = 4\quad}$$

$$8x - 3y \qquad = 41$$

Step 3: Combine the derived equations into a system and solve:

$$4x + 3y = 7$$

$$(+)\frac{8x - 3y = 41}{12x \qquad = 48}$$

$$x = 4$$

Solve for y:

$$4x + 3y = 7$$

$$x = 4: \quad 4(4) + 3y = 7$$

$$3y = -9$$

$$y = -3$$

Step 4: Solve for z:

$$x - 3y - 2z = 11$$

$$\left.\begin{matrix} x = 4 \\ y = -3 \end{matrix}\right\} \quad (4) - 3(-3) - 2z = 11$$

$$4 + 9 - 2z = 11$$

$$-2z = -2$$

$$z = 1$$

The ordered triple, $(4, -3, 1)$, is the solution.

3. Solve: $\begin{cases} x - y = 3 & \text{(A)} \\ 2x - z = 1 & \text{(B)} \\ y + 3z = 8 & \text{(C)} \end{cases}$

SOLUTION The basic strategy is to reduce the system from one of three variables to an equivalent system of two equations in two variables.

Other selections are possible, but here we elect to solve equation (A) for y in terms of x; then substitute for y in equation (C) to produce an equation in x and z:

$$x - y = 3 \qquad \text{(A)}$$

$$y = x - 3$$

Substitute for y in equation (C).

$$y + 3z = 8 \qquad \text{(C)}$$

$$y = x - 3: \quad x - 3 + 3z = 8$$

$$x + 3z = 11 \qquad \text{(D)}$$

Now we can combine derived equation (D) and original equation (B) into a system of two linear equations in the same two variables and solve.

$$\begin{cases} 2x - z = 1 & \text{(B)} \\ x + 3z = 11 & \text{(D)} \end{cases}$$

By Program **9.3**,

$$2x - z = 1 \qquad\qquad 2x - z = 1$$
$$-2 \cdot (x + 3z = 11) \quad \overset{\Rightarrow}{(+)} \frac{-2x - 6z = -22}{-7z = -21}$$
$$z = 3$$

Substitute 3 for z in equation (B) to obtain x:

$$2x - z = 1 \qquad \text{(B)}$$
$$z = 3: \quad 2x - (3) = 1$$
$$2x = 4$$
$$x = 2$$

Substitute 3 for z in equation (C) to obtain y:

$$y + 3z = 8 \qquad \text{(C)}$$
$$z = 3: \quad y + 3(3) = 8$$
$$y + 9 = 8$$
$$y = -1$$

The solution is the ordered triple, $(2, -1, 3)$.

EXERCISE 9-6

Solve.

1. $\begin{cases} x - y + z = 2 \\ 2x + y + z = 7 \\ x - 2y + 2z = 3 \end{cases}$ 2. $\begin{cases} 2x + y + z = 1 \\ x - y - z = 2 \\ x + 2y - z = 2 \end{cases}$

3. $\begin{cases} 2x + y - 2z = 4 \\ 3x - y + 4z = 4 \\ 3x + 2y + 2z = 0 \end{cases}$ 4. $\begin{cases} x - y - 2z = 3 \\ 3x - z = -2 \\ 2x + 3y + z = 1 \end{cases}$

5. $\begin{cases} x - 3y - 3z = -2 \\ 3x - 2y + 2z = -3 \\ 2x + y - z = 5 \end{cases}$ 6. $\begin{cases} 2x - 3y + z = 3 \\ 2x - 4y + 3z = -2 \\ x - 5y - z = 1 \end{cases}$

7. $\begin{cases} x + y + z = 6 \\ 2x + 3y - z = 11 \\ 3x - y + 2z = 9 \end{cases}$ 8. $\begin{cases} 2a + 3b + c = 6 \\ a - 2b + 3c = -3 \\ 3a + b - c = 8 \end{cases}$

$$9. \begin{cases} 2x - 3y - z = 4 \\ x - 5y + z = -4 \\ 3x - 4y + 2z = -8 \end{cases}$$

$$10. \begin{cases} 3x - y + z = 7 \\ 4x - 7y + 2z = 10 \\ 2x + 3y - 3z = 1 \end{cases}$$

$$11. \begin{cases} a - 2b + c = 3 \\ 3a - b + 2c = 3 \\ 2a + b + c = 0 \end{cases}$$

$$12. \begin{cases} 3x - y + 3z = 1 \\ x + y - z = 1 \\ x - y + 2z = 1 \end{cases}$$

$$13. \begin{cases} a + b = 1 \\ b + c = 9 \\ a + c = -6 \end{cases}$$

$$14. \begin{cases} x - y = 2 \\ y + z = -1 \\ x + z = 1 \end{cases}$$

$$15. \begin{cases} x + 2z = 7 \\ 2y - 3z = -5 \\ 3x + y = 5 \end{cases}$$

$$16. \begin{cases} x + 2y = 4 \\ 2x + z = 7 \\ y - 2z = -5 \end{cases}$$

$$17. \begin{cases} \dfrac{1}{2}a - \dfrac{3}{4}b = 0 \\ 2b - \dfrac{1}{3}c = \dfrac{5}{6} \\ \dfrac{2}{3}a - \dfrac{1}{2}c = \dfrac{7}{12} \end{cases}$$

$$18. \begin{cases} \dfrac{2x}{3} - y = -\dfrac{5}{6} \\ \dfrac{2y}{5} - \dfrac{z}{2} = -\dfrac{13}{15} \\ \dfrac{4x}{3} + \dfrac{z}{4} = -\dfrac{1}{2} \end{cases}$$

$$19. \begin{cases} a - b + c - d = 2 \\ a + b - c + d = 6 \\ a - b - c - d = -2 \\ a + b + c - d = 8 \end{cases}$$

$$20. \begin{cases} 2a + b + c - d = 9 \\ a - 2b + c - d = 0 \\ a + 2b - c + d = 6 \\ a + b - 2c + d = 3 \end{cases}$$

Hint: Reduce to lower-order systems.

7. APPLICATIONS

The most difficult part in applying mathematics, usually, is translating the situation and circumstances of a problem into appropriate mathematical symbolism. Often the translating is easier when several variables are used instead of one.

To illustrate, consider this simple problem, solved first by a single equation in a single variable—as in Chapter 5—and then by a system of equations in two variables.

The sum of two numbers is 15. The larger is four times the smaller. What are the numbers?

using one equation

Let x = larger number; then $15 - x$ = smaller number.

Equation:

$$4(15 - x) = x$$

(i.e., four times the smaller number equals the larger)

Solving,

$$60 - 4x = x$$

$$60 = 5x$$

$$12 = x$$

The larger number, x, is 12; the smaller, $15 - x$, is 3.

using two equations

Let x = larger number and y = smaller number. From the statement of the problem,

(i) The sum of the two numbers is 15; so

$$x + y = 15$$

(ii) The larger number is four times the smaller; so

$$x = 4y$$

The system:

$$\begin{cases} x + y = 15 \\ x = 4y \end{cases}$$

By Program **9.2**,

$$x + y = 15$$

$$x = 4y: \quad (4y) + y = 15$$

$$5y = 15$$

$$y = 3$$

Solving for x,

$$x = 4y$$

$$y = 3: \quad x = 4(3)$$

$$x = 12$$

The two numbers are 12 and 3.

Solving a system of equations is seldom easier than solving a single equation. But translating from fact to algebra is often easier when a system of equations is used. The most critical point to remember is that as many *independent* equations as there are variables must be developed from the given data.

1. The perimeter of a rectangular lot is 370 m. Its length is 85 m greater than its width. What are the dimensions of the lot?

SOLUTION Let x = length and y = width, as in Figure 9.7.

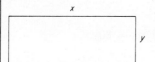

Figure 9.7

One equation (length is 85 m greater than width):

$$x = y + 85$$

Another equation (using the perimeter formula, $P = 2l + 2w$):

$$370 = 2x + 2y$$

Combining the equations,

$$\begin{cases} x - y = 85 \\ 2x + 2y = 370 \end{cases}$$

By Program **9.3**,

$$\begin{array}{cc} 2 \cdot (x - y = 85) \\ 2x + 2y = 370 \end{array} \Rightarrow \begin{array}{r} 2x - 2y = 170 \\ 2x + 2y = 370 \\ \hline (+) \quad 4x \quad\quad = 540 \\ x = 135 \end{array}$$

Solving for y,

$$x - y = 85$$

$$x = 135: \quad (135) - y = 85$$

$$50 = y$$

Result: The length is 135 m, and the width is 50 m. Check this result back in the *statement* of the problem.

2. Two children sit 10 ft apart on opposite sides of a seesaw (see Figure 9.8). One weighs 80 lb and the other 120 lb. How far from the fulcrum is each child if they balance the seesaw?

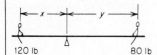

120 lb   80 lb   **Figure 9.8**

Let x = distance from fulcrum to 120-lb child and
y = distance from fulcrum to 80-lb child

One equation (distance between children is 10 ft):

$$x + y = 10$$

Another equation (moments balance):

$$120x = 80y$$

The system:

$$\begin{cases} x + y = 10 \\ 120x - 80y = 0 \end{cases}$$

By Program **9.2**,

$$x = 10 - y$$

$$120(\,10 - y\,) - 80y = 0$$

$$1200 - 120y - 80y = 0$$

$$1200 = 200y$$

$$6 = y$$

Solving for x,

$$x + y = 10$$

$$y = 6: \quad x + (6) = 10$$

$$x = 4$$

Result: The seesaw will balance when the 120-lb child sits 4 ft from the fulcrum and the 80-lb child 6 ft from the fulcrum.

3.   The sum of $4000 is invested in two funds, one drawing annual interest at 7% and the other at 9%. The combined yield is $310 per year. How much is invested at each rate?

SOLUTION Let x = amount invested at 7%, and
y = amount invested at 9%.

Then

$$x + y = 4000$$

and

$$0.07x + 0.09y = 310$$

are two independent equations in two variables based upon the given data; expressed as a system:

$$\begin{cases} x + y = 4000 \\ 0.07x + 0.09y = 310 \end{cases}$$

We may solve the system by substitution,

$$x = 4000 - y$$

$$x = 4000 - y: \quad 0.07(4000 - y) + 0.09y = 310$$

$$280 - 0.07y + 0.09y = 310$$

$$-0.07y + 0.09y = 310 - 280$$

$$0.02y = 30$$

$$y = 1500$$

Solving for x,

$$x + y = 4000$$

$$y = 1500: \quad x + 1500 = 4000$$

$$x = 2500$$

Result: $2500 is invested at 7% and $1500 at 9% to produce an annual return of $310.

4. The sum of the angle measures of every triangle in a plane is 180°. Determine the size of the angles of a triangle in which the largest angle is five times the smallest and the smallest is 40° less than the middle-sized angle.

SOLUTION Let a = largest angle, and
b = middle-sized angle, and
c = smallest angle

One equation (sum of the angles):

$$a + b + c = 180$$

Another equation (largest angle is five times the smallest):

$$a = 5c$$

A third equation (mid-sized angle is 40° larger than the smallest):

$$b = c + 40$$

The system:

$$\begin{cases} a + b + c = 180 \\ a = 5c \\ b = c + 40 \end{cases}$$

Solving the system:

(i) Substitute $5c$ for a in the top equation to derive an equation in b and c:

$$a + b + c = 180$$

$$a = 5c: \quad (5c) + b + c = 180$$

$$b + 6c = 180$$

(ii) Combine the derived equation with the bottom equation of the system into a system of two equations in two variables, b and c:

$$\begin{cases} b + 6c = 180 \\ b = c + 40 \end{cases}$$

(iii) Using Program **9.2**,

$$b + 6c = 180$$

$$b = c + 40: \quad (c + 40) + 6c = 180$$

$$40 + 7c = 180$$

$$7c = 140$$

$$c = 20$$

Solving for b,

$$b = c + 40$$

$$c = 20: \quad b = (20) + 40$$

$$b = 60$$

Solving for a,

$$a + b + c = 180$$

$$\left. \begin{array}{l} b = 60 \\ c = 20 \end{array} \right\} \quad a + (60) + (20) = 180$$

$$a = 180 - 60 - 20$$

$$a = 100$$

Result: The angles of the triangle are $100°$, $60°$, and $20°$.

EXERCISE 9-7

Use systems of equations to solve the problems.

1. The sum of two numbers is 6. Their difference is 1. What are the numbers?
2. The difference between two numbers is 0, and their sum is 32. What are the numbers?
3. The sum of two numbers is 18. Three times the smaller number is 1 less than twice the larger. What are the numbers?

4. The sum of the digits of a two-digit number is 12. The tens digit is 4 more than the ones digit. What is the number?

5. The sum of two numbers is 5. When one of the numbers is doubled and added to the other, the sum is 3. What are the numbers?

6. A rectangle is 12 m longer than it is wide and has a perimeter of 44 m. What are its dimensions?

7. A child has 50¢ in seven coins, all nickels and dimes. How many of each coin does she have?

8. Tickets for a benefit show were $2 for adults and 75¢ for children. Receipts were $66 for 58 paid admissions. How many of each type of ticket were sold?

9. The length of a rectangle exceeds its width by 8 ft, and the perimeter of the rectangle is 76 ft. Find the dimensions of the rectangle.

10. During one month the charge for 25 daily papers and 5 Sunday papers was $4. During the next month the paper bill was $3.90 for 27 daily and 4 Sunday papers. What is the unit cost for daily and Sunday papers?

11. Between two cities an 8-min phone call costs $3.15. A 10-min call costs $3.85. Find the basic rate for the first 3 minutes and the overtime rate for each additional minute.

12. A chemistry student wishes to make 100 cc of 27% acid solution by combining quantities of 20% and 30% solutions. How many cubic centimeters of each will be necessary?

13. Pete and Al work together and complete a job in 4 days. If Pete worked alone, it would take him twice the time that it would take Al, working alone, to finish the job. How long would it take each, working alone, to complete the job?

14. A fraction is equivalent to $\frac{3}{8}$. If 5 is subtracted from the numerator, it will be $\frac{1}{4}$ the denominator. Name the fraction.

15. An airplane flew 900 mi, with the wind, in 4 hr. On the return flight, against the same wind, the trip took 7.2 hr. What was the speed of the airplane in still air and the speed of the wind?

16. To mail a package requires $1.13 in postage. The sender uses 11 stamps, all 8¢ and 13¢ stamps. How many of each did she put on the package?

17. A club earned $345 by selling 200 tickets to a charity football game. The club received $1 for each general admission ticket sold and $2 for each reserved-seat ticket. How many of each type of ticket must have been sold?

18. The sum of three numbers is 3. The second is four times the first, and the sum of the first and third is −1. What are the numbers?

19. The perimeter of a triangle is 30 m. The longest side is 4 m greater than the shortest side. The shortest side is 2 m less than the next larger side. What are the dimensions of the triangle?

20. There are 13 coins in a box—pennies, nickels, and dimes. They have a combined value of 48¢. If there are four times as many pennies as nickels, how many of each coin is in the box?

21. A, B, and C have a total weight of 23 g. They balance when distributed as shown in the diagram. If A is $1\frac{1}{2}$ times as heavy as B, determine the individual weights of A, B, and C.

22. The sum of the digits of a three-digit number is 13. The sum of the ones and tens digits is 1 more than the hundreds digit, and the hundreds digit is three times the ones digit. What is the number?

8. GRAPHING SYSTEMS OF INEQUALITIES★

Solutions of inequalities has taken on new importance in recent years among the applications of mathematics. As with systems of equations, the solution of a system of inequalities may be represented by the intersection of the graphs of the inequalities that make up the system.

As we noted in Section 1 of this chapter, when the system consists of equations in two variables, the solution of the system is represented by the coordinates of a point. When the system consists of inequalities in two variables, the solution set is represented by a plane region. (Recall Section 10, Chapter 8.)

EXAMPLES

1. Solve: $\begin{cases} x + y < 4 \\ x - y > 3 \end{cases}$

SOLUTION The graph of $x + y < 4$ is shown in Figure 9.9 as a tinted region; the graph of $x - y > 3$ is shown as a shaded region. The intersection of these two regions (a shaded-tinted region) represents the solution set of the system.

★ The content of this section, in more basic courses of study, may be postponed without significant effect upon the remaining parts of the book.

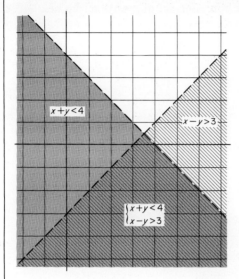

$x+y<4$

$x-y>3$

$\begin{cases} x+y<4 \\ x-y>3 \end{cases}$

Figure 9.9

2. Solve: $\begin{cases} x \ge 3 \\ x - 2y > 2 \end{cases}$

SOLUTION | The graph of $x \ge 3$ is shown in Figure 9.10 as a tinted region, including the bounding set of points, because $x \ge 3$ means $x > 3$ or $x = 3$.

The graph of $x - 2y > 2$ is shown as a shaded region, excluding the bounding set of points, because the graph of $x - 2y = 2$ is not included in that of $x - 2y > 2$.

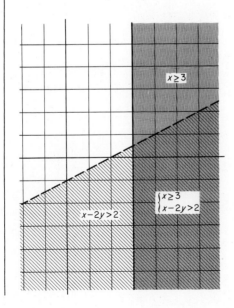

$x \ge 3$

$\begin{cases} x \ge 3 \\ x-2y>2 \end{cases}$

$x-2y>2$

Figure 9.10

The graph of the solution set of the system is that set of points in the shaded-tinted region, *including* the bounding set of points at the left, and *excluding* the bounding set of points above.

3. Solve: $\begin{cases} x - y \le 5 \\ x - y \ge 1 \end{cases}$

SOLUTION | The graphs of the inequalities of the system are shown in Figure 9.11. The respective bounding sets of points are included. The graph of the solution set of the system is the set of points in the shaded-tinted band, including the bounding points.

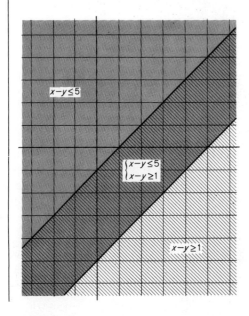

Figure 9.11

| EXERCISE 9-8

Solve the systems graphically.

1. $\begin{cases} x + y < 3 \\ x - y < 1 \end{cases}$

2. $\begin{cases} x + y < 6 \\ x - y < 2 \end{cases}$

3. $\begin{cases} 2x - y > 4 \\ x - y > 1 \end{cases}$

4. $\begin{cases} 2x + 3y \ge 10 \\ x - 3y > 5 \end{cases}$

5. $\begin{cases} 3x - 2y < 8 \\ 2x + y > 3 \end{cases}$

6. $\begin{cases} 3x - 2y > 8 \\ 2x + y < 3 \end{cases}$

7. $\begin{cases} 4x + y < 0 \\ 2x - y \le -6 \end{cases}$

8. $\begin{cases} 2x + 3y \le 5 \\ y > 3 \end{cases}$

9. $\begin{cases} x - y > 4 \\ x - y < 1 \end{cases}$

10. $\begin{cases} 2x - y \le 3 \\ 2x - y \ge 5 \end{cases}$

334 Chapter 9 SYSTEMS OF EQUATIONS AND INEQUALITIES

The solution set of each of the following systems is represented by the region common to all inequalities in the system. Graph the solution set.

11. $\begin{cases} x \geq 1 \\ y \geq 0 \\ x - y \geq 0 \end{cases}$ 12. $\begin{cases} x > 1 \\ y < 2 \\ x - y < 0 \end{cases}$ 13. $\begin{cases} x \leq 1 \\ y \leq 3 \\ 2x + 3y \leq 0 \end{cases}$ 14. $\begin{cases} x > -2 \\ y < 1 \\ 2x - y < 2 \end{cases}$

Determine the coordinates of the points of the triangular region defined by:

15. The system of inequalities of Exercise 13.

16. The system of inequalities of Exercise 14.

Write a system of inequalities to define a triangular region having the given points as vertices.

17. $(3, 3)$, $(6, 6)$, $(3, 6)$ 18. $(5, -2)$, $(-2, -2)$, $(5, 5)$

REVIEW

PART A

Answer True or False.

1. If the equations of a system have the same slope, the system will not have a single solution.

2. To solve a system of equations by graphing is to employ an approximate method.

3. A system of two linear equations is called independent when the graphs of the equations are parallel lines.

4. When the graphs of the equations of a system coincide, the system is said to be dependent.

5. If two independent equations of a system are added together term by term, the resulting equation has the same solution set as the two equations of the system.

6. In the determinant $\begin{vmatrix} a & b \\ c & d \end{vmatrix}$, the elements a and d are in the principal diagonal.

7. Solution by Cramer's Rule requires knowledge of determinants.

8. Every system of three linear equations in three variables has a single solution.

9. To solve problems using a system of equations requires as many independent equations as there are variables.

10. The solution set of a system of two inequalities in two variables consists of the solutions that each of the inequalities have in common.

PART B

Apply the slope-intercept analysis to determine if the system is independent, inconsistent, or dependent.

1. $\begin{cases} x - y = 3 \\ x - 2y = 1 \end{cases}$

2. $\begin{cases} x - y = 2 \\ x - y = 4 \end{cases}$

3. $\begin{cases} y = \dfrac{1}{2}x + 1 \\ y = \dfrac{5}{10}x + 1 \end{cases}$

4. $\begin{cases} 2x - 7y = 1 \\ 7x - 2y = 1 \end{cases}$

5. $\begin{cases} 2x - 3y = 0 \\ -3y = 0 \end{cases}$

6. $\begin{cases} y = \dfrac{3}{4}x - 2 \\ y = -\dfrac{3}{4}x + 2 \end{cases}$

7. $\begin{cases} \dfrac{x}{6} - \dfrac{y}{4} = \dfrac{1}{3} \\ -\dfrac{x}{3} + \dfrac{y}{2} = \dfrac{1}{3} \end{cases}$

8. $\begin{cases} 2y - 3 = 3x \\ 3x = 3 \end{cases}$

9. $\begin{cases} 3x - 2y = 1 \\ 2x - 3y = 1 \end{cases}$

10. $\begin{cases} 2y = 3x - 6 \\ \dfrac{x}{2} - \dfrac{y}{3} = 1 \end{cases}$

Solve the system by graphing.

11. $\begin{cases} 2x + 4y = -1 \\ 4x + 5y = 4 \end{cases}$

12. $\begin{cases} 3x + 2y = 11 \\ 2x + 3y = 4 \end{cases}$

13. $\begin{cases} 2x - 3y = 5 \\ 6y - 4x = -10 \end{cases}$

14. $\begin{cases} x - y = -2 \\ 3x - y = 1 \end{cases}$

15. $\begin{cases} 3x - 2y + 4 = 0 \\ -6x + 4y + 7 = 0 \end{cases}$

16. $\begin{cases} 11x + 2y + 6 = 0 \\ y + 3 = 0 \end{cases}$

Solve the system by either substitution or addition.

17. $\begin{cases} 4x - 3y = -5 \\ 3x - 5y = -1 \end{cases}$

18. $\begin{cases} 2x + 3y = 2 \\ 3x - y = 14 \end{cases}$

19. $\begin{cases} 4x - 7y = -32 \\ 7x + 3y = 66 \end{cases}$

20. $\begin{cases} 4x + 3y = 4 \\ 6x - 9y = -3 \end{cases}$

21. $\begin{cases} 3x + 2y + 9 = 0 \\ x + 3 = 0 \end{cases}$

22. $\begin{cases} 4x - 7y + 2 = 0 \\ 8x - 6y = 8y \end{cases}$

23. $\begin{cases} \dfrac{3}{4}x - \dfrac{2}{3}y = -2\dfrac{1}{4} \\ \dfrac{1}{2}x + \dfrac{3}{2}y = 4\dfrac{1}{3} \end{cases}$

24. $\begin{cases} \dfrac{2}{3}x + \dfrac{3}{4}y = 8 \\ \dfrac{1}{3}x + \dfrac{3}{8}y = 4 \end{cases}$

25. $\begin{cases} \dfrac{x - 3}{3} - \dfrac{y + 2}{2} = 1 \\ \dfrac{x + 3}{2} - \dfrac{y + 1}{3} = 4 \end{cases}$

26. $\begin{cases} \dfrac{x - 4}{4} + \dfrac{y + 1}{3} = -\dfrac{5}{6} \\ \dfrac{x + 2}{5} + \dfrac{y + 1}{3} = \dfrac{2}{3} \end{cases}$

27. $\begin{cases} 2x - y + z = -1 \\ x - 2y + 2z = -5 \\ x + 3y + z = 6 \end{cases}$ **28.** $\begin{cases} 2a - b + c = 9 \\ a - 2b + 3c = 15 \\ 3a + 2b + c = 5 \end{cases}$

29. $\begin{cases} 6x + 2y + 3z = 5 \\ x - 3y - z = -20 \\ 2x - 4y + 2z = -12 \end{cases}$ **30.** $\begin{cases} 4x - \dfrac{y}{2} + z = 4 \\ x + 2y - z = 17 \\ 3x + 4y - \dfrac{z}{3} = 27 \end{cases}$

31. $\begin{cases} 2y - z = 9 \\ x + 3y = 12 \\ 2x + z = -1 \end{cases}$ **32.** $\begin{cases} 2a - c = -6 \\ 3b + a = 3 \\ 4b - c = 8 \end{cases}$

Evaluate each determinant.

33. $\begin{vmatrix} 2 & 0 \\ 1 & 3 \end{vmatrix}$ **34.** $\begin{vmatrix} -2 & -3 \\ 1 & 4 \end{vmatrix}$ **35.** $\begin{vmatrix} 6 & -1 \\ -1 & 5 \end{vmatrix}$

36. $\begin{vmatrix} -9 & 0 \\ 5 & 0 \end{vmatrix}$ **37.** $\begin{vmatrix} 1 & -n \\ m & 2 \end{vmatrix}$ **38.** $\begin{vmatrix} p & -4 \\ q & 3 \end{vmatrix}$

Solve by Cramer's Rule.

39. Problem 17   **40.** Problem 12   **41.** Problem 13

42. Problem 22   **43.** Problem 5   **44.** Problem 2

45. Problem 7   **46.** Problem 8   **47.** Problem 23

48. Problem 24

49. The difference between two numbers is 3. Four times the smaller is 3 more than twice the larger. What are the numbers?

50. A fraction is equal to $\dfrac{5}{8}$. The difference between its numerator and denominator is $10\dfrac{1}{2}$. Name the fraction.

51. Admission charges for a church benefit were 50¢ for children and 80¢ for adults. Total receipts from the 160 paid admissions were $117.50. How many of each kind of ticket must have been sold?

52. A chemist wishes to produce 400 cc of 45% alcohol solution by using quantities of 70% solution and 30% solution. How much of each should he use?

53. An office manager bought two kinds of letterheads, one costing $10 per box and the other $8 per box. In all, she paid $116 for 13 boxes. How many of each kind did she buy?

54. With the river's current, a girl can swim 9 mi in 2 hr. Against the current she can do only $1\dfrac{1}{2}$ mph. What is the speed of the current?

55. The sum of the measures of the angles of a triangle is 180°. The smallest angle is half the largest, and 20° less than the other angle. What are the sizes of the angles of the triangle?

56. The sum of the digits of a three-digit integer is 9. The tens' digit is one more than the hundreds' digit, and twice the ones' digit. Name the number.

Graph the solution set for each system.

57. $\begin{cases} x + y \geq 3 \\ x - y \geq 1 \end{cases}$

58. $\begin{cases} x + y < 3 \\ x - y > 2 \end{cases}$

59. $\begin{cases} 2x + 3y < -7 \\ x - 4y > 2 \end{cases}$

60. $\begin{cases} 2x - y \leq 1 \\ x - y \geq 2 \end{cases}$

61. $\begin{cases} x - y > 3 \\ x - y < 1 \end{cases}$

62. $\begin{cases} x + y < 1 \\ x + y > 4 \end{cases}$

63. $\begin{cases} x \geq 1 \\ y \geq 3 \\ x - y \leq 0 \end{cases}$

64. $\begin{cases} x \geq 1 \\ y < 2 \\ x - y < 3 \end{cases}$

10 LOGARITHMS

1. EXPONENTS AND LOGARITHMS

Before the invention of calculators, mathematicians developed a set of "auxiliary numbers" called **logarithms** to help with the tedious work of computing products, quotients, powers, and roots. These numbers continue to be of interest today because of their many applications to science, business, and technology.

Logarithms, which are essentially exponents, have both the properties and the operational characteristics of exponents. Recall from Chapter 6 this relationship and terminology:

$$\text{exponent} \searrow$$
$$2^4 = 16$$
$$\text{base} \nearrow \qquad \nwarrow \text{equivalent}$$
$$\qquad\qquad \text{numerical value}$$

In words, "The fourth power of 2 is 16."

339

This same relationship can also be expressed in another way.

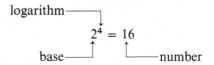

In words, "Four is the logarithm of 16 when the base is 2" or, more simply, "The log(arithm) of 16, base 2, is 4."

Following are more instances, in both forms, drawn from the table of powers of 2.

2^6	2^5	2^4	2^3	2^2	2^1	2^0	2^{-1}	2^{-2}	2^{-3}	2^{-4}
64	32	16	8	4	2	1	$\dfrac{1}{2}$	$\dfrac{1}{4}$	$\dfrac{1}{8}$	$\dfrac{1}{16}$

In exponential form	*In logarithmic form*
$2^5 = 32$	The logarithm of 32, base 2, is 5.
$2^2 = 4$	The logarithm of 4, base 2, is 2.
$2^{-3} = \dfrac{1}{8}$	The logarithm of $\dfrac{1}{8}$, base 2, is -3.

In general,

> If $a^L = n$, where a denotes any positive number other than 1, the exponent L is called the **logarithm of n to the base a**; symbolically, $\mathbf{log}_a\, \mathbf{n} = \mathbf{L}$.

For a given base, then, *the logarithm of a number is the exponent* to which the base must be raised in order to produce that number. Consequently, the two mathematical sentences

$$\log_a n = L \quad \text{and} \quad a^L = n$$

are equivalent to one another, provided that a is a positive number other than 1.

EXAMPLES

1. Write each in logarithmic form.

(a) $2^7 = 128$
(b) $3^4 = 81$
(c) $5^{-2} = 0.04$

SOLUTION (a) $\log_2 128 = 7$
(b) $\log_3 81 = 4$
(c) $\log_5 0.04 = -2$

2. Write each in exponential form.

SOLUTION (a) $\log_3 243 = 5$
(b) $\log_{10} 100 = 2$
(c) $\log_a 25 = 2$

(a) $3^5 = 243$
(b) $10^2 = 100$
(c) $a^2 = 25$

3. Which replacement for the variable will make the sentence true:
(a) $\log_4 16 = L$; (b) $\log_a 125 = 3$; (c) $\log_4 n = 0.5$?

SOLUTION (a) $\log_4 16 = L$: $4^L = 16$
                           $4^2 = 16$
                           $L = 2$
(b) $\log_a 125 = 3$: $a^3 = 125$
                           $5^3 = 125$
                           $a = 5$
(c) $\log_4 n = 0.5$: $4^{0.5} = n$
                           $4^{1/2} = n$
                           $2 = n$

EXERCISE 10-1

Express each in logarithmic form.

1. $2^3 = 8$	**2.** $10^3 = 1000$	**3.** $3^5 = 243$
4. $4^3 = 64$	**5.** $16^{1/4} = 2$	**6.** $36^{0.5} = 6$
7. $7^{-2} = \dfrac{1}{49}$	**8.** $2^{-2} = 0.25$	**9.** $3^{-1} = 0.33$
10. $27^{0.33} = 3$	**11.** $10^{-2} = \dfrac{1}{100}$	**12.** $4^{-3} = \dfrac{1}{64}$
13. $8^{-1/3} = \dfrac{1}{2}$	**14.** $37^0 = 1$	

Express each in exponential form.

15. $\log_4 64 = 3$	**16.** $\log_3 81 = 4$	**17.** $\log_{0.1} 0.001 = 3$
18. $\log_{0.5} 0.125 = 3$	**19.** $\log_8 2 = 0.33$	**20.** $\log_{10} 0.01 = -2$

21. $\log_{10} 2 = 0.301$ **22.** $\log_{10} 5 = 0.699$ **23.** $\log_{10} 17 = 1.230$

24. $\log_a 8 = 3$ **25.** $\log_2 p = 3$ **26.** $\log_{10} 17 = x$

Which replacement for the variable in each of the following will make the sentence true?

27. $\log_3 81 = L$ **28.** $\log_2 32 = L$ **29.** $\log_a 8 = 3$

30. $\log_a 16 = 2$ **31.** $\log_5 n = 2$ **32.** $\log_4 n = -2$

33. $\log_a 6 = 0.5$ **34.** $\log_4 \dfrac{1}{4} = L$ **35.** $\log_8 4 = L$

36. $\log_a 216 = 3$ **37.** $\log_5 n = 3$ **38.** $\log_3 \dfrac{1}{27} = L$

39. $\log_a 0.64 = 2$ **40.** $\log_5 625 = L$ **41.** $\log_{11} n = -2$

42. $\log_{12} n = 0.5$ **43.** $\log_b 3 = 1$ **44.** $\log_2 n = -5$

45. $\log_2 x = -4$ **46.** $\log_{17} M = 0$ **47.** $\log_a 4 = \dfrac{1}{2}$

48. $\log_x 32 = \dfrac{5}{2}$ **49.** $\log_b 8 = 1.5$ **50.** $\log_x 8 = -1.5$

2. LOGARITHM TABLE

Our Hindu–Arabic system of number notation is said to be a decimal system because it has 10 as its base number (*decem* is the Latin word for ten). Logarithms that have 10 as a base are called **common logarithms**. Systems of logarithms having other numbers as base are not only feasible but also sometimes preferable.

Because we limit our discussion in most of this chapter to common logarithms, we adopt the usual practice of dropping the base subscript (as used in the notation of the previous section) when the base is 10.

Thus "$\log n$" means "$\log_{10} n$."

Exponential form	*Logarithmic form*
$0.01 = 10^{-2}$	$\log 0.01 = -2$
$0.1 = 10^{-1}$	$\log 0.1 = -1$
$1 = 10^0$	$\log 1 = 0$
$10 = 10^1$	$\log 10 = 1$
$100 = 10^2$	$\log 100 = 2$

The equivalent exponential and logarithmic forms of several numbers in ordered sequence are displayed above. From this display we can infer that common logarithms for numbers between 1 and 10 are numbers between 0 and 1.

Such tables as Table II on pages 430 and 431, partially reproduced in Figure 10.1, provide approximate common logarithms for numbers between 1 and 10.

N	0	1	2	3	4	5	6	7	8	9
10	0.0000	0.0043	0.0086	0.0128	0.0170	0.0212	0.0253	0.0294	0.0334	0.0374
11	0.0414	0.0453	0.0492	0.0531	0.0569	0.0607	0.0645	0.0682	0.0719	0.0755
12	0.0792	0.0828	0.0865	0.0899	0.0934	0.0969	0.1004	0.1038	0.1072	0.1106
13	0.1139	0.1173	0.1206	0.1239	0.1271	0.1303	0.1335	0.1367	0.1399	0.1430
14	0.1461	0.1492	0.1523	0.1553	0.1584	0.1614	0.1644	0.1673	0.1703	0.1732
15	0.1761	0.1790	0.1818	0.1847	0.1875	0.1903	0.1931	0.1959	0.1987	0.2014
16	0.2041	0.2068	0.2095	0.2122	0.2148	0.2175	0.2201	0.2227	0.2253	0.2279
17	0.2304	0.2330	0.2355	0.2380	0.2405	0.2430	0.2455	0.2480	0.2504	0.2529
18	0.2553	0.2577	0.2601	0.2625	0.2648	0.2672	0.2695	0.2718	0.2742	0.2765
19	0.2788	0.2810	0.2833	0.2856	0.2878	0.2900	0.2923	0.2945	0.2967	0.2989
20	0.3010	0.3032	0.3054	0.3075	0.3096	0.3118	0.3139	0.3160	0.3181	0.3201
21	0.3222	0.3243	0.3263	0.3284	0.3304	0.3324	0.3345	0.3365	0.3385	0.3404
22	0.3424	0.3444	0.3464	0.3483	0.3502	0.3522	0.3541	0.3560	0.3579	0.3598
23	0.3617	0.3636	0.3655	0.3674	0.3692	0.3711	0.3729	0.3747	0.3766	0.3784
24	0.3802	0.3820	0.3838	0.3856	0.3874	0.3892	0.3909	0.3927	0.3945	0.3962
25	0.3979	0.3997	0.4014	0.4031	0.4048	0.4065	0.4082	0.4099	0.4116	0.4133
26	0.4150	0.4166	0.4183	0.4200	0.4216	0.4232	0.4249	0.4265	0.4281	0.4298

Figure 10.1

EXAMPLES

1. Find the value of log 1.92, using Table II.

SOLUTION The first two digits of 1.92, "19," identify the row in Table II (Figure 10.1); the third digit, "2," identifies the column in which the logarithm is to be found. Thus

$$\log 1.92 \approx 0.2833$$

Note: $\approx$ means "is approximately equal to."

2. Find the value of log 1.72.

SOLUTION The value of log 1.72 is found at the intersection of row 17 and column 2. Thus

$$\log 1.72 \approx 0.2355$$

3. From Figure 10.1,

$$\log 2.01 \approx 0.3032$$
$$\log 1.89 \approx 0.2765$$
$$\log 2.3 = \log 2.30 \approx 0.3617$$
$$\log 2 = \log 2.00 \approx 0.3010$$
$$\log 2.55 \approx 0.4065$$

Within our decimal system of number notation it is possible to "name" a number in a wide variety of ways. One is to express a number as a product in which one factor is *some number between 1 and 10* and the other factor is *some integral power of 10.*

For instance,

$$128 = 1.28 \times 10^2 \qquad\qquad 3 = 3 \times 10^0$$
$$3542 = 3.542 \times 10^3 \qquad\qquad 0.6 = 6 \times 10^{-1}$$
$$71{,}000 = 7.1 \times 10^4 \qquad\qquad 0.481 = 4.81 \times 10^{-1}$$
$$15 = 1.5 \times 10^1 \qquad\qquad 0.0075 = 7.5 \times 10^{-3}$$

Numbers expressed in this special factor form are said to be in *scientific notation.* (Recall Program **6.1**, page 175.)

Using scientific notation, we can extend the use of Table II to the writing of approximate logarithms for many other numbers. For example, the common logarithm for 12.8 can be produced by reasoning as follows:

$$\log 1.28 \approx 0.1072$$

or

$$1.28 \approx 10^{0.1072}$$

and

$$12.8 = 1.28 \times 10^1$$
$$\approx 10^{0.1072} \times 10^1$$
$$\approx 10^{1 + 0.1072}$$
$$\approx 10^{1.1072}$$

Thus $\log 12.8 \approx 1.1072$.

Similarly, we can develop the common logarithm for 128, using Table II and the laws of exponents.

$$128 = 1.28 \times 100 = 1.28 \times 10^2$$
$$\approx 10^{0.1072} \times 10^2$$
$$\approx 10^{2 + 0.1072}$$
$$\approx 10^{2.1072}$$

Thus $\log 128 \approx 2.1072$.

EXAMPLES

1. $2.34 \approx 10^{0.3692}$; so, $\log 2.34 \approx 0.3692$.

$23.4 = 2.34 \times 10^1 \approx 10^{0.3692} \times 10^1 = 10^{1.3692}$;

so, $\qquad \log 23.4 \approx 1.3692$.

$234 = 2.34 \times 10^2 \approx 10^{0.3692} \times 10^2 = 10^{2.3692}$;

so, $\qquad \log 234 \approx 2.3692$.

2. $1.61 \approx 10^{0.2068}$;   so, $\log 1.61 \approx 0.2068$.

$16.1 \approx 10^{1.2068}$;   so, $\log 16.1 \approx 1.2068$.

$161 \approx 10^{2.2068}$;   so, $\log 161 \approx 2.2068$.

$16,100 \approx 10^{4.2068}$;   so, $\log 16,100 \approx 4.2068$.

EXERCISE 10-2

Express the following in scientific notation.

1. 4000	**2.** 350	**3.** 1836
4. 4,832,000	**5.** 3,000,000	**6.** 4275
7. 0.83	**8.** 0.624	**9.** 0.0004
10. 0.00000074	**11.** 0.0006325	**12.** 8.42

13. If $10^{0.4} = 2.51$, what would be the proper exponent, p, for each of the following?

(a) $10^p = 25.1$ (b) $10^p = 251$ (c) $10^p = 0.251$
(d) $10^p = 0.0251$ (e) $10^p = 2510$

Use Table II to locate the proper exponent (logarithm) having 10 as the base.

14. 3.86	**15.** 7.34	**16.** 5.03	**17.** 8.66
18. 4.1	**19.** 8	**20.** 6.03	**21.** 60.3
22. 300	**23.** 3000	**24.** 682	**25.** 0.631
26. 0.0631	**27.** 0.00631		

3. CHARACTERISTIC AND MANTISSA

The numeral for a common logarithm may be thought to consist of two parts, separated by a decimal point. The part to the left of the decimal point is called the **characteristic**; the part to the right of the decimal point is called the **mantissa**. For example:

$$\overset{\text{characteristic}}{\log 213 = \underset{\text{mantissa}}{2.3284}}$$

$$\overset{\text{characteristic}}{\log 1.8 = \underset{\text{mantissa}}{0.2553}}$$

In general, for any positive number N:

- The mantissa of its common logarithm is related to the sequence of digits in the decimal numeral for N.
- The characteristic of its common logarithm is related to the location of the decimal point in the decimal numeral for N.

1. When the sequence of digits for N is the same—regardless of the location of the decimal point—the mantissa is the same:

$$2.13 \approx 10^{0.3284}; \quad \text{so,} \quad \log 2.13 \approx 0.\underline{3284}$$

$$21.3 = 2.13 \times 10 \approx 10^{0.3284} \times 10^1 = 10^{1.3284};$$

$$\text{so,} \quad \log 21.3 \approx 1.\underline{3284}$$

$$213 = 2.13 \times 10^2 \approx 10^{0.3284} \times 10^2 = 10^{2.3284};$$

$$\text{so,} \quad \log 213 \approx 2.\underline{3284}$$

$$2130 = 2.13 \times 10^3 \approx 10^{0.3284} \times 10^3 = 10^{3.3284};$$

$$\text{so,} \quad \log 2130 \approx 3.3284$$

Notice also that the characteristic is 1 less than the number of digits to the left of the decimal point in the decimal expression for N; e.g., in $\log 213 \approx 2.3284$, there are *three* digits to the left of the decimal point in 213, and the characteristic of its logarithm is *two*.

2. $\log 1.28 \approx 0.1072$
   $\log 12.8 \approx 1.1072$
   $\log 128 \approx 2.1072$
   $\log 12{,}800 \approx 4.1072$

Common logarithms for numbers between 0 and 1 follow the same pattern as that for numbers greater than 1. For example:

$$2.13 \approx 10^{0.3284}; \qquad \text{so,} \quad \log 2.13 \approx 0.3284$$

$$0.213 = 2.13 \times 10^{-1} \approx 10^{-1+0.3284}; \quad \text{so,} \quad \log 0.213 \approx -1 + 0.3284$$

$$0.0213 = 2.13 \times 10^{-2} \approx 10^{-2+0.3284}; \quad \text{so,} \quad \log 0.0213 \approx -2 + 0.3284$$

$$0.00213 = 2.13 \times 10^{-3} \approx 10^{-3+0.3284}; \quad \text{so,} \quad \log 0.00213 \approx -3 + 0.3284$$

We could write

$$\log 0.213 \text{ as } -0.6716 \, (= -1 + 0.3284)$$

$$\log 0.0213 \text{ as } -1.6716 \, (= -2 + 0.3284)$$

$$\log 0.00213 \text{ as } -2.6716 \, (= -3 + 0.3284)$$

and so on. But that makes pencil-and-paper computation with logarithms cumbersome. There are advantages to expressing these negative characteristics as binomials—the difference between some number and a multiple of ten.

For instance, the negative characteristic -1 may be expressed equivalently as $9 - 10$, the characteristic -2 as $8 - 10$, -3 as $7 - 10$, and so on. The characteristic of a very small number might be -13, which may be written as $7 - 20$. The negative logarithms displayed in the previous paragraph, then, may also be expressed as follows:

$-1 + 0.3284$ as $9.3284 - 10$ (which is the equivalent of -0.6716)

$-2 + 0.3284$ as $8.3284 - 10$ (which is the equivalent of -1.6716)

$-3 + 0.3284$ as $7.3284 - 10$ (which is the equivalent of -2.6716)

and we may write

$$\log 0.213 \approx 9.3284 - 10$$

$$\log 0.0213 \approx 8.3284 - 10$$

$$\log 0.00213 \approx 7.3284 - 10$$

$$\log 0.000213 \approx 6.3284 - 10$$

By expressing logarithms in this way, the same mantissa will correspond to the same sequence of digits in the number's decimal numeral, regardless of where the decimal point happens to occur. This binomial alternative has great utility when tables of logarithms are used, as we do here. However, for those using computers or calculators with logarithmic tables built in, the binomial notation is unnecessary.

PROGRAM 10.1

To determine the common logarithm for a number whose decimal contains three significant digits*:

Step 1: Express the number in scientific notation; the exponent to the 10-factor is the characteristic.

Step 2: Locate the mantissa of the desired logarithm in Table II at the intersection of the *row* corresponding to the first two significant digits of the decimal and the *column* corresponding to its third digit.

Step 3: Combine the characteristic and mantissa found in Steps 1 and 2 for the desired logarithm.

EXAMPLES

1. Determine the common logarithm for 17.9.

SOLUTION

Step 1: 17.9, in scientific notation, is 1.79×10^1. The characteristic is 1.

Step 2: In 17.9, the first two significant digits are 17, the third, 9. The mantissa will be found in Table II at the intersection of the row labeled 17 and the column headed by 9. Thus the desired mantissa is 0.2529.

Step 3: The common logarithm for 17.9 is 1.2529.

* It may be useful to review the footnote on page 175 relating to significant digits.

2. Determine the common logarithm for 0.00234.

SOLUTION   *Step 1:* 0.00234, in scientific notation, is 2.34×10^{-3}. The characteristic is -3, or $7 - 10$.

Step 2: In 0.00234, the first two significant digits are 23, the third, 4. The mantissa will be found in Table II at the intersection of row 23 and column 4: 0.3692.

Step 3: The common logarithm for 0.00234 is $7.3692 - 10$.

3. Determine the common log for 0.021.

SOLUTION   *Step 1:* $0.021 = 2.1 \times 10^{-2}$; characteristic is -2, or $8 - 10$.

Step 2: In 0.021, the first two significant digits are 21, the third is assumed to be 0. The mantissa is found at the intersection of row 21 and column 0: 0.3222.

Step 3: $\log 0.021 \approx 8.3222 - 10$.

4. Determine the common log for 102,000.

SOLUTION   *Step 1:* $102,000 = 1.02 \times 10^5$; characteristic: 5.

Step 2: Mantissa at row 10, column 2: 0.0086.

Step 3: $\log 102,000 \approx 5.0086$.

| **EXERCISE 10-3**

Use Table II to determine the common logarithm for each of the following.

1. 321	**2.** 847	**3.** 63	**4.** 807
5. 42	**6.** 5.6	**7.** 0.37	**8.** 8.42
9. 5000	**10.** 0.031	**11.** 0.0052	**12.** 0.0257
13. 100	**14.** 797	**15.** 0.999	**16.** 0.001
17. 3040	**18.** 0.0502	**19.** 162,000	**20.** 5,000,000
21. 0.00032	**22.** 0.00427	**23.** 5.0600	**24.** 0.00001

4. ANTILOGARITHMS

Determining the logarithm of a number—that is, "Given N, find $\log N$"—can be diagrammed as

$$N \xrightarrow[\text{Table II}]{\text{via}} \log N$$

The inverse of this operation

$$N \xleftarrow[\text{Table II}]{\text{via}} \log N$$

which is to say, "Given log N, find N," is referred to as determining the **anti-logarithm**. Briefly, if the logarithm of N is L, then N is the *antilogarithm* of L.

One procedure for determining the antilogarithm of a specific logarithm is to reverse Program **10.1** (To find the logarithm of a number). Say that we wish to determine the antilogarithm (or "antilog," for short) for 2.1818. We start with the last step, Step 3, of Program **10.1** and identify the characteristic and mantissa:

$$2.\underbrace{1818}$$

characteristic ————┘ └———— mantissa

Next, go to Step 2 of the program and locate the mantissa in Table II at the intersection of row 15 and column 2. The mantissa 0.1818, then, represents the number 1.52.

Finally, go to Step 1 and recognize that the characteristic, 2, represents the exponent of the 10-factor when the number that we seek is expressed in scientific notation—that is, 10^2.

Thus the number that has 2.1818 for a common logarithm is

$$1.52 \times 10^2 = 152$$

The following program yields the same result, as is demonstrated in Example 1, but is more efficient. It bypasses the scientific notation stage and goes directly to the standard decimal for the antilogarithm.

PROGRAM 10.2

To determine an antilogarithm, using a four-place table of mantissas:

Step 1:   Locate the mantissa of the logarithm in the table. The first two significant digits of the desired antilogarithm will be those of the row containing the mantissa, and the third significant digit will be that of the column containing the mantissa.

Step 2:   Place the decimal point temporarily (in the form of a caret, $\wedge$) after the first significant digit in the antilogarithm; then

(a) If the characteristic of the logarithm is positive, move the decimal point in the antilogarithm the same number of places to the right as given by the characteristic. (It may be necessary to annex zeros.)

(b) If the characteristic of the logarithm is negative, move the decimal point in the antilogarithm the same number of places to the left as given by the absolute value of the characteristic. (It may be necessary to annex zeros.)

1. Determine the antilogarithm of 2.1818.

SOLUTION | *Step 1:* The mantissa is 0.1818. In the table (see Figure 10.1), 0.1818 is in row 15 and column 2. The sequence of significant digits in the antilog of 2.1818 is 152.

Step 2: Temporarily place the decimal point ($_\wedge$) after the first digit: 1$_\wedge$52. The characteristic of the logarithm under consideration is positive: 2. Move the decimal point two places to the right: 1$_\wedge$5 2. The antilogarithm of 2.1818 is 152.

2. Determine the antilogarithm of 8.2405 − 10.

SOLUTION | *Step 1:* The mantissa is 0.2405. In the table (Figure 10.1) it is in row 17, column 4. The sequence of significant digits in the desired antilogarithm is 174.

Step 2: Temporarily place the decimal point ($_\wedge$) after the first significant digit: 1$_\wedge$74. The characteristic is negative: (8 − 10); its absolute value is $|(8 − 10)| = |−2| = 2$. Move the decimal point two places to the left; it will be necessary to annex one zero: 01$_\wedge$74. The antilogarithm of 8.2405 − 10 is 0.0174.

Note: With the scientific notation stage included, log $N = 8.2405 − 10$.
Mantissa: 0.2405; equivalent to 1.74.
Characteristic: 8 − 10; equivalent to −2.
Scientific notation: $N = 1.74 \times 10^{-2} = 0.0174$.

3. Determine the number whose logarithm is 4.0792.

SOLUTION | *Step 1:* Mantissa 0.0792 is in row 12, column 0. The sequence of significant digits will be 120.

Step 2: The characteristic is 4. The decimal point is located four places to the right after the first digit: 12000. The number whose logarithm is 4.0792 is 12,000.

4. Determine the number whose logarithm is 8.2833 − 20.

SOLUTION | *Step 1:* Mantissa 0.2833 is in row 19, column 2. The sequence of significant digits will be 192.

Step 2: The characteristic is negative, 8 − 20; its absolute value is 12. The decimal point is located 11 places before the first significant digit; 0.00000000000192 is the number whose logarithm is 8.2833 − 20.

Note: In such cases, it is preferable to express the number in scientific notation:

$$0.00000000000192 = 1.92 \times 10^{-12}.$$

Use Table II to determine the antilogarithm.

1. 1.4150	**2.** 0.3096	**3.** 2.5315
4. 9.5658 − 10	**5.** 8.6010 − 10	**6.** 9.6474 − 10
7. 4.8710	**8.** 3.8825	**9.** 9.8287 − 10
10. 7.9009 − 10	**11.** 2.2253	**12.** 0.0043
13. 8.7966 − 10	**14.** 6.9335 − 10	**15.** 5.3160
16. 0.5705	**17.** 9.6758 − 10	**18.** 6.9542
19. 3.9015	**20.** 0.8041	**21.** 1.7396
22. 8.0934 − 20	**23.** 2.3560 − 10	**24.** 9.4099 − 10

5. INTERPOLATION

Up to this point the examples and exercises have been selected to fit precisely the mantissa entries of Table II. In practice that will not always be the case. If we want the mantissa for a four-digit number, or the antilogarithm of a logarithm whose mantissa is not found directly in the table, we may elect one of several alternatives:

1. Take the nearest entry in the tables as sufficiently approximate, or
2. Consult a more extensive table of mantissas, or
3. Use a process called **interpolation**—the computation of an intermediate number between two stated or tabulated numbers.

Because the logarithmic function is nonlinear, our linear interpolation method will be approximate. The error is minimal, however, when the interval between the entries is as small as those we will be using.

Basic to interpolation as it applies to logarithms is the fact that

---

If a, b, c are positive numbers and a < b < c, (i.e., b is between a and c), then

$$\log a < \log b < \log c$$

From this property and what we already know about mantissas of logarithms, it follows that if the logarithms of *a*, *b*, and *c* are ordered in a certain way, then so too are their respective mantissas. To illustrate, from Table II we may note that

$$\log 32.3 \approx 1.5092$$

$$\log 32.4 \approx 1.5105$$

$$\log 32.5 \approx 1.5119$$

Clearly,

$$32.3 < 32.4 < 32.5$$

and

$$\log 32.3 < \log 32.4 < \log 32.5$$

because

$$1.5092 < 1.5105 < 1.5119$$

and the respective mantissas have the same order:

$$0.5092 < 0.5105 < 0.5119$$

Suppose now that we wish to determine the logarithm of 5.176, a four-digit number. As noted at the beginning of this section, we could elect to use the nearest mantissa in Table II. That would be 0.7143, the mantissa for 5.18. Or we may elect to consult a five-place table of mantissas in which we would find 0.71391 to be the mantissa for 5.176.

Or we may elect to use Table II, a four-place table of mantissas, and interpolate as follows.

Identify in Table II the mantissas for 5.170 and 5.180 (which are the same as the mantissas for 5.17 and 5.18). In keeping with the generalization above about the order of numbers, their respective logarithms and mantissas, we know that the mantissa for 5.176 will be a number between the mantissas for 5.170 and 5.180, because $5.170 < 5.176 < 5.180$.

The following display summarizes the data. The letter c represents a "correction" that is to be added to the smaller tabulated mantissa to produce the desired mantissa, m.

$$\begin{array}{cc} \underline{N} & \text{Mantissa} \\ \end{array}$$

$$0.010 \left[0.006 \left[\begin{array}{cc} 5.170 & 0.7135 \text{ (listed)} \\ 5.176 & m \end{array} \right] c \right] 0.0008$$

$$\begin{array}{cc} 5.180 & 0.7143 \text{ (listed)} \end{array}$$

The differences, 0.006 and 0.010 in the display, are presumed to be in the same ratio as the differences c and 0.0008. Equating the two ratios creates a proportion

$$\frac{c}{0.0008} = \frac{0.006}{0.010}$$

which in turn leads to the equation

$$(0.010)(c) = (0.006)(0.0008)$$

which can be solved:

$$c = \frac{(0.006)(0.0008)}{0.010}$$

$$= 0.00048$$

$$\approx 0.0005 \quad \text{(rounded to four places)}$$

Because we are working with a four-place table of mantissas, c is rounded to four places: 0.0005. That number when added to the smaller tabulated mantissa yields the desired interpolated mantissa. Thus

$$m \approx 0.7135 + c$$

$$\approx 0.7135 + 0.0005$$

$$\approx 0.7140$$

Consequently, by Program **10.1**,

$$\log 5.175 \approx \underset{\text{characteristic}}{0}.\underset{\text{mantissa}}{7140}$$

Note: The decimal burden in computing the correction, c, can be eased somewhat by simplifying the ratio on the right to a fraction with a denominator of 1. In this case

$$\frac{c}{0.0008} = \frac{0.006}{0.010} = \frac{0.6}{1}$$

Then

$$(c)(1) = (0.0008)(0.6)$$

$$= 0.00048$$

EXAMPLES

1. Use Table II to write the logarithm for 74.43.

SOLUTION | According to Step 1, Program **10.1**: 74.43 in scientific notation is 7.443×10^1; the characteristic is 1. The mantissa: Table II lists mantissas for 7.44 and 7.45 (i.e., 7.440 and 7.450), but not for 7.443. We may interpolate to determine the mantissa for 7.443.

$$0.010 \left[0.003 \begin{bmatrix} \underline{N} & \text{Mantissa} \\ 7.440 & 0.8716 \\ 7.443 & m \end{bmatrix} c \right] 0.0006$$
$$7.450 \quad 0.8722$$

We compute c by solving the proportion

$$\frac{c}{0.0006} = \frac{0.003}{0.010} = \frac{0.3}{1}$$

$$(1)(c) = (0.3)(0.0006)$$

$$c = 0.00018$$

$$\approx 0.0002 \quad \text{(rounded to four places)}$$

Adding the correction to the smaller mantissa, we obtain the mantissa, m, for 74.43:

$$m \approx 0.8716 + c$$

$$\approx 0.8716 + 0.0002$$

$$\approx 0.8718$$

Finally, combining the characteristic, 1, with the interpolated mantissa, we have

$$\log 74.43 \approx 1.8718$$

2. Interpolate to find the logarithm for 166.4.

SOLUTION $166.4 = 1.664 \times 10^2$; characteristic 2. The mantissa is interpolated from Table II:

$$\begin{array}{cc} N & \text{Mantissa} \\ \hline \end{array}$$

$$0.010 \begin{bmatrix} 0.004 \begin{bmatrix} 1.660 & 0.2201 \\ 1.664 & m \end{bmatrix} c \\ 1.670 & 0.2227 \end{bmatrix} 0.0026$$

The proportion:

$$\frac{c}{0.0026} = \frac{0.004}{0.010} = \frac{0.4}{1}$$

$$c = (0.4)(0.0026)$$

$$= 0.00104$$

$$\approx 0.0010 \quad \text{(rounded to four places)}$$

The interpolated mantissa for 1.664 is

$$m \approx \text{mantissa for } 1.660 + c$$

$$\approx 0.2201 + 0.0010$$

$$\approx 0.2211$$

Combining the characteristic (2) and mantissa (0.2211), we have

$$\log 166.4 \approx 2.2211$$

3. Write the common logarithm for 0.02347.

SOLUTION $0.02347 = 2.347 \times 10^{-2}$; characteristic -2. Mantissa:

$$\begin{array}{cc} N & \text{Mantissa} \\ \hline \end{array}$$

$$0.010 \begin{bmatrix} 0.007 \begin{bmatrix} 2.340 & 0.3692 \\ 2.347 & m \end{bmatrix} c \\ 2.350 & 0.3711 \end{bmatrix} 0.0019$$

Computing the correction, c:

$$\frac{c}{0.0019} = \frac{0.007}{0.010} = \frac{0.7}{1}$$

$$c = (0.7)(0.0019)$$

$$= 0.00133$$

$$\approx 0.0013$$

The mantissa, m, is the mantissa for 2.340 plus the correction, c:

$$m \approx 0.3692 + 0.0013$$

$$\approx 0.3705$$

Combining mantissa (0.3705) and characteristic (-2 or $8 - 10$), we obtain

$$\log 0.02347 \approx 8.3705 - 10$$

As we shall see in the next several sections, when we compute with logarithms we will often need to find the antilogarithm of a logarithm having an unlisted mantissa. Again, we might elect to use the nearest tabulated mantissa, or consult a more extensive table of mantissas, or improve the accuracy of the four-place table by interpolation. The following relates to the latter alternative.

Suppose that we wished to determine the antilog of 2.1538; that is, given $\log n = 2.1538$, determine n.

The mantissa, 0.1538, is not listed in Table II. The two listed mantissas nearest to 0.1538 are 0.1523 and 0.1553, which have 1.42 and 1.43, respectively, for antilogarithms.

$$0.01 \begin{bmatrix} c \begin{bmatrix} 1.42 & 0.1523 \text{ (listed)} \\ n & 0.1538 \text{ (unlisted)} \end{bmatrix} 0.0015 \\ 1.43 & 0.1553 \text{ (listed)} \end{bmatrix} 0.0030$$

$$\underline{N} \qquad \underline{\text{Mantissa}}$$

In the display above, n represents the antilogarithm for the unlisted mantissa, and can be obtained by adding the "correction," c, to the smaller listed antilogarithm.

Using proportion to compute c, we have

$$\frac{c}{0.01} = \frac{0.0015}{0.0030} = \frac{15}{30} = \frac{1}{2}$$

$$c = \frac{1}{2}(0.01)$$

$$= 0.005$$

Adding c to the smaller listed antilogarithm yields n, the interpolated antilogarithm for 0.1538:

$$n \approx 1.42 + c$$

$$\approx 1.42 + 0.005$$

$$\approx 1.425$$

Thus "1425" is the sequence of digits for all antilogarithms that have 0.1538 as mantissa. In particular, the antilogarithm of 2.1538 (the characteristic is 2) is

$$1.425 \times 10^2 = 142.5$$

Or, in terms of Step 2(a), Program **10.2**:

$$1 \; 4 \; 2 \, . \, 5$$

EXAMPLES

1. Determine the antilogarithm of 1.2344.

SOLUTION | The mantissa, 0.2344, is not listed in Table II. The two nearest listed mantissas with their corresponding antilogarithms are

$$
\begin{array}{cc}
\underline{N} & \text{Mantissa} \\
\end{array}
$$

$$
0.01 \left[c \begin{bmatrix} 1.71 & 0.2330 \text{ (listed)} \\ n & 0.2344 \text{ (unlisted)} \\ 1.72 & 0.2355 \text{ (listed)} \end{bmatrix} 0.0014 \right] 0.0025
$$

First we compute c:

$$\frac{c}{0.01} = \frac{0.0014}{0.0025} = \frac{14}{25}$$

$$c = \frac{(0.01)(14)}{25}$$

$$= 0.0056$$

$$\approx 0.006 \quad \text{(rounded to one significant digit)}$$

Then use c to determine n:

$$n \approx 1.71 + c$$

$$\approx 1.71 + 0.006$$

$$\approx 1.716$$

The digit sequence for logarithms having 0.2344 as mantissa is "1716." Therefore, the antilogarithm of 1.2344 (characteristic 1, mantissa 0.2344) is $1 \; 7 \; 1$ 6, or 17.16.

2. Determine the antilogarithm for $7.5722 - 10$.

SOLUTION The mantissa is 0.5722. From Table II:

$$
\begin{array}{cc}
N & \text{Mantissa} \\
\end{array}
$$

$$
0.01 \left[c \left[\begin{array}{cc} 3.73 & 0.5717 \text{ (listed)} \\ n & 0.5722 \text{ (unlisted)} \end{array} \right] 0.0005 \right] 0.0012
$$

$$
\begin{array}{c}
3.74 \quad 0.5729 \text{ (listed)}
\end{array}
$$

$$
\frac{c}{0.01} = \frac{0.0005}{0.0012} = \frac{5}{12}
$$

$$
c = (0.01)\left(\frac{5}{12}\right)
$$

$$
= 0.004167
$$

$$
\approx 0.004 \quad \text{(rounded to one significant digit)}
$$

$$
n \approx 3.73 + c
$$

$$
\approx 3.73 + 0.004
$$

$$
\approx 3.734
$$

The digit sequence for the antilogarithm of $7.5722 - 10$ is "3743." The characteristic is $7 - 10$, or -3. Thus

$$
0\ 0\ 3_{\wedge}7\ 3\ 4
$$

and the antilogarithm for $7.5722 - 10$ is 0.003734.

3. Find n if $\log n = 8.6680$.

SOLUTION From Table II:

$$
\begin{array}{cc}
N & \text{Mantissa} \\
\end{array}
$$

$$
0.01 \left[c \left[\begin{array}{cc} 4.65 & 0.6675 \\ n & 0.6680 \end{array} \right] 0.0005 \right] 0.0009
$$

$$
\begin{array}{c}
4.66 \quad 0.6684
\end{array}
$$

$$
\frac{c}{0.01} = \frac{0.0005}{0.0009} = \frac{5}{9}
$$

$$
c = (0.01)\left(\frac{5}{9}\right)
$$

$$
c = 0.0056
$$

$$
\approx 0.006 \quad \text{(rounded, one significant digit)}
$$

The digit sequence for numbers having 0.6680 as mantissa is "4656." In particular, if $\log n = 8.6680$, then $n = 4.656 \times 10^8$.

EXERCISE 10-5

Determine the common logarithm for each.

1. 4.625	**2.** 0.2074	**3.** 74.37
4. 0.05603	**5.** 523.4	**6.** 67.63
7. 0.006403	**8.** 7855	**9.** 424,700
10. 0.07523	**11.** 0.9771	**12.** 8.794×10^{-6}

Determine the antilogarithm for each.

13. 0.6174	**14.** $9.6558 - 10$	**15.** 2.9104
16. $8.4096 - 10$	**17.** 4.6844	**18.** 2.5012
19. 1.8604	**20.** $9.8304 - 10$	**21.** $7.2307 - 10$
22. $3.5605 - 10$	**23.** 5.9348	**24.** 9.7751

6. PRODUCTS AND QUOTIENTS BY LOGARITHMS

Paralleling the Laws of Exponents (Section 6, Chapter 6) are comparable Laws of Logarithms. In this section we discuss two of the Laws by which we may use logarithms to compute products and quotients.

Recall Law of Exponents I:

$$a^m \cdot a^n = a^{m+n}$$

and note the similarity to

Logarithm Law I

The logarithm of a product of positive numbers is equal to the sum of the logarithms of those numbers.

Consider the product $A \cdot B$.

Let $\log A = x$ and $\log B = y$.

Then $A = 10^x$ and $B = 10^y$ and $A \cdot B = 10^{x+y}$.

Thus

$$\log (A \cdot B) = x + y \quad \text{or} \quad \log (A \cdot B) = \log A + \log B$$

By extension,

$$\log (A \cdot B \cdot C \cdot \cdots \cdot N) = \log A + \log B + \log C + \cdots + \log N$$

PROGRAM 10.3

To compute the product of several positive numbers by logarithms:

Step 1: Determine the logarithm for each of the numbers.

Step 2: Add the logarithms of Step 1.

Step 3: Determine the antilogarithm of the result of Step 2 for the desired product.

EXAMPLES

1. Compute the product 3.62×47.9 by logarithms.

SOLUTION *Step 1:* $\log 3.62 \approx 0.5587$; $\log 47.9 \approx 1.6803$

Step 2:

$$
\begin{array}{rr}
\log 3.62 \approx & 0.5587 \\
\log 47.9 \approx & 1.6803 \\
\hline
\log (3.62 \times 47.9) \approx & 2.2390
\end{array}
$$

$(+)$

Step 3: The antilogarithm of 2.2390 is 173.4. Therefore, 3.62×47.9 is 173.4 (approximately).

2. Find the product of 0.00732×0.0167.

SOLUTION $\log (0.00732 \times 0.0167) = \log 0.00732 + \log 0.0167$

$$
\begin{array}{rl}
\log 0.00732 \approx & 7.8645 - 10 \\
\log 0.0167 \approx & 8.2227 - 10 \\
\hline
\log (0.00732 \times 0.0167) \approx & 16.0872 - 20 = 6.0872 - 10
\end{array}
$$

$(+)$

$$\text{antilog of } 6.0872 - 10 \approx 0.0001222$$

Thus

$$0.00732 \times 0.0167 \approx 0.0001222$$

SOLUTION **3.** Find the product of $(3.624) \times (-3.71) \times (0.0046)$ by logarithms.

Our definition of logarithm precludes logarithms of negative numbers. However, we can use logarithms to compute a product involving one or more negative numbers. We do so by computing the product of the factors as though they were all positive numbers, and then make the product negative if the number of negative factors involved is odd or positive if the number of negative factors is even.

$$(3.624) \times (-3.71) \times (0.0046) = -[(3.624) \times (3.71) \times (0.0046)]$$

$$\log [3.624 \times 3.71 \times 0.0046] = \log 3.624 + \log 3.71 + \log 0.0046$$

$$
\begin{array}{rl}
\log 3.624 \approx & 0.5592 \\
\log 3.71 \approx & 0.5694 \\
\log 0.0046 \approx & 7.6628 - 10 \\
\log \text{product} \approx & \!\!\!(+)\overline{8.7914 - 10}
\end{array}
$$

$$\text{antilog } 8.7914 - 10 \approx 0.06186$$

Thus

$$(3.624)(-3.71)(0.0046) \approx -0.06186$$

Logarithm Law II relates to quotients of numbers, and it parallels Law of Exponents II:

$$a^m \div a^n = a^{m-n}$$

Logarithm Law II
The logarithm of a quotient of two positive numbers is equal to the logarithm of the dividend less the logarithm of the divisor.

Consider the quotient $A \div B$.
Let $\log A = x$ and $\log B = y$.
Then $A = 10^x$ and $B = 10^y$ and $A \div B = 10^x \div 10^y = 10^{x-y}$.
Thus

$$\log (A \div B) = x - y \quad \text{or} \quad \log (A \div B) = \log A - \log B$$

PROGRAM 10.4

To compute the quotient of two positive numbers by logarithms:

Step 1: Determine the logarithms for the dividend and divisor.

Step 2: Subtract the logarithm of the divisor from the logarithm of the dividend.

Step 3: Determine the antilogarithm of the result of Step 2 for the desired quotient.

EXAMPLES

1. Determine the quotient, $86.3 \div 0.427$, by logarithms.

SOLUTION   *Step 1:* log 86.3 $\approx$ 1.9360   (dividend)

log 0.427 $\approx$ 9.6304 − 10   (divisor).

Step 2: Subtract (here it is necessary to adjust the log of the dividend, 1.9360, to an equivalent 11.9360 − 10):

$$\begin{array}{r} \log 86.3 \approx \quad 11.9360 - 10 \\ \log 0.427 \approx \quad 9.6304 - 10 \\ \hline \log (86.3 \div 0.427) \approx \;^{(-)}\; \overline{2.3056} \end{array}$$

Step 3: Antilog of 2.3056 $\approx$ 202.1. Therefore $86.3 \div 0.427 \approx 202.1$.

2. Determine the quotient $0.00426 \div 0.0798$.

SOLUTION   log $(0.00426 \div 0.0798)$ = log 0.00426 − log 0.0798

$$\begin{array}{rll} \log 0.00426 \approx & 7.6294 - 10 = & 17.6294 - 20 \\ \log 0.0798 \approx \;^{(-)}& 8.9020 - 10 = \;^{(-)}& 8.9020 - 10 \\ \hline \log \text{quotient} \approx & & 8.7274 - 10 \end{array}$$

antilog 8.7274 − 10 $\approx$ 0.05339

Thus

$$0.00426 \div 0.0798 \approx 0.05339$$

3. Compute by logarithms: $\dfrac{463 \times 2.18}{0.674}$.

SOLUTION $\log \dfrac{463 \times 2.18}{0.674} = \log (463 \times 2.18) - \log 0.674$

$$= \log 463 + \log 2.18 - \log 0.674$$

$$\begin{array}{rll} \log 463 \approx & 2.6656 \\ \log 2.18 \approx \;^{(+)}& 0.3385 \\ \hline \log \text{numerator} \approx & 3.0041 & = \quad 13.0041 - 10 \\ \log 0.674 \approx \;^{(-)}& 9.8287 - 10 = \;^{(-)}& 9.8287 - 10 \\ \hline \log \dfrac{463 \times 2.18}{0.674} \approx & & 3.1754 \end{array}$$

antilog 3.1754 $\approx$ 1498

Therefore

$$\frac{463 \times 2.18}{0.674} \approx 1498$$

Use logarithms to compute.

1. 14 × 36   **2.** 17 × 360   **3.** 325 × 1.6

4. 428 × 0.632   **5.** 0.637 × 0.142   **6.** 0.0032 × 6.78

7. 0.0243 × 0.725   **8.** (−3.62) × (−0.043)

9. 526 × 0.000327   **10.** 426.5 × 817   **11.** 3.624 × 67.85

12. 0.6387 × 0.4321   **13.** (−6.328) × (4.26)   **14.** 0.836 × 0.09799

15. 327 × 491 × 632   **16.** 524 × 0.623 × 0.042

17. (−4.32) × (8.43) × (−0.6745)   **18.** 0.0326 × 0.0478 × 0.832

19. 3.246 × 49.37 × 0.03265 × 48.21

20. 0.004263 × 0.07327 × 0.0004263 × 0.0003271

21. 448 ÷ 56   **22.** 36.2 ÷ 1.78   **23.** 4.29 ÷ 2.63

24. 8360 ÷ 51.7   **25.** 67.3 ÷ 205   **26.** (−8.37) ÷ (54.3)

27. 0.427 ÷ 0.372   **28.** 0.62 ÷ 0.078   **29.** 0.426 ÷ 1.63

30. (−527) ÷ (−0.0263)   **31.** 6.87 ÷ 6.93

32. 0.0032 ÷ 0.0627   **33.** (−8.037) ÷ (0.426)   **34.** 84.2 ÷ 5936

35. 1 ÷ 0.8625   **36.** 1 ÷ 4.284   **37.** 42,880 ÷ 67.3

38. $\dfrac{32.6 \times 42.9}{5.26}$ **39.** $\dfrac{(0.0632) \times (-0.0427)}{(-63.2)}$

40. $\dfrac{32.7 \times 4.635}{37.93 \times 5.26}$

7. POWERS AND ROOTS BY LOGARITHMS

Logarithm Law III relates to powers, and it parallels Law of Exponents III:

$$(a^n)^m = a^{(m)(n)}$$

Logarithm Law III

The logarithm of a power of a positive number is equal to the product of the exponent and the logarithm of the number.

Consider A^p.

Let $\log A = x$.

Then $A = 10^x$ and $A^p = (10^x)^p = 10^{px}$.

Thus

$$\log (A^p) = px \quad \text{or} \quad \log (A^p) = p \log A$$

PROGRAM 10.5

To compute the expansion of a given power of a positive number by logarithms:

Step 1: Determine the logarithm for the number to be raised to the given power.

Step 2: Multiply the logarithm of Step 1 by the exponent of the given power.

Step 3: Determine the antilogarithm of the result of Step 2 for the desired expansion.

EXAMPLES

1. Compute the expansion of $(8.62)^3$.

SOLUTION

Step 1: log 8.62 ≈ 0.9355

Step 2: Multiply log 8.62 by the exponent 3:

$$\begin{array}{r} \log 8.62 \approx \quad 0.9355 \\ (\times)\ \underline{\qquad 3} \\ \log (8.62)^3 \approx \quad 2.8065 \end{array}$$

Step 3: Antilog 2.8065 ≈ 640.4. Therefore, $(8.62)^3$ ≈ 640.4.

Note: By arithmetic, $(8.62)^3 = 640.503928$. Mantissas are rounded numbers, and rounding errors are magnified when raising to higher powers.

2. Expand $(0.073)^6$.

SOLUTION

$\log (0.073)^6 = 6 \log (0.073)$

$$\begin{array}{r} \log 0.073 \approx \quad 8.8633 - 10 \\ (\times)\ \underline{\qquad\qquad 6} \\ \log (0.073)^6 \approx \quad 53.1798 - 60 = 3.1798 - 10 \end{array}$$

$$\text{antilog } 3.1798 - 10 \approx 0.0000001513$$

Therefore,

$$(0.073)^6 \approx 0.0000001513, \quad \text{or} \quad 1.513 \times 10^{-7}$$

3. Expand $(-3.62)^5$.

SOLUTION

$(-3.62)^5 = -(3.62)^5$

(The number of factors, 5, is odd; the expansion is a negative number.)

$\log (3.62)^5 = 5 \log 3.62$

$\log 3.62 \approx \quad 0.5587$

$\log (3.62)^5 \approx (\times) \dfrac{5}{2.7935}$

$\text{antilog } 2.7935 \approx 621.6$

Thus

$$(-3.62)^5 \approx -621.6$$

In an exponential sense, roots and powers are essentially the same. That is, $\sqrt[n]{a}$ (the nth root of a) is the same as $a^{1/n}$. Thus Logarithm Law III applies equally for raising to a power and extracting a root.

EXAMPLES

1. Compute $\sqrt[5]{27.4}$.

SOLUTION $\quad \sqrt[5]{27.4} = (27.4)^{1/5}$

Using Program **10.5**:

Step 1: $\log 27.4 \approx 1.4378$

Step 2: $\left(\dfrac{1}{5}\right)(\log 27.4) \approx \dfrac{1.4378}{5} \approx 0.2876$

Step 3: Antilog $0.2876 \approx 1.939$. Therefore, $\sqrt[5]{27.4} \approx 1.939$.

2. Compute $\sqrt[3]{0.0261}$.

SOLUTION $\quad \sqrt[3]{0.0261} = (0.0261)^{1/3}$

$\log \sqrt[3]{0.0261} = \left(\dfrac{1}{3}\right)(\log 0.0261) = \dfrac{\log 0.0261}{3} \approx \dfrac{8.4166 - 10}{3}$

Note: To divide $8.4166 - 10$ by 3 would yield a quotient of $2.8055 - 3\frac{1}{3}$, which expresses the correct negative logarithm for $\sqrt[3]{0.0261}$. But in this form we cannot readily find its antilogarithm from Table II. The difficulty can be avoided if we replace the logarithm binomial, *before dividing*, with an equivalent binomial whose second term is ten times the divisor. So we replace $8.4166 - 10$ with the equivalent $28.4166 - 30$ and then divide by 3.

$\log \sqrt[3]{0.0261} = \dfrac{\log 0.0261}{3} \approx \dfrac{8.4166 - 10}{3} = \dfrac{28.4166 - 30}{3}$

$\approx 9.4722 - 10$

364 Chapter 10 LOGARITHMS

Antilog $9.4722 - 10$ is 0.2966 (approximately). Thus $\sqrt[3]{0.0261} \approx 0.2966$.

Note: The two logarithm expressions, $9.4722 - 10$ and $2.8055 - 3\frac{1}{3}$, are, in fact, equivalent, as can be shown.

$$9.4722 - 10: \quad \begin{array}{r} -10.0000 \\ +\ 9.4722 \\ \hline -\ 0.5278 \end{array} \qquad 2.8055 - 3\frac{1}{3}: \quad \begin{array}{r} -3.3333 \\ +2.8055 \\ \hline -0.5278 \end{array}$$

The antilogarithm of the former $(9.4722 - 10)$ can be found directly from Table II, whereas the latter $\left(2.8055 - 3\frac{1}{3}\right)$ cannot.

3. Simplify $\left[\dfrac{(3.27)(62)}{0.0183}\right]^{2/3}$.

SOLUTION $\qquad \log\left[\dfrac{(3.27)(62)}{0.0183}\right]^{2/3} = \dfrac{2}{3}\,(\log 3.27 + \log 62 - \log 0.0183)$

$$\begin{array}{rl}
\log 3.27 \approx & 0.5145 \\
\log 62 \approx & 1.7924 \\
\log \text{numerator} \approx\ (+) & \overline{12.3069 - 10} \\
\log 0.0183 \approx & 8.2625 - 10 \\
\log [\ \ ] \approx\ (-) & \overline{4.0444}
\end{array}$$

$$\log [\ \ ]^{2/3} \approx \frac{2}{3} \times 4.0444$$

$$\approx 2.6963$$

antilog $2.6963 \approx 496.9$

Thus

$$\left[\frac{(3.27)(62)}{0.0183}\right]^{2/3} \approx 496.9$$

| EXERCISE 10-7

Use logarithms to compute.

1. $(67)^3$	**2.** $(1.08)^7$	**3.** $(42.7)^4$
4. $(0.67)^5$	**5.** $(0.0426)^2$	**6.** $(4.12)^{10}$
7. $(0.00632)^3$	**8.** $(-8.42)^3$	**9.** $(4.37)^{4.1}$
10. $(6.275)^2$	**11.** $(-0.09874)^4$	**12.** $(3.67 \times 49.2)^3$
13. $(0.0637)^3 \times (0.0548)^4$	**14.** $(18.37 \times 0.0625)^6$	
15. $(18.37)^6 \times (0.0625)^6$	**16.** $(3.72)^2 \times (46.3)^3 \times (0.0379)^4$	

17. $(-3.62)^4 \times (0.4875)^5$ 18. $\left(\dfrac{3.62}{48.2}\right)^3$

19. $\left(\dfrac{0.000627}{684}\right)^4$ 20. $\left[\dfrac{32.65 \times 87.21}{(-62.4)}\right]^3$

21. $\sqrt{729}$ 22. $\sqrt{2.25}$ 23. $\sqrt{13.69}$

24. $\sqrt{0.0484}$ 25. $\sqrt[3]{512}$ 26. $\sqrt[3]{9261}$

27. $\sqrt{63.74}$ 28. $\sqrt[3]{7.986}$ 29. $\sqrt[3]{0.062}$

30. $\sqrt[3]{0.00274}$ 31. $\sqrt[5]{4.24}$ 32. $\sqrt[3]{-3.682}$

33. $\sqrt[7]{0.0263}$ 34. $\sqrt{3.26 \times 5.71}$ 35. $\sqrt[3]{0.026 \times 0.384}$

36. $\sqrt[5]{0.00627 \times 0.03814}$ 37. $\sqrt{\dfrac{32.7}{48.5}}$

38. $\left(\dfrac{0.673}{0.284}\right)^{-1/2}$ 39. $\sqrt[3]{\dfrac{-1.03}{0.427}}$ 40. $\sqrt[3]{\dfrac{0.042 \times (3.26)^2}{1.932}}$

41. $\left[\dfrac{3.62 \times 1.48}{6.21}\right]^{2/3}$ 42. $\left[\dfrac{0.032 \div 146.9}{0.002}\right]^{3/4}$

43. $\left[\dfrac{6.21 \times 10^5}{48.3}\right]^{-2/5}$ 44. $\left[\dfrac{3.841 \times 6.257}{4.32 \times 0.0006}\right]^{-3/4}$

The formula

$$A = P(1 + r)^n$$

gives the accumulated value, A, of P dollars invested at interest rate r (expressed as a decimal), compounded n times.

45. If $1000 is invested at 8%, compounded annually, what will be the accumulated value of the investment after 4 years?

46. If $2500 is invested at 7%, compounded annually, how much will the investment amount to after 10 years?

47. Same as Problem 45, except that the compounding takes place quarterly (i.e., 2% per quarter, 16 quarters).

48. How much should be put on deposit in a bank which pays 6% compounded annually in order to have $1000 available in the account at the end of 5 years?

49. Had the Indians who reportedly sold Manhattan for $24 in 1626 put the money into a bank account that paid 5% interest, compounded annually, how much would the account be worth by 1980?

8. EXPONENTIAL EQUATIONS

Equations in which the variable occurs in the position of the exponent are called exponential equations. Often the use of logarithms is necessary to solve such equations, as we shall see in the latter part of this section. However, when it is possible to express both members as convenient powers of the same base, the need for logarithms can be avoided. Program **10.6** provides the method. It is based upon the fact that

$$\text{If } n^x = n^y, \text{ then } x = y. \ (n \neq 0, 1)$$

PROGRAM 10.6

To solve an exponential equation (Method I):

Step 1: Express both members of the equation as powers having the same base.

Step 2: Equate the exponents of the members of the equation of Step 1 and solve; the solution set of this equation is also the solution set of the given exponential equation.

Step 3: Check by substituting the solution(s) obtained in Step 2 for the variable in the given exponential equation.

EXAMPLES

1. Solve the equation $2^{x+1} = 16$.

SOLUTION *Step 1:* Since $16 = 2^4$, both members of the equation can be expressed as powers of 2:

$$2^{x+1} = 2^4$$

Step 2: Equate exponents and solve:

$$x + 1 = 4$$
$$x = 3$$

Step 3: Check by substituting the solution obtained (3) for the variable in the given exponential equation:

$$2^{x+1} = 16$$
$$x = 3: \quad 2^{(3)+1} \overset{?}{=} 16$$
$$2^4 = 16$$

Therefore, the solution of $2^{x+1} = 16$ is 3.

2. Solve the equation $3^{x-7} = 27^x$.

SOLUTION *Step 1:* Since $27 = 3^3$, then $27^x = (3^3)^x = 3^{3x}$; the given equation may be expressed as

$$3^{x-7} = 3^{3x}$$

Step 2:
$$x - 7 = 3x$$
$$-7 = 3x - x$$
$$-\frac{7}{2} = x$$

Step 3: Check:
$$3^{x-7} = 27^x$$

$$x = -\frac{7}{2}: \quad 3^{(-7/2)-7} \overset{?}{=} 27^{-7/2}$$

$$3^{-21/2} \overset{?}{=} 27^{-7/2}$$

$$3^{-21/2} \overset{?}{=} (3^3)^{-7/2}$$

$$3^{-21/2} = 3^{-21/2}$$

Therefore, $3^{x-7} = 27^x$ has one solution: $-\frac{7}{2}$.

3. Solve $5^{x^2-4} = \left(\dfrac{1}{125}\right)^x$.

SOLUTION

$$5^{x^2-4} = \left(\frac{1}{125}\right)^x$$

$$5^{x^2-4} = [(125)^{-1}]^x$$

$$5^{x^2-4} = 5^{-3x}$$

$$x^2 - 4 = -3x$$

$$x^2 + 3x - 4 = 0$$

$$(x + 4)(x - 1) = 0$$

$$x = -4 \qquad x = 1$$

Check:

$$5^{x^2-4} = \left(\frac{1}{125}\right)^x$$

$$x = -4: \quad 5^{(-4)^2-4} \overset{?}{=} \left(\frac{1}{125}\right)^{-4} \qquad\qquad x = 1: \quad 5^{(1)^2-4} \overset{?}{=} \left(\frac{1}{125}\right)^1$$

$$5^{16-4} \overset{?}{=} \left(\frac{1}{5^3}\right)^{-4} \qquad\qquad\qquad 5^{1-4} \overset{?}{=} \left(\frac{1}{5^3}\right)^1$$

$$5^{12} = (5^{-3})^{-4} \qquad\qquad\qquad\qquad 5^{-3} = 5^{-3}$$

The solution set of $5^{x^2-4} = \left(\dfrac{1}{125}\right)^x$ is $\{-4, 1\}$.

When the numbers of an exponential equation do not conveniently reduce to powers of some base, as they do in the foregoing examples, a more comprehensive method is available, outlined in Program **10.7**. It is based upon the fact that

> *If $a = b$, then* $\log a = \log b$.

PROGRAM 10.7

To solve an exponential equation (Method II):

Step 1:   Express each member of the equation as an equivalent product, quotient, power, or root (not sum or difference).

Step 2:   Equate the logarithms of each member of Step 1.

Step 3:   Solve the logarithmic equation of Step 2; the solution set of this equation will also be the solution set of the original exponential equation.

Step 4:   Check by substituting the solution(s) obtained in Step 3 for the variable in the given equation.

EXAMPLES

1.   Solve $3^x = 5 - 1$.

SOLUTION   *Step 1:* Simplify the right member:

$$3^x = 5 - 1$$
$$3^x = 4$$

Step 2: Equate the logarithms of each member:

$$\log 3^x = \log 4$$

Step 3: Solve for x:

$$x \log 3 = \log 4$$
$$x = \frac{\log 4}{\log 3} \approx \frac{0.6021}{0.4771} \approx 1.262$$

Note: We can compute the quotient 1.262 by dividing 0.6021 by 0.4771, by using arithmetic or by logarithms.

$$
\begin{array}{rl}
\log 0.6021 \approx & 9.7797 - 10 \\
\log 0.4771 \approx & 9.6786 - 10 \\
\log \text{quotient} \approx & (-)\overline{0.1011}
\end{array}
$$

$$\text{antilog } 0.1011 \approx 1.262$$

Since 0.6021 is log 4, then log 0.6021 may be thought of as a "log of a log," or log (log 4), or simply "log log 4."

Step 4: Check:

$$3^x = 5 - 1$$

$$3^x = 4$$

$$x = 1.262: \quad 3^{1.262} \stackrel{?}{=} 4$$

$$(10^{0.4771})^{1.262} \stackrel{?}{=} 10^{0.6021}$$

$$10^{0.6021} = 10^{0.6021}$$

2. Solve $5^{x-2} = 3^2 + 4$.

SOLUTION *Step 1:* Simplify the right member:

Note: $\log (3^2 + 4) \neq \log 3^2 + \log 4$.

$$5^{x-2} = 3^2 + 4$$

$$5^{x-2} = 9 + 4$$

$$5^{x-2} = 13$$

Step 2: Equate the logarithms of each member:

$$\log 5^{x-2} = \log 13$$

Step 3: Solve the logarithmic equation of Step 2:

$$(x - 2)(\log 5) = \log 13$$

$$x \log 5 - 2 \log 5 = \log 13$$

$$x = \frac{\log 13 + 2 \log 5}{\log 5}$$

$$x \approx \frac{1.1139 + 2(0.6990)}{0.6990}$$

$$x \approx \frac{2.5119}{0.6990}$$

$$x \approx 3.593$$

Note: If the value of x is computed by means of four-place tables, it is necessary to round off the numerator 2.5119 to 2.512 (four places) before entering Table II for the appropriate mantissa.

Step 4: Check:

$$5^{x-2} = 3^2 + 4$$

$$x = 3.593: \quad 5^{(3.593)-2} \stackrel{?}{=} 3^2 + 4$$

$$5^{1.593} \stackrel{?}{=} 13$$

$$\text{antilog } [1.593(\log 5)] \stackrel{?}{=} 13$$

$$\text{antilog } [1.1135] \stackrel{?}{=} 13$$

$$12.99 \approx 13$$

(The discrepancy is due to rounding errors and the approximate nature of logarithms.)

3. Solve $0.6^x = 83.2^{2x-5}$.

SOLUTION The *logarithm* of 0.6 is negative. It is advisable to express the logarithm first in the usual binomial form $(9.7782 - 10)$ and then as an equivalent negative number: -0.2218.

$$0.6^x = (83.2)^{2x-5}$$

$$\log 0.6^x = \log (83.2)^{2x-5}$$

$$x \log 0.6 = (2x - 5)(\log 83.2)$$

$$x(9.7782 - 10) \approx 2x \log 83.2 - 5 \log 83.2$$

$$x(-0.2218) \approx 2x(1.9201) - 5(1.9201)$$

$$-0.2218x \approx 3.8402x - 9.6005$$

$$-4.0620x \approx -9.6005$$

$$x \approx \frac{-9.6005}{-4.0620}$$

$$x \approx 2.364$$

Check:

$$0.6^x = 83.2^{2x-5}$$

$$x = 2.364: \quad 0.6^{2.364} \stackrel{?}{=} 83.2^{2(2.364)-5}$$

$$0.6^{2.364} \stackrel{?}{=} 83.2^{-0.272}$$

$$2.364 \log 0.6 \stackrel{?}{=} -0.272 \log 83.2$$

$$(2.364)(-0.2218) \stackrel{?}{=} -0.272(1.9201)$$

$$-0.524 \approx -0.522$$

(The discrepancy is due to rounding errors and the approximate nature of logarithms.)

Solve and verify.

1. $2^{x+2} = 32$ **2.** $3^{x+1} = 243$ **3.** $3^{x-1} = \dfrac{1}{81}$

4. $2^{2x-1} = \dfrac{1}{128}$ **5.** $3^{2x} = 81^2$ **6.** $5^{x-2} = 25^x$

7. $4^{x-1} = \left(\dfrac{1}{16}\right)^x$ **8.** $9^{x+1} = \left(\dfrac{1}{27}\right)^x$ **9.** $(\sqrt{5})^{x-1} = 25^x$

10. $(\sqrt{2})^{x-4} = (\sqrt[3]{2})^{x+1}$ **11.** $\dfrac{(\sqrt{2})^{x-1}}{2} = (\sqrt[3]{2})^x$

12. $3^{x^2-4} = \left(\dfrac{1}{27}\right)^x$ **13.** $4^{x^2} = \dfrac{8^x}{2}$ **14.** $(27)^{x^2-2x} = \dfrac{(81)^{2x}}{(9)^4}$

15. $\dfrac{(5)^{x^2}}{(25\sqrt{5})^{x+2}} = 5(\sqrt{5^5})^x$ **16.** $\sqrt{10^{x^2}} = 10^x$

Solve.

17. $8.3^x = 68.89$ **18.** $94^x = 4.547$ **19.** $44^x = 6.633$

20. $7^x = 192$ **21.** $10^x = 3.82$ **22.** $4^{x+1} = 3$

23. $3^x = 5^{x+2}$ **24.** $7^x = 3^{2x-4}$ **25.** $0.5^x = 26^{x-2}$

26. $16^{x^2} = 85$ **27.** $2^{3x} = 3^{2x+1}$ **28.** $3^x = \pi$

29. $(3.75)^x = (4.73)^{x-1}$ **30.** $3^{x+2} = 2^{2x-1}$ **31.** $3^{2x+1} = 5^{x+1}$

32. $5^{x+2} = 4^{2x-1}$ **33.** $16^{3x-1} = (2^7)(4.2)^x$ **34.** $\dfrac{3^{2x}}{2} = \dfrac{5^4}{3}$

35. $\dfrac{5^{x+2}}{5} = \dfrac{2^{2x}}{2}$ **36.** $\begin{cases} 3^{x+y} = 1000 \\ 3^{x-y} = 100 \end{cases}$

37. Recall the compound interest formula, $A = P(1 + r)^n$ (see Exercise 10-7, page 366). At a rate of 6%, compounded annually, about how long will it take for an initial investment of $1000 to grow to $1500?

38. Using the formula of Problem 37, how long to the nearest year will it take an initial investment to double if the rate of interest is 6%, compounded annually?

39. If a radioactive substance decomposes according to the formula, $y = x(0.98)^t$, in which y milligrams (mg) of an original x mg remains after t centuries, how long will it take for half of a given original amount to decompose? (That is, what is the "half-life" of the substance?)

40. Using the formula of Problem 39, how many centuries will it take for an original 4 mg to reduce to 1 mg?

9. CHANGE OF BASE★

As noted in the beginning of the chapter, any positive number other than 1 may be used as a base for logarithms. Because of the decimal nature of our Hindu–Arabic system of number notation, the choice of ten as a base has many computational advantages.

There is another frequently used base that has many other advantages, particularly at levels of mathematics beyond the present. That base is an irrational number, commonly referred to as e. Like any other irrational number, e can be expressed only approximately in decimal notation: $2.7183\ldots$.

Tables are readily available for common logarithms (base 10) and for natural logarithms (base e). But that is not so for other bases, even though there are occasions when logarithms to bases other than 10 or e are needed. A simple formula helps to bridge the gap, and may be derived as follows.

Let

$$x = \log_b N$$

which may also be expressed as

$$b^x = N$$

Now if we use an arbitrary number, a, as the base and equate the respective logarithms of each member to that base, we obtain

$$\log_a b^x = \log_a N$$

which may be expressed alternatively as

$$x \log_a b = \log_a N$$

Solving for x:

$$x = \frac{\log_a N}{\log_a b}$$

Coming back to our original assumption, that $x = \log_b N$, we have in the solution for x immediately above the makings of a formula for changing the base of a logarithm:

$$\log_b N = \frac{\log_a N}{\log_a b}$$

When tables for common logarithms are readily available, as in our case, the formula simplifies to

$$\log_b N = \frac{\log_{10} N}{\log_{10} b} = \frac{\log N}{\log b}$$

In words: To compute the logarithm of a given number, N, to a given base, b, divide the common logarithm of N by the common logarithm of b.

★ The content of this section, in more basic courses of study, may be postponed without significant effect upon the remaining parts of the book.

EXAMPLES

1. Determine $\log_5 3$.

SOLUTION $\quad \log_5 3 = \dfrac{\log 3}{\log 5}$

$\qquad \approx \dfrac{0.4771}{0.6990}$

$\qquad \approx 0.6825$

2. Find $\log_2 0.004$.

SOLUTION $\quad \log_2 0.004 = \dfrac{\log 0.004}{\log 2}$

$\qquad \approx \dfrac{7.6021 - 10}{0.3010}$

$\qquad \approx \dfrac{-2.3979}{0.3010}$

$\qquad \approx -7.966$

3. Determine the logarithm of 7 to the base e.

SOLUTION $\quad \log_e 7 = \dfrac{\log 7}{\log e} \approx \dfrac{\log 7}{\log 2.718}$

$\qquad \approx \dfrac{0.8451}{0.4343}$

$\qquad \approx 1.946$

Note: The symbol "ln" is widely accepted to mean "$\log_e$"; thus, $\log_e 7 = \ln 7 \approx 1.946$.

EXERCISE 10-9

Determine the logarithms.

1. $\log_3 262$ **2.** $\log_5 182$ **3.** $\log_2 14.3$

4. $\log_4 4.68$ **5.** $\log_e 1.87$ **6.** $\log_e 2.31$

7. $\log_5 0.62$ **8.** $\log_4 0.87$ **9.** $\log_3 0.0042$

10. $\log_7 0.0341$ **11.** $\log_e 0.067$ **12.** $\log_e 0.0083$

13. $\log_{2.1} 13.2$ **14.** $\log_{3.5} 971$ **15.** $\log_{6.8} 0.00432$

16. $\log_{2.7} 0.00032$

PART A

Answer True or False.

1. Logarithms are essentially exponents.
2. If $\log_{10} 6 = z$, then $10^z = 6$.
3. 120,000, written in scientific notation, would be expressed as 1.2×10^5.
4. If $10^p = 360$, then $2 < p < 3$.
5. The common logarithms for the numbers 23.45 and 2.345 have the same mantissa.
6. The common logarithms for all numbers between 28 and 64 have the same characteristic.
7. If the logarithm of 12 is x, then the antilogarithm of x is 12.
8. Both 0.308 and 0.038 have the same significant digits.
9. If $a < b$, then $\log a < \log b$.
10. If $a^x = b$, then $x = \log b - \log a$.

PART B

Express in logarithmic form.

1. $0.5^2 = 0.25$
2. $0.5^{-2} = 4$
3. $2^6 = 64$
4. $81^{0.5} = 9$
5. $64^{0.67} = 16$
6. $8^{-0.33} = 0.5$

Express in exponential form.

7. $\log_9 81 = 2$
8. $\log_8 2 = \frac{1}{3}$
9. $\log_{16} 2 = 0.25$
10. $\log_{10} 6.3 = 0.799$
11. $\log_{10} 2.6 = 0.415$
12. $\log_a 64 = 6$

What replacement for the variable will make the sentence true?

13. $\log_6 36 = L$
14. $\log_a 2 = 0.25$
15. $\log_8 n = 4$
16. $\log_a 0.001 = -3$
17. $\log_{16} n = -2.25$
18. $\log_{100} 1000 = L$

Express in scientific notation.

19. 3864
20. 30,006
21. 40,000,000
22. 0.042
23. 0.000789
24. 0.00000000000042

From Table II, what replacement for p will make the following true?

25. $10^p = 3.75$
26. $10^p = 4$
27. $10^p = 5.09$
28. $10^p = 460$
29. $10^p = 0.46$
30. $10^p = 0.0046$

Use Table II to write the appropriate common logarithm.

31. 863
32. 0.0082
33. 0.000001
34. 426,000
35. 31,800
36. 0.5757

37. 283.4   **38.** 0.006764   **39.** 3.397×10^{-6}

40. 8,995,000

Use Table II to write the appropriate antilogarithm.

41. 0.8675   **42.** $8.8035 - 10$ **43.** 2.1139

44. $3.5024 - 10$ **45.** $7.0253 - 20$ **46.** $7.8865 - 10$

47. 0.9144   **48.** 4.6250   **49.** 1.8750

50. 3.5791   **51.** $6.8003 - 20$ **52.** 7.3504

Compute, using logarithms.

53. 427×0.697 **54.** $(0.0035) \times (-4.26)$ **55.** 0.00823×0.0647

56. $824 \times 3.62 \times 87.41$ **57.** $(-2.38) \times (-0.063) \times (0.4278)$

58. $83.62 \times 42.51 \times 3.879$ **59.** $0.0001046 \times 0.003275$

60. $0.00362 \times 0.005714 \times 0.0000632 \times (-0.004621)$

61. $4320 \div 61.8$ **62.** $(-3.94) \div (54.8)$

63. $(0.073) \div (0.0846)$ **64.** $(-428) \div (-0.0367)$

65. $0.00476 \div 0.0832$ **66.** $1 \div 3.762$

67. $\dfrac{(-0.0432) \times (6.814)}{(-82.6)}$ **68.** $\dfrac{32.75 \times 106.2}{0.0361 \times 0.04287}$

69. $(56.8)^4$ **70.** $(0.0682)^3$ **71.** $(0.0003714)^5$

72. $(-3.62)^3$ **73.** $(4.26 \times 3.81)^2$ **74.** $(-6.37 \times 0.0428)^3$

75. $(-8.37)^3 \times (4.26)^2$ **76.** $\left(\dfrac{0.00624 \times 0.8365}{0.0724}\right)^4$

77. $\sqrt{8.365}$ **78.** $\sqrt[3]{0.0487}$

79. $(0.0632)^{1/5}$ **80.** $\sqrt[5]{0.0624 \times 0.0381}$

81. $(6.742)^{-2/3}$ **82.** $\sqrt[3]{\dfrac{4.62}{37.5}}$

83. $\sqrt[4]{\dfrac{(4.06)^2 \times (3.62)^3}{(-1.8)^2}}$ **84.** $\left[\dfrac{(3.07)^3 \times (2.634)^{1/2}}{(8.62)^4}\right]^{-1/2}$

Solve and verify.

85. $2^{x+2} = 128$ **86.** $3^{x-1} = \dfrac{1}{243}$ **87.** $3^{3x} = 27^3$

88. $9^{x-1} = \left(\dfrac{1}{81}\right)^x$ **89.** $(\sqrt{3})^{x-1} = 27^x$ **90.** $\dfrac{(\sqrt{3})^{x+1}}{3} = (\sqrt[3]{3})^x$

91. $(9)^{x^2} = \dfrac{3^x}{3^{-1}}$ **92.** $\dfrac{(3)^{x^2}}{3(\sqrt{3})^{3x}} = (27)^3(3\sqrt{3})^x$

Solve for x.

93. $6.8^x = 46.24$ **94.** $86^x = 9.274$ **95.** $10^x = 4.61$

96. $5^x = 3^{x+2}$ **97.** $0.5^x = 26^{x+2}$ **98.** $5^{3x} = 7^{2x-1}$

99. $(53.1)^{1-x} = 129.7$ **100.** $5^{x+2} = 3^{2x+1}$ **101.** $16^{3x+1} = 2^7(4.2)^x$

102. $\dfrac{5^{x+3}}{5} = \dfrac{3^{2x+2}}{3}$

Write the logarithm to the base indicated.

103. $\log_3 56$

104. $\log_5 29$

105. $\log_4 0.0062$

106. $\log_3 0.0163$

107. $\log_e 1.57$

108. $\log_e 2.68$

109. $\log_e 0.09425$

110. $\log_e 0.00006213$

111. A \$100,000 certificate of deposit (CD), gaining interest at 8%, compounded quarterly (i.e., 2% per quarter), will be worth how much at maturity, 6 years from now?

112. How much must a person invest in a certificate of deposit, paying 9% interest, compounded annually, if she wishes her investment to grow to \$8000 in 5 years?

113. An investor is looking for an opportunity to invest \$25,000 with the expectation that it will double in 6 years. What annual compound interest rate must he find?

114. How many years will it take for \$100 to grow to \$250 in a savings account for which the interest rate is 6%, compounded quarterly?

115. City planners expect the population p of a city to grow according to the formula, $p = 150,000(1.09)^{t/20}$, where t is the number of years after 1980. If the formula proves to be accurate, what will be the population in 1988?

116. The growth of 60,000 bacteria in a certain culture is a function of time, t, measured in hours, and defined by $N = 60,000(2.7)^{0.3t}$. How many bacteria will be present in the culture after 7 hours?

117. How long will it take the bacteria in Problem 116 to grow to 1 million?

SUPPLEMENTARY UNIT: FUNCTIONS

1. CONCEPT OF A FUNCTION

Consider these situations, each of which may be expressed by an equation in two variables:

- In a state that charges a 4% sales tax, the amount of tax charged *depends* upon the amount of the purchase.

  If x = number of dollars per purchase, and

  y = amount of sales tax to be charged, then

  $$y = 0.04x$$

  expresses the relationship between the tax to be charged and the amount of the purchase.

378

- For an examination, the number of tests needed *corresponds* to the number of examinees.

$$\text{If } x = \text{number of examinees, and}$$

$$y = \text{number of tests needed, then}$$

$$y = x$$

expresses the relationship between needed tests and the number of examinees.

- The weight of a bucket of water varies as the number of gallons it contains.

$$\text{If } x = \text{number of gallons of water in the bucket, and}$$

$$62.5 = \text{weight in pounds of a gallon of water, and}$$

$$2 = \text{weight in pounds of the bucket when empty, and}$$

$$y = \text{total weight of the bucket of water, then}$$

$$y = 62.5x + 2$$

expresses the relationship between the total weight and the amount of water in the bucket.

Each of the three equations above defines a **function**—a rule that assigns to each *input* number exactly one *output* number. In each instance the input number was designated by x (amount of purchase, number of examinees, gallons of water). The variable that represents an input number in a function is called the **independent variable**.

In each of the three examples, the output number was designated by y (amount of sales tax to be charged, number of tests needed, weight of the bucket and its contents). The variable that represents the output number in a function is called the **dependent variable**. In effect, the output number depends upon the input number. Thus *the dependent variable is said to be a function of the independent variable.*

The set of all possible input numbers is called the **domain** of the function. The set of all possible output numbers is called the **range** of the function.

EXAMPLES

1. The equation

$$y = 2x - 3$$

defines a function in which the independent variable is x and the dependent variable is y. Determine the value of y for each x when the domain of x is the set $\{-2, 3, 8\}$.

SOLUTION

$$y = 2x - 3$$

When $x = -2$:

$$y = 2(-2) - 3 = -7$$

When $x = 3$:

$$y = 2(3) - 3 = 3$$

When $x = 8$:

$$y = 2(8) - 3 = 13$$

In brief, the set $\{-7, 3, 13\}$ is the range of the function defined by the equation, $y = 2x - 3$, when the domain is the set $\{-2, 3, 8\}$.

2.   The equation, $y = 2x + 1$, defines a function in x. Graph the function when the domain of the independent variable, x, is the set of nonnegative real numbers.

SOLUTION   Apply Program **8.2** (To graph an equation in two variables). Using only non-negative real numbers for x, we obtain the graph shown in Figure A.1.

$$y = 2x + 1$$

x	0	1	2
y	1	3	5

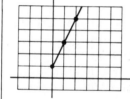

Figure A.1

Clearly, the range of the function defined by $y = 2x + 1$ is the set of real numbers *equal to or greater than* 1, when the domain is the set of nonnegative real numbers.

Not all equations in x and y define functions, however. For instance

$$y^2 = x$$

does not define a function in x. The reason: for each positive input number, x, there will be *two* output numbers. When x is replaced by 4, say, two output numbers, y, will result: -2 and $+2$:

$$y^2 = x$$

$$x = 4: \quad y^2 = 4$$

$$y = \pm 2$$

This result is contrary to a basic characteristic of a function, namely that for each input number there is to be *exactly* one output number.

EXAMPLE

Which of these two equations (x the independent variable) defines a function?

(a) $y^2 = 4x$

(b) $y = 4x^2$

SOLUTION   Often a quick way to determine whether or not an equation defines a function is to sketch its graph.

(a) $y^2 = 4x$

x	0	1	2
y	0	± 2	$\pm\sqrt{8}$

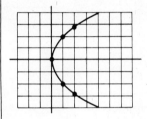

Figure A.2

(b) $y = 4x^2$

x	0	1	-1
y	0	4	4

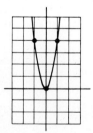

Figure A.3

Note that in the case of (a) $y^2 = 4x$, Figure A.2, except for the vertex, $(0, 0)$, each abscissa (x value) has associated with it two different ordinates (y values). So $y^2 = 4x$ does not define a function in x because at least one input number generates more than one output number.

In the case of (b) $y = 4x^2$, Figure A.3, each abscissa has associated with it one and only one ordinate. In this instance, y is indeed a function of x.

The analysis developed in the foregoing example may be reduced to the following technique: Visualize a vertical line moving left to right across the graph of an equation. If in its sweep the visualized line *never* intersects the graph at more than one point, then the equation defines a function.

EXAMPLES

Use the vertical line test to decide which of these graphs define a function.

1.
Function

2.
Function

3.
Not a function

4.
Function

5.
Not a function

6.
Not a function

EXERCISE A-1

For the function defined by the equation (*consider x to be the independent variable*), *compute the output numbers for the given set of input numbers.*

1. $y = 3x$ Input: $\{-2, 0, 3, 5\}$

2. $y = \dfrac{1}{2}x$ Input: $\{4, 3, 0, -2\}$

3. $y = 3x + 5$ Input: $\{2, 0, -1, -5\}$

4. $y = 6 - x$ Input: $\{0, -3, -5, -8\}$

5. $y = \dfrac{2}{x - 3}$ Input: $\{-2, 0, 1, 2\}$

6. $y = \dfrac{x - 1}{3}$ Input: $\{-2, 0, 1, 4\}$

7. $y = |3x - 2|$ Input: $\{2, 0, -1, -2\}$

8. $y = \sqrt{x - 1}$ Input: $\{1, 5, 10, 17\}$

9. $y = \dfrac{x}{\sqrt{x^2 + 1}}$ Input: $\{0, -1, 1, 3\}$

10. $y = 2|x - 4|$ Input: $\{-3, -1, 0, 6\}$

Sketch the graph of each equation for which the domain is the set of real numbers; decide whether or not the equation may be used to define y as a function of x.

11. $y = 3x - 2$	12. $y = 6 - 2x$	13. $x + 2y = 1$
14. $3y - 2x = 0$	15. $y = |3x - 2|$	16. $x = |2y|$

17. $y^2 = x + 1$ **18.** $x^2 = y + 1$ **19.** $x^2 + y^2 = 4$

20. $x^2 - 4y^2 = 16$ **21.** $xy = 2$ **22.** $x^2 + 4y^2 = 16$

Describe the range of the function defined by the equation (x the independent variable). Unless otherwise indicated, the domain is the set of real numbers.

23. $y = 3x - 7$ **24.** $y = -2x + 3$ **25.** $y = 2x^2 - 1$

26. $y = -2x^2 + 1$ **27.** $y = |-x + 1|$ **28.** $y = -|x + 1|$

29. $y = 3x - 7$ Domain: $x > 2$ **30.** $y = -2x + 3$ Domain: $x < 3$

2. FUNCTIONAL NOTATION

Functions are usually designated by letters, such as f, g, h, F, G, and so on (the most frequent being f). When a function, say f, is defined by an equation, say $y = 2x - 3$, the notation

$$f(x) = 2x - 3$$

is likely to be used.

The symbol "$f(x)$" is read "f of x," as in "f of x equals two x minus three," shown above.

Note: $f(x)$ does *not* mean "f times x."

Thus when an input number, x, is selected from the domain of the function, $f(x)$ represents the corresponding output number. To illustrate, if

$$f(x) = 2x - 3$$

then

$$f(-2) = 2(-2) - 3 = -7$$

$$f(3) = 2(3) - 3 = 3$$

$$f(8) = 2(8) - 3 = 13$$

That is, for the function $f(x) = 2x - 3$:

- input -2 for x; output -7, which is $f(-2)$
- input 3 for x; output 3, which is $f(3)$
- input 8 for x; output 13, which is $f(8)$

In effect, $f(x)$ operates in the same way as y in $y = 2x - 3$, except the $f(x)$ notation more clearly identifies the functional relationship.

EXAMPLES

1. If $f(x) = 2x + 3$, what is (a) $f(4)$? (b) $f(-2)$?

SOLUTION Given: $$f(x) = 2x + 3.$$

(a) $$f(4) = 2(4) + 3$$
$$= 8 + 3$$
$$= 11$$

(b) In the beginning it may be helpful to write the given function first with ( ) in place of the variable—for instance,

$$f(\ ) = 2(\ ) + 3$$

followed by insertion of the variable replacement.

$$f(-2) = 2(-2) + 3$$
$$= -4 + 3$$
$$= -1$$

2. Given $f(x) = 3x^2 - 2$; determine $f(0); f(2); f(-2)$.

SOLUTION
$$f(x) = 3x^2 - 2$$
$$f(0) = 3(0)^2 - 2 = -2$$
$$f(2) = 3(2)^2 - 2 = 10$$
$$f(-2) = 3(-2)^2 - 2 = 10$$

3. If $f(x) = 2x^2 - 1$, find $f(a + 2)$.

SOLUTION
$$f(x) = 2x^2 - 1$$
$$f(\ ) = 2(\ )^2 - 1$$
$$f(a + 2) = 2(a + 2)^2 - 1$$
$$= 2(a^2 + 4a + 4) - 1$$
$$= 2a^2 + 8a + 8 - 1$$
$$= 2a^2 + 8a + 7$$

As mentioned previously, functions may be designated by letters other than f. Also variables may be expressed by letters other than x. For instance, $g(x)$ names a function g whose independent variable is x. An example would be

$$g(x) = x^2 + 3x - 1$$

Similarly, we may have a function q with independent variable b, as expressed by

$$q(b) = b^2 - 4$$

(read "q of b equals b-squared minus 4").

Then

$$q(0) = (0)^2 - 4$$
$$= -4$$
$$q(-3) = (-3)^2 - 4$$
$$= 9 - 4$$
$$= 5$$

and so on.

EXAMPLES

1. If $g(x) = 2x + 3$ and $f(x) = -x^2$, find (a) $g(2) + f(2)$; (b) $f(-2) - g(3)$.

SOLUTION (a) $g(x) = 2x + 3$ $\qquad$ $f(x) = -x^2$
$\qquad\quad$ $g(2) = 2(2) + 3$ $\qquad$ $f(2) = -(2)^2$
$\qquad\qquad\;\; = 7$ $\qquad\qquad\quad\;\; = -4$

Thus

$$g(2) + f(2) = 7 - 4$$
$$= 3$$

(b) $f(-2) = -(-2)^2 = -4$
$\quad\; g(3) = 2(3) + 3 = 6 + 3 = 9$

So

$$f(-2) - g(3) = (-4) - (9)$$
$$= -4 - 9$$
$$= -13$$

2. If $F(x) = g(x) - p(x)$ and $g(x) = x^2 - 1$ while $p(x) = x - 3$, find $F(3)$.

Note: In this case, $g(x)$ and $p(x)$ are *combined* to produce a new function, $F(x)$.

SOLUTION One approach:

$$F(x) = g(x) - p(x)$$
$$F(3) = g(3) - p(3)$$
$$\qquad\qquad \hookrightarrow p(3) = 3 - 3 = 0$$
$$\qquad \hookrightarrow g(3) = (3)^2 - 1 = 9 - 1 = 8$$
$$= 8 - 0$$
$$= 8$$

Another approach: Express $F(x)$ in terms of the variable, simplify, then substitute 3 for the variable.

$$F(x) = g(x) - p(x)$$
$$= (x^2 - 1) - (x - 3)$$
$$= x^2 - 1 - x + 3$$
$$= x^2 - x + 2$$
$$F(3) = (3)^2 - (3) + 2$$
$$= 9 - 3 + 2$$
$$= 8$$

3.　If $f(x) = 6x$ and $k(x) = 2 - 3x$, determine $f[k(x)]$, which is read "f of k of x."

SOLUTION　$f[k(x)] = f[2 - 3x]$

and

$$f[2 - 3x] = 6(2 - 3x) = 12 - 18x$$

Thus

$$f[k(x)] = 12 - 18x$$

　　　Unless otherwise stated, we shall consider the domain of the variable (input numbers) to be the set of all real numbers except those that produce function values that are imaginary numbers. In such cases we say the function is not defined for those input numbers. For instance, if

$$f(x) = \sqrt{x}$$

then the function is not defined for negative numbers; it *is* defined only for $x \geq 0$.
　　　Also excluded are numbers that produce 0 denominators. For instance, if

$$f(x) = \frac{1}{x}$$

then $f(0) = \frac{1}{0}$, which is meaningless. The function *is* defined for all real numbers except 0.

EXERCISE A-2

If $f(x) = x^2 - x + 2$, *find*

1. $f(2)$　　　　**2.** $f(5)$　　　　**3.** $f(0)$　　　　**4.** $f(-3)$

If $g(a) = 2a^2 - 1$, find

5. $g(-1)$ **6.** $g(0)$ **7.** $g\left(\dfrac{1}{2}\right)$ **8.** $g\left(\dfrac{4}{3}\right)$

9. $g(p)$ **10.** $g(-p)$ **11.** $g(a - 3)$ **12.** $g(a^2 + 1)$

Determine the function values indicated.

13. $F(m) = \dfrac{m - 3}{2m^2}$; $F(2)$; $F(3)$

14. $P(t) = \dfrac{2t - 1}{6 - 3t}$; $P(0)$; $P(-1)$

15. $G(x) = \dfrac{2x^2}{3x - 1}$; $G\left(\dfrac{1}{2}\right)$; $G\left(\dfrac{1}{3}\right)$

16. $h(y) = \dfrac{2y - 3}{y + 6}$; $h\left(\dfrac{3}{2}\right)$; $h(-6)$

17. $f(x) = |2x - 1|$; $f(-2)$; $f(0)$

18. $p(x) = \dfrac{2}{|x|}$; $p(-4)$; $p(\sqrt{3})$

19. $s(x) = \dfrac{3}{\sqrt{x}}$; $s(9)$; $s(-4)$

20. $p(m) = \dfrac{\sqrt{x + 3}}{\sqrt{x}}$; $p(0)$; $p(-1)$

If $f(x) = x + 2$ and $g(x) = 2x - 3$, express more simply.

21. $3f(-1)$[i.e., $3 \cdot f(-1)$]   **22.** $f(2) - g(1)$
23. $2f(0) - 3g(1)$ **24.** $f(2) \cdot g(2)$

25. $\dfrac{f(3)}{g(3)}$ **26.** $f(x + a) - f(a)$ **27.** $f[g(2)]$

28. $g[f(2)]$ **29.** $g[2f(x)]$ **30.** $f[g(x + 1)]$

If $f(x) = x - 3$ and $p(x) = 2x^2 + x - 3$, for what numbers will the functions be undefined?

31. $A(x) = \dfrac{p(x)}{f(x)}$ **32.** $B(x) = \dfrac{f(x)}{p(x)}$

33. $C(x) = \sqrt{f(x)}$ **34.** $D(x) = \sqrt{p(x)}$

Example: Express the area of a square as a function of its edge dimension, x. Answer: $A = f(x) = x^2$.

35. Express the area of a rectangle as a function of its length in inches, x, if its width is 8 in.
36. Express the perimeter of a square as a function of its edge dimension, x.
37. Express the perimeter of the rectangle of Exercise 35 as a function of x.

38. Express the area of a triangle of base 20 and altitude h as a function of its altitude.

39. Express the area of a circle as a function of its radius.

40. Express the radius of a circle as a function of its circumference.

41. Express the area of a circle as a function of its circumference.

42. Express the area of a triangle as a function of its altitude, h, if the altitude is 3 in. longer than its base.

43. A salesman's earnings per month, $f(x)$, are a function of the number of sales, x, he produces, as defined by $f(x) = x^2 - 3x - 9$. How many sales must he make before earning his first dollar? (*Hint*: Substitute integral values for x.)

44. A manufacturer can earn a profit of p dollars on a shipment of x items as defined by the function, $p = f(x) = x^2 - 4x - 12$. How many items must be in a shipment to produce a profit in excess of $100?

45. If the function, $f(x) = 1380 - |4x - 96|$ expresses the weekly profit from the sale of x widgets, what sales volume will produce the maximum profit per week?

REVIEW

PART A

Answer True or False.

1. In a function, the same output numbers may have different input numbers, but not the other way around.

2. The function symbol "$f(x)$" means "f times x."

3. If $g(a) = 2a^2 - 4$, then $g(2) = 2(2)^2 - 4$.

4. When a function is defined by an equation in two variables, say p and q, in which p represents the input numbers, then q is the dependent variable.

PART B

Determine the function values indicated.

1. $f(x) = 3x^3 - 2x + 1; f(2); f(0)$

2. $p(x) = -4x + 3x^3 + 7; p(10); p(-1)$

3. $g(x) = |2x - 3|; g(0); g(-6)$

4. $f(c) = |c^2 - 2c - 1|; f(2); f(-2)$

5. $f(x) = \sqrt{x^2 - x - 1}; f(0); f(3)$

6. $g(n) = \sqrt[3]{n^2 + n - 8}; g(0); g(-2)$

7. $f(a) = \dfrac{a^2 - 1}{|a + 2|}; f(3); f(1)$

8. $g(a) = \dfrac{|a^2 - a|}{a + 4}; g(2); g(-2)$

9. $f(b) = \dfrac{b^2 - 1}{b + 1}$; $f(a - 1)$; $f(a + 1)$

10. $m(n) = \dfrac{n^2 - 2n}{|n + x|}$; $m(2 - x)$; $m(x - 1)$

If $f(x) = 2x - 1$ and $g(x) = x^2 - 4$, express more simply.

11. $f(3) - 3g(3)$

12. $\dfrac{f(-1)}{g(2)}$

13. $f[g(2)]$

14. $g[f(3)]$

15. $g[f(p - 1)]$

16. $f[g(a + 2)]$

For what numbers will the function be defined?

17. $p(x) = \sqrt{x^2 - 3}$

18. $f(a) = \sqrt{4 - a^2}$

19. $f(x) = \dfrac{3}{|x - 1|}$

20. $g(x) = \dfrac{|x^2 - 2x + 1|}{-3}$

21. $F(x) = \sqrt{2x^2 - 5x - 3}$

22. $G(x) = \sqrt{12 + 5x - 2x^2}$

Which of these equations define a function in x, assuming that no restrictions are placed on the variables?

23. $x^2 - y^2 = 16$

24. $3x - 2y = 5$

25. $xy = 9$

26. $x = |2y - 1|$

27. $x^2 = y + 1$

28. $x + 3y^2 = 2$

29. Express the area of a rectangle 5 ft long as a function of its width in feet, x.

30. Express the perimeter of an equilateral triangle as a function of the measure of its side, m.

31. An assembly line in a factory can produce up to 8 items per day. Daily overhead costs per item is expressed by the function $f(n) = 2n^2 - 24n + 113$, where n is the number of items produced that day. What should be the daily production goal if minimizing overhead is the only consideration?

32. A desk manufacturer finds that his cost per desk (c) is a function of the number of desks (x) that his shop can produce in a day. The governing equation, he has found, is $c = x^2 - 6x + 10$. How many desks should he produce each day so that the cost of each is at a minimum?

SUPPLEMENTARY UNIT: PROGRESSIONS

B

1. SEQUENCES

A set of numbers arranged in some order, so there is a first number, a second number, a third number, and so on, is called a **sequence**. For example:

$$2, 4, 6, 8, 10, \ldots$$

$$3, 5, 4, 6, 5, 7, 6, 8$$

are sequences. The numbers of a sequence are referred to as **terms**. When there is a last term, the sequence is said to be **finite**; when there is not a last term, the sequence is said to be **infinite**.

A common type of notation for a sequence is

$$a_1, a_2, a_3, a_4, \ldots, a_n, a_{n+1}, \ldots$$

where a_1 is the first term;

a_2 is the second term;

a_3 is the third term;

$\vdots$

a_n is the nth term, called the **general term**;

a_{n+1} is the next term after the general term; and so on.

There may or may not be a pattern among the terms of a sequence. For the first sequence above,

$$2, 4, 6, 8, 10, \ldots$$

the pattern is obvious: add 2 to each term to get the next term. In the second instance,

$$3, 5, 4, 6, 5, 7, 6, 8$$

there is also a pattern, but it is not so obvious: alternatively, add 2 and subtract 1, starting with 3 and stopping at 8.

The pattern of a sequence may be symbolized by the general term. For instance,

$$a_n = \frac{n-3}{2}$$

will generate the sequence

$$-1, -\frac{1}{2}, 0, \frac{1}{2}, 1, 1\frac{1}{2}, \ldots$$

as the numbers 1, 2, 3, 4, 5, 6, ... are substituted for n. Thus, if

$$a_n = \frac{n-3}{2}$$

then

$$a_1 = \frac{1-3}{2} = \frac{-2}{2} = -1$$

$$a_2 = \frac{2-3}{2} = \frac{-1}{2} = -\frac{1}{2}$$

$$a_3 = \frac{3-3}{2} = \frac{0}{2} = 0$$

$$a_4 = \frac{4-3}{2} = \frac{1}{2}$$

etc.

Note: In some presentations, function notation as discussed in Supplementary Unit A is used instead of the notation employed here. That is:

$$a_n = a(n) = \frac{n-3}{2}$$

$$a_1 = a(1) = \frac{1-3}{2} = -1$$

$$a_2 = a(2) = \frac{2-3}{2} = -\frac{1}{2}$$

etc.

In this instance,

$$a(n) = \frac{n-3}{2}$$

is said to define a **sequence function** in which the domain is the set of natural numbers, and the range is the set

$$\{-1, -\tfrac{1}{2}, 0, \tfrac{1}{2}, \ldots\}$$

EXAMPLES

1. Generate the sequence for which $a_n = n + 5$, for $n = 3, 4, 5, 6$.

SOLUTION $a_n = n + 5$

$a_3 = 3 + 5 = 8$

$a_4 = 4 + 5 = 9$

$a_5 = 5 + 5 = 10$

$a_6 = 6 + 5 = 11$

The desired sequence is 8, 9, 10, 11.

2. Express the sequence for which the general term is 2^n, for $0 \le n \le 7$, $n \in$ {integers}.

SOLUTION The sequence: $2^0, 2^1, 2^2, 2^3, 2^4, 2^5, 2^6, 2^7$ or 1, 2, 4, 8, 16, 32, 64, 128.

A **series** is the indicated sum of the terms of a sequence, expressed in expanded form. For the sequence that has a_n as its general term, the related series would be

$$a_1 + a_2 + a_3 + a_4 + \cdots + a_n + \cdots$$

When there is a last term to the series, the series is said to be **finite**. When there is no last term, the series is **infinite**.

A **partial sum** of a series is the sum of a finite number of consecutive terms of the series, starting with the first term. The first, second, third, and nth partial sums of the series above would be

$$S_1 = a_1$$
$$S_2 = a_1 + a_2$$
$$S_3 = a_1 + a_2 + a_3$$
$$S_n = a_1 + a_2 + \cdots + a_n$$

1. Compute the sum of the first three terms of a sequence defined by $a_n = 2n - 1$.

SOLUTION S_3 (third partial sum) $= a_1 + a_2 + a_3$, where $a_n = 2n - 1$.

$$a_1 = 2(1) - 1 = 1$$
$$a_2 = 2(2) - 1 = 3$$
$$a_3 = 2(3) - 1 = 5$$

So $S_3 = 1 + 3 + 5 = 9$.

2. Compute the sixth partial sum of the sequence for which the general term is $n^2 - n$ (i.e., $a_n = n^2 - n$).

SOLUTION

n	$n^2 - n$	a_n
1	$1^2 - 1$	0
2	$2^2 - 2$	2
3	$3^2 - 3$	6
4	$4^2 - 4$	12
5	$5^2 - 5$	20
6	$6^2 - 6$	30

Thus $S_6 = 0 + 2 + 6 + 12 + 20 + 30 = 70$.

EXERCISE B-1

For each sequence, add three more terms to continue the pattern.

1. 15, 18, 21, ___, ___, ___

2. 10, 30, 50, ___, ___, ___

3. 16, 11, 6, ___, ___, ___

4. 96, 85, 74, ___, ___, ___

5. 1, 3, 9, 27, ___, ___, ___

6. $1, \dfrac{1}{2}, \dfrac{1}{4}, \dfrac{1}{8},$ ___, ___, ___

7. 2, -4, 8, ___, ___, ___, 128

8. 1, -10, ___, -1000, ___, ___

Write the first four terms of a sequence for which the general term is given.

9. $n + 3$

10. $n + 2$

11. $3n - 5$

12. $4n - 8$

13. $6 - 3n$

14. $7 - 2n$

15. $2n^2$

16. $n^3 - 1$

17. $\dfrac{n + 2}{n}$

18. $\dfrac{3n - 1}{n}$

19. $\dfrac{n - 2}{3n}$

20. $\dfrac{2 - n}{2n}$

Compute the indicated partial sum for the series that has the given general term, a_n, n being the number of the term.

21. S_4; $a_n = 2n$

22. S_5; $a_n = n + 1$

23. S_6; $a_n = 2n - 3$

24. S_5; $a_n = 3 - n$

25. $S_4; a_n = n^2 - 6$ **26.** $S_3; a_n = 2n^2 - 1$

27. $S_4; a_n = 8 - 3n^2$ **28.** $S_3; a_n = n - n^2$

29. $S_3; a_n = \dfrac{3n - 2}{2n + 1}$ **30.** $S_5; a_n = \dfrac{n^2 + 1}{n - 1}$

2. ARITHMETIC PROGRESSION

An **arithmetic progression** is a sequence in which each term after the first is obtained by adding a constant to the preceding term in the sequence. This constant is referred to as the **common difference**.

EXAMPLES

1. 2, 5, 8, 11, 14 is an arithmetic progression having a common difference of 3.
2. $-4, -5, -6, -7$ is an arithmetic progression with a common difference of -1.
3. 2.7, 3.8, 4.9, 6.0, 7.1 is an arithmetic progression with a common difference of 1.1.
4. $\sqrt{2}, 2\sqrt{2}, 3\sqrt{2}, 4\sqrt{2}$ is an arithmetic progression with a common difference of $\sqrt{2}$.
5. $\dfrac{1}{2}, \dfrac{1}{4}, 0, -\dfrac{1}{4}, -\dfrac{1}{2}$ is an arithmetic progression with a common difference of $-\dfrac{1}{4}$.

If we let a_1 represent the first term of an arithmetic progression and d the common difference, then

the first term, a_1, is a_1;

the second term, a_2, is $a_1 + d$;

the third term, a_3, is $(a_1 + d) + d$, or $a_1 + 2d$;

the fourth term, a_4, is $(a_1 + 2d) + d$, or $a_1 + 3d$;

the fifth term, a_5, is $(a_1 + 3d) + d$, or $a_1 + 4d$;

the sixth term, a_6, is $a_1 + 5d$;

$\quad\quad\quad \vdots$

the twentieth term, a_{20}, is $a_1 + 19d$;

etc.

Note that the coefficient of d in the progression above is always one less than the number of the term. In general

$$a_n = a_1 + (n - 1)d$$

represents any specified term (the nth term) in an arithmetic progression in which the first term is a_1 and the common difference is d.

EXAMPLES

1. Find the seventh term of an arithmetic progression in which the first term is 11 and the common difference is 4.

SOLUTION   The seventh term may be designated as a_7. We use

$$a_n = a_1 + (n - 1)d$$

as a formula and substitute for the variables to find a_7:

$$
\left.
\begin{array}{l}
n = 7 \\
a_1 = 11 \\
d = 4
\end{array}
\right\}
\quad
\begin{array}{l}
a_7 = 11 + (7 - 1)(4) \\
a_7 = 11 + (6)(4) \\
a_7 = 35
\end{array}
$$

The seventh term is 35.
The progression: 11, 15, 19, 23, 27, 31, **35**.

2. Find the 135th term of an arithmetic progression in which $a_1 = \dfrac{1}{2}, d = -\dfrac{1}{4}$.

SOLUTION   The general term: $a_n = a_1 + (n - 1)d$. In this progression, $a_1 = \dfrac{1}{2}, d = -\dfrac{1}{4}$, $n = 135$. Thus

$$a_{135} = \frac{1}{2} + (135 - 1)\left(-\frac{1}{4}\right)$$

$$= \frac{1}{2} + (134)\left(-\frac{1}{4}\right)$$

$$= \frac{1}{2} - 33\frac{1}{2}$$

$$= -33$$

3. Find the 13th term of the arithmetic progression, 7, 17, 27, ...

SOLUTION   Here $a_1 = 7, n = 13, d = 17 - 7 = 10$, and

$$a_n = a_1 + (n - 1)d$$

Therefore,

$$a_{13} = 7 + (13 - 1)(10)$$

$$= 7 + 120$$

$$= 127$$

The 13th term in the progression is 127.

The general-term formula can often be used to complete the data for an arithmetic progression, as the examples below illustrate. Note that if three of the variables are replaced by numbers in the formula

$$u_n = a_1 + (n - 1)d$$

the value of the fourth variable is determined.

EXAMPLES

1. What is the first term of an eight-term arithmetic progression in which the common difference is 6 and the 8th term is 17?

SOLUTION Here $a_1 = ?$, $a_n = 17$, $n = 8$, and $d = 6$. Using the general-term formula,

$$a_n = a_1 + (n - 1)d$$

we develop an equation in which the variable is a_1, and solve for it.

$$17 = a_1 + (8 - 1)(6)$$

$$17 = a_1 + 42$$

$$-25 = a_1$$

Thus the first term of the given progression is -25.

2. What is the common difference in a ten-term arithmetic progression in which the first term is 0.32 and the last term is 0.59?

SOLUTION The "last term" in this instance is a_{10}. Thus $a_1 = 0.32$, $a_{10} = 0.59$, $n = 10$, and $d = ?$. Using the general-term formula,

$$a_n = a_1 + (n - 1)d$$

we substitute and solve for d:

$$0.59 = 0.32 + (10 - 1)d$$

$$0.59 - 0.32 = 9d$$

$$0.27 = 9d$$

$$0.03 = d$$

3. How many terms are there in an arithmetic progression in which the first and last terms are $8\frac{1}{4}$ and $12\frac{1}{2}$, respectively, and in which the common difference is $\frac{1}{8}$?

396 Appendix B SUPPLEMENTARY UNIT: PROGRESSIONS

Here $a_1 = 8\frac{1}{4}$, $a_n = 12\frac{1}{2}$, $d = \frac{1}{8}$, and $n = $?

$$a_n = a_1 + (n - 1)d$$

$$12\frac{1}{2} = 8\frac{1}{4} + (n - 1)\left(\frac{1}{8}\right)$$

$$12\frac{1}{2} - 8\frac{1}{4} = \frac{n - 1}{8}$$

$$4\frac{1}{4} = \frac{n - 1}{8}$$

$$34 = n - 1$$

$$35 = n$$

The **means** of a progression are the terms that occur between any two given terms in a sequence (called the **extremes**). To determine a certain number of means between two given numbers in an arithmetic progression, we may use the general-term formula, find the appropriate d value, and then compute the desired terms.

EXAMPLES

1. Insert three arithmetic means between 6 and 41.

SOLUTION The resulting arithmetic progression will be a five-term sequence: 6, —, —, —, 41. So $n = 5$, $a_1 = 6$, $a_5 = 41$, and $d = $?

We use the general-term formula, substitute, and solve for d:

$$a_n = a_1 + (n - 1)d$$

$$41 = 6 + (5 - 1)d$$

$$41 = 6 + 4d$$

$$35 = 4d$$

$$8\frac{3}{4} = d$$

The five-term arithmetic progression is: 6, $14\frac{3}{4}$, $23\frac{1}{2}$, $32\frac{1}{4}$, 41; the desired means are $14\frac{3}{4}$, $23\frac{1}{2}$, $32\frac{1}{4}$.

2. Insert one arithmetic mean between 86 and 70.

SOLUTION The resulting arithmetic progression will be: 86, —, 70.

Thus $a_1 = 86$, $n = 3$, $a_n = 70$, and $d = $?

$$a_n = a_1 + (n - 1)d$$
$$70 = 86 + (3 - 1)d$$
$$70 = 86 + 2d$$
$$70 - 86 = 2d$$
$$-\frac{16}{2} = d$$
$$-8 = d$$

The three-term arithmetic progression is: 86, 78, 70; the desired mean is 78.

Note: The "average" of 86 and 70 is

$$\frac{86 + 70}{2} = \frac{156}{2} = 78$$

Thus the mean or middle term of a three-term arithmetic progression, a, $a + d$, $a + 2d$, is the term $a + d$, which we can also determine by "averaging" the first and third terms, a and $a + 2d$:

$$\text{"average"} = \frac{(a) + (a + 2d)}{2} = \frac{a + a + 2d}{2} = \frac{2a + 2d}{2}$$
$$= a + d$$

EXERCISE B-2

Add three more terms to each of the sequences so that the whole sequence forms an arithmetic progression.

1. 19, 25, 31, ...
2. 6, -2, -10, ...
3. 106, 54, 2, ...
4. $\frac{1}{2}, \frac{5}{6}, 1\frac{1}{6}, \ldots$
5. $\frac{2}{3}, \frac{5}{12}, \frac{1}{6}, \ldots$
6. 0.04, 0.017, ...

Find the term indicated in parentheses for each progression.

7. 5, 11, 17, ..., (20th)
8. 17, 13, 9, ..., (16th)
9. 4, $6\frac{1}{2}$, 9, ..., (21st)
10. $3\frac{1}{2}$, 1, $-1\frac{1}{2}$, ..., (17th)
11. $\frac{1}{4}, \frac{1}{6}, \frac{1}{12}, \ldots,$ (25th)
12. 0.08, 0.22, 0.36, ..., (11th)
13. $\sqrt{2}$, 0, $-\sqrt{2}$, ..., (18th)
14. 1.314, 1.248, 1.182, ..., (26th)

In the following, a_1 represents the first term of an arithmetic progression, a_n the nth term, n the number of terms, and d the common difference. Compute the missing datum.

15. $a_1 = 2$, $d = 3$, $a_n = 38$, $n = $?
16. $a_1 = -93$, $a_n = 97$, $n = 20$, $d = $?

17. $a_n = 1\frac{1}{2}, n = 11, d = -\frac{3}{4}, a_1 = ?$

18. $a_1 = -\frac{1}{3}, d = -\frac{1}{3}, a_n = -3, n = ?$

19. $d = \frac{3}{8}, n = 17, a_n = 6\frac{3}{4}, a_1 = ?$

20. $a_n = 36.6, a_1 = 1.4, n = 12, d = ?$

21. $d = -0.2, a_1 = 4.3, a_n = 1.1, n = ?$

22. $a_1 = -3.6, d = -0.4, a_n = -8.8, n = ?$

23. $n = 14, a_n = -8\frac{13}{21}, a_1 = \frac{2}{3}, d = ?$

24. $a_1 = x, d = x - 2a, a_n = 8x - 14a, n = ?$

25. Insert three arithmetic means between 3 and 27.

26. Insert four arithmetic means between $\frac{2}{3}$ and $3\frac{1}{6}$.

27. Insert three arithmetic means between 7 and -37.

28. Insert four arithmetic means between 0.04 and -0.09.

What replacement for x will make the three-term sequence an arithmetic progression?

29. $3x, 4x + 2, 8x - 2$ **30.** $\frac{1}{x}, x - 2, 1 + \frac{2}{x}$

3. SUM OF TERMS IN ARITHMETIC PROGRESSION

A general formula for the sum of the terms of an arithmetic progression (i.e., an arithmetic series) may be developed in this way. Let S_n represent the sum of n terms in arithmetic sequence in which a_1 is the first term, d the common difference, and a_n the final term. Then

$$S_n = a_1 + (a_1 + d) + (a_1 + 2d) + (a_1 + 3d) + \cdots + (a_n - 2d) + (a_n - d) + a_n$$

If we now add the reverse of the expression for S_n above to itself, we obtain

$$
\begin{aligned}
S_n &= a_1 + (a_1 + d) + (a_1 + 2d) + \cdots + (a_n - 2d) + (a_n - d) + a_n \\
(+) \quad S_n &= a_n + (a_n - d) + (a_n - 2d) + \cdots + (a_1 - 2d) + (a_1 - d) + a_1 \\
\hline
2S_n &= (a_1 + a_n) + (a_1 + a_n) + (a_1 + a_n) + \cdots + (a_1 + a_n) + (a_1 + a_n)
\end{aligned}
$$

Since there are n terms in the sequence, there will be n addends of $(a_1 + a_n)$ in the combined sum above; that is,

$$2S_n = n(a_1 + a_n)$$

or

$$S_n = \frac{n}{2}(a_1 + a_n)$$

EXAMPLES

1. Find the sum of the six terms of the arithmetic progression: 8, 12, 16, 20, 24, 28.

SOLUTION   In this progression, $n = 6$, $a_1 = 8$, and $a_n = 28$. Using the formula

$$S_n = \frac{n}{2}(a_1 + a_n)$$

we obtain

$$S_6 = \frac{6}{2}(8 + 28)$$

$$= 3(36)$$

$$= 108$$

Therefore, $8 + 12 + 16 + 20 + 24 + 28 = 108$, which can be verified by adding the terms.

2. Find the sum of the first one hundred counting numbers.

SOLUTION   Here $a_1 = 1$, $n = 100$, and $a_n = 100$.

$$S_n = \frac{n}{2}(a_1 + a_n)$$

$$S_{100} = \frac{100}{2}(1 + 100)$$

$$= 50(101)$$

$$= 5050$$

An alternative formula for the sum of n terms of an arithmetic progression can be obtained by substituting the equivalent for a_n into $S_n = \frac{n}{2}(a_1 + a_n)$:

$$\begin{cases} S_n = \frac{n}{2}(a_1 + a_n) \\ a_n = a_1 + (n - 1)d \end{cases}$$

$$S_n = \frac{n}{2}[a_1 + \boxed{a_1 + (n - 1)d}]$$

$$= \frac{n}{2}[2a_1 + (n - 1)d]$$

The five parts of an arithmetic progression that we have been representing by S_n, a_1, a_n, n, and d are called the **elements** of an arithmetic progression. If *any three* of these five elements are known, by appropriate use of the formulas

$$a_n = a_1 + (n-1)d, \quad S_n = \frac{n}{2}(a_1 + a_n), \quad \text{and} \quad S_n = \frac{n}{2}[2a_1 + (n-1)d]$$

the remaining two elements can be determined.

EXAMPLES

1. The first term of an arithmetic progression is 12, the last is 18, and the sum is 75. Find the common difference and number of terms in the progression.

SOLUTION Use $S_n = \frac{n}{2}(a_1 + a_n)$ to find n:

$$\left. \begin{array}{l} S_n = 75 \\ a_1 = 12 \\ a_n = 18 \end{array} \right\}$$

$$S_n = \frac{n}{2}(a_1 + a_n)$$

$$75 = \frac{n}{2}(12 + 18)$$

$$75 = 15n$$

$$5 = n$$

Use $a_n = a_1 + (n-1)d$ to find d:

$$a_n = a_1 + (n-1)d$$

$$18 = 12 + (5-1)d$$

$$18 - 12 = 4d$$

$$\frac{3}{2} = d$$

2. In an arithmetic progression the sum is 585, the first term is -3, and the common difference is 6. Find the number of terms and the last term.

SOLUTION Here $S_n = 585$, $a_1 = -3$, and $d = 6$. To find n, we use

$$S_n = \frac{n}{2}[2a_1 + (n-1)d]$$

Then

$$585 = \frac{n}{2}[2(-3) + (n-1)6]$$

$$585 = \frac{n}{2}[-6 + 6n - 6]$$

$$585 = \frac{n}{2}[6n - 12]$$

$$585 = 3n^2 - 6n$$

$$0 = 3n^2 - 6n - 585$$

$$0 = n^2 - 2n - 195$$

$$0 = (n - 15)(n + 13)$$

$$n - 15 = 0 \qquad n + 13 = 0$$

$$n = 15 \qquad\qquad n = -13$$

An arithmetic progression of -13 terms makes no sense; so we reject the solution, -13, and accept 15 as the number of terms in the progression. To find the value of the last term, we substitute 15 for n in

$$a_n = a_1 + (n-1)d$$

$$a_{15} = -3 + (15 - 1)6$$

$$= -3 + (14)(6)$$

$$= 81$$

3. The third term of an eight-term arithmetic progression is 8; the seventh term is -4. Find the remaining elements of this progression.

SOLUTION | Solution of a system of equations is involved here. If the third term is 8 and the seventh term is -4, the progression may be shown as

$$\underline{\quad}, \underline{\quad}, 8, \underline{\quad}, \underline{\quad}, \underline{\quad}, -4, \underline{\quad}$$

Now consider 8 to be the third term of a three-term progression, and -4 as the seventh term of a seven-term progression. Both progressions have the same first term, a_1, and the same common difference, d. Using the general-term formula

$$a_n = a_1 + (n-1)d$$

we obtain the pair of equations

$$\begin{cases} 8 = a_1 + (3-1)d \\ -4 = a_1 + (7-1)d \end{cases}$$

or, more simply,

$$\begin{cases} a_1 + 2d = 8 \\ a_1 + 6d = -4 \end{cases}$$

Using Program **9.3**, we add the negative of the bottom equation to the top equation, and solve for d:

$$-4d = 12$$

$$d = -3$$

Since the third term is $a_1 + 2d = 8$, and $d = -3$, then

$$a_1 + 2(-3) = 8$$

$$a_1 = 14$$

The eighth or final term of the progression must be

$$a_n = a_1 + (n-1)d$$

$$a_8 = 14 + (8-1)(-3)$$

$$= 14 + (7)(-3)$$

$$= -7$$

Finally, to find S_8:

$$S_n = \frac{n}{2}(a_1 + a_n)$$

$$S_8 = \frac{8}{2}[14 + (-7)]$$

$$= 4(7)$$

$$= 28$$

EXERCISE B-3

Some of the elements of an arithmetic progression, a_1, n, d, a_n, and S_n are given. Find the remaining elements.

1. $a_1 = 7$, $n = 12$, $a_n = 40$

2. $a_1 = 7$, $d = 4$, $a_n = 79$

3. $n = 10$, $d = 2$, $a_n = 25$

4. $a_1 = 3$, $n = 14$, $d = 5$

5. $a_1 = 10$, $a_n = 74$, $S_n = 714$

6. $a_1 = 21$, $n = 24$, $S_n = 2160$

7. $n = 22$, $d = -5$, $S_n = -869$

8. $n = 38$, $a_n = 46$, $S_n = 342$

9. $d = \frac{1}{2}$, $a_n = 0$, $S_n = -\frac{105}{2}$

10. $a_1 = 6.1$, $d = -0.3$, $S_n = 64.6$

11. The second term of a seven-term arithmetic progression is 8, and the sixth term is 20. Determine the remaining elements.

12. A charity raffle involves 60 tickets numbered consecutively, in multiples of 5, from 5 to 300, inclusive. If each purchaser pays in cents the number on the ticket he draws, how much will be realized from the raffle?

13. A woman plans to save $100 this year and, in each succeeding year, $50 more than the year before. If she follows this plan, how much will she have accumulated at the end of 15 years?

14. A regular savings program of $500 deposited at the beginning of each year at 5% simple interest will accumulate to what amount at the end of the 14th year?

15. An object falls 16 ft during the first second, 48 ft during the second, 80 ft during the third, and so on.

(a) How far does it fall during the twentieth second?
(b) How far will it have fallen after twenty seconds?
(c) How long will it take to drop 22,500 ft?

16. The ancient Greeks were interested in "triangular numbers," the first four of which were represented as ., .·., .·.·., .·.·.·. (1, 3, 6, 10). How many dots would there be in the representation of:

(a) the eighth triangular number?
(b) the forty-third triangular number?

4. GEOMETRIC PROGRESSION

A **geometric progression** is a sequence in which each term after the first is obtained by *multiplying* the preceding term by a constant. This constant is referred to as the **common ratio**.

EXAMPLES

1. 1, 2, 4, 8, 16, 32 is a geometric progression having a common ratio of 2.

2. $-3, 6, -12, 24, -48$ is a geometric progression with a common ratio of -2.

3. 6, 0.6, 0.06, 0.006, 0.0006 is a geometric progression with a common ratio of 0.1.

4. $\sqrt{2}, 2, 2\sqrt{2}, 4, 4\sqrt{2}, 8$ is a geometric progression with a common ratio of $\sqrt{2}$.

5. $-\dfrac{1}{5}, -\dfrac{1}{25}, -\dfrac{1}{125}, -\dfrac{1}{625}$ is a geometric progression with a common ratio of $\dfrac{1}{5}$.

If we denote the first term of a geometric progression by a_1, and the common ratio by r, then

the first term, a_1, is a_1;
the second term, a_2, is $r \cdot a_1$, or $a_1 r$;
the third term, a_3, is $r(a_1 r)$, or $a_1 r^2$;
the fourth term, a_4, is $r(a_1 r^2)$, or $a_1 r^3$;
the fifth term, a_5, is $a_1 r^4$;
$$\vdots$$
the twentieth term, a_{20}, is $a_1 r^{19}$;
etc.

Note that the exponent of the r-factor is always one less than the number of the term. In general

$$a_n = a_1 r^{n-1}$$

which represents any specified term (the nth) in a geometric progression in which the first term is a_1 and the common ratio is r.

EXAMPLES

1. Find the fifth term of a geometric progression in which the first term is 3 and the common ratio is 2.

SOLUTION The fifth term may be designated as a_5. We may use the general-term expression as a formula

$$a_n = a_1 r^{n-1}$$

substitute 3 for a_1 and 2 for r, and solve:

$$a_5 = (3)(2)^{5-1}$$
$$= (3)(2)^4$$
$$= (3)(16)$$
$$= 48$$

Thus 48 is the fifth term of a geometric progression in which the first term is 3 and the common ratio is 2:

$$3, 6, 12, 24, \mathbf{48}$$

2. Find the sixth term of the geometric progression: $-2, \frac{1}{2}, -\frac{1}{8}, \cdots$.

SOLUTION Here $a_1 = -2$, $n = 6$, and the common ratio, r, is found by dividing one of the terms by the term before it in the sequence $\left[\text{e.g., } r = \frac{1}{2} \div (-2) = -\frac{1}{4} \right]$. The sixth term (a_6) of the progression can be found by substituting in the general-term formula.

$$a_n = a_1 r^{n-1}$$

$$\left.\begin{array}{c} a_1 = -2 \\ n = 6 \\ r = -\dfrac{1}{4} \end{array}\right\}$$

$$a_6 = (-2)\left(-\frac{1}{4}\right)^{6-1}$$

$$= (-2)\left(-\frac{1}{4}\right)^{5}$$

$$= (-2)\left(-\frac{1}{1024}\right)$$

$$= \frac{1}{512}$$

The progression is: $-2, \dfrac{1}{2}, -\dfrac{1}{8}, \dfrac{1}{32}, -\dfrac{1}{128}, \mathbf{\dfrac{1}{512}}$.

3. Find the tenth term of the geometric progression in which the first term is 1.3 and the common ratio is 1.1.

SOLUTION Here $a_1 = 1.3$, $n = 10$, and $r = 1.1$; because of the high power to which one factor must be raised, computing a_{10} is best done by logarithms:

$$a_n = a_1 r^{n-1}$$

$$a_{10} = (1.3)(1.1)^{10-1}$$

$$= (1.3)(1.1)^{9}$$

$$\log 1.1 \approx \qquad 0.0414$$

$$9 \log 1.1 \approx \overset{(\times)}{} \frac{9}{0.3726}$$

$$\log 1.3 \approx \frac{0.1139}{}$$

$$\log a_{10} \approx \overset{(+)}{} \frac{}{0.4865}$$

$$a_{10} = 3.066 \quad \text{(approximately)}$$

When there is incomplete information about a finite geometric progression, the general-term formula may often be used to complete that information. The following examples illustrate.

EXAMPLES

1. What is the first term of a six-term geometric progression in which the ratio is $\sqrt{3}$ and the sixth term is 27?

Here $a_1 = ?$, $a_n = 27$, $r = \sqrt{3}$, and $n = 6$. We find a_1 by:

$$a_n = a_1 r^{n-1}$$

$$a_6 = a_1 r^{6-1}$$

$$27 = a_1 (\sqrt{3})^{6-1}$$

$$27 = a_1 (\sqrt{3})^5$$

$$\frac{27}{9\sqrt{3}} = a_1$$

$$\sqrt{3} = a_1$$

2.   What is the common ratio in an eight-term geometric progression in which the fourth term is 1 and the eighth term is $\dfrac{625}{1296}$?

SOLUTION   The sequence may be represented in this way:

$$\underline{\quad}, \underline{\quad}, \underline{\quad}, 1, \underline{\quad}, \underline{\quad}, \underline{\quad}, \frac{625}{1296}$$

The common ratio r, which must be common throughout, can be found if we consider a part of the progression, namely, the last five terms, as a progression

$$1, \underline{\quad}, \underline{\quad}, \underline{\quad}, \frac{625}{1296}$$

in which $a_1 = 1$, $a_n = \dfrac{625}{1296}$, $n = 5$. Substituting these data in the general-term formula allows us to solve for r:

$$a_n = a_1 r^{n-1}$$

$$\frac{625}{1296} = (1)(r)^{5-1}$$

$$\frac{625}{1296} = r^4$$

$$\frac{5^4}{6^4} = r^4$$

$$\frac{5}{6} = r$$

3. How many terms are there in a geometric progression in which the first and last terms are 16 and $\dfrac{1}{64}$, respectively, and for which the common ratio is $\dfrac{1}{2}$?

SOLUTION Here $a_1 = 16$, $a_n = \dfrac{1}{64}$, $r = \dfrac{1}{2}$; n may be found in this way:

$$a_n = a_1 r^{n-1}$$

$$\frac{1}{64} = (16)\left(\frac{1}{2}\right)^{n-1}$$

$$\frac{1}{64 \times 16} = \left(\frac{1}{2}\right)^{n-1}$$

$$\frac{1}{2^6 \times 2^4} = \left(\frac{1}{2}\right)^{n-1}$$

$$\left(\frac{1}{2}\right)^{10} = \left(\frac{1}{2}\right)^{n-1}$$

By Program **10.6**:

$$10 = n - 1$$

$$11 = n$$

Terms that occur between two extreme terms in a sequence so that all of the terms of the sequence form a geometric progression are called the **geometric means** of the progression. We may generate these interior terms by determining the common ratio of the progression, and then use it to produce the terms.

EXAMPLES

1. Insert two geometric means between 7 and 189.

SOLUTION The resulting geometric progression will be a four-term sequence:

$$7, \text{\_\_\_}, \text{\_\_\_}, 189.$$

Thus, $a_1 = 7$, $a_n = 189$, and $n = 4$; r may be found by:

$$a_n = a_1 r^{n-1}$$

$$189 = 7(r)^{4-1}$$

$$27 = r^3$$

$$3 = r$$

The four-term geometric progression, then, is 7, 21, 63, 189; the desired means are 21 and 63.

2. Insert three geometric means between 11 and 891.

SOLUTION | The resulting geometric progression will have five terms; so $n = 5$, $a_1 = 11$, and $a_n = 891$.

$$a_n = a_1 r^{n-1}$$
$$891 = 11(r)^{5-1}$$
$$81 = r^4$$
$$+3, -3 = r$$

Thus there are two geometric progressions that fit the given data. One has a common ratio of $+3$: ($11, \textbf{33}, \textbf{99}, \textbf{297}, 891$), and the other a common ratio of -3: ($11, -\textbf{33}, \textbf{99}, -\textbf{297}, 891$). The respective geometric means are shown in boldface.

EXERCISE B-4

Add three more terms to each of the sequences so that the whole sequence forms a geometric progression.

1. $2, 6, 18, \ldots$ **2.** $12, 6, 3, \ldots$ **3.** $1, -3, 9, \ldots$

4. $\dfrac{3}{4}, -\dfrac{3}{8}, \dfrac{3}{16}, \ldots$ **5.** $-3, 0.6, -0.12, \ldots$ **6.** $3, 3\sqrt{2}, 6, \ldots$

Find the term indicated in parentheses for the given geometric progression.

7. $1, 3, 9, \ldots,$ (6th) **8.** $3, 6, 12, \ldots,$ (7th)

9. $\dfrac{1}{2}, 1, 2, \ldots,$ (10th) **10.** $\dfrac{1}{2}, -\dfrac{1}{10}, \dfrac{1}{50}, \ldots,$ (8th)

11. $\sqrt{2}, -2, 2\sqrt{2}, \ldots,$ (11th) **12.** $\sqrt[3]{2}, \sqrt[3]{4}, 2, \ldots,$ (15th)
13. $i, -1, -i, 1, i, \ldots,$ (16th) **14.** $5, -1, 0.2, \ldots,$ (9th)

In the following, a_1 represents the first term of a geometric progression, n is the number of terms, a_n the last term, and r the common ratio.

15. $a_n = 48, a_1 = 3, r = 2, n = ?$

16. $a_n = 243, a_1 = 3, n = 5, r = ?$

17. $a_n = -1250, n = 5, r = 5, a_1 = ?$

18. $a_n = 81, a_1 = \dfrac{1}{3}, r = 3, n = ?$

19. $a_n = -2916, a_1 = -4, n = 7, r = ?$

20. $a_n = -\dfrac{1}{32}, n = 8, a_1 = 4, r = ?$

21. $a_n = 24.3, n = 6, r = 3, a_1 = ?$

22. $a_n = \dfrac{1}{972}, a_1 = \dfrac{3}{4}, r = -\dfrac{1}{3}, n = ?$

23. $a_n = -162, a_1 = 2\sqrt{3}, n = 8, r = ?$

24. $a_n = 64, a_1 = \sqrt[3]{2}, r = \sqrt[3]{2}, n = ?$

25. Insert three geometric means between 4 and 324.

26. Insert four geometric means between 160 and 5.

27. Insert four geometric means between $-\dfrac{2}{3}$ and $5\dfrac{1}{16}$.

28. Insert five geometric means between 5 and 625.

What replacement for x will make the three-term sequence a geometric progression?

29. $x - 2, 2x + 1, 7x - 4$.

30. $3x - 1, 5x + 1, 10x + 2$

5. SUM OF TERMS IN GEOMETRIC PROGRESSION

The sum of the first n terms in a geometric progression, S_n, in which the first term is a_1 and the common ratio is r, may be written

$$S_n = a_1 + a_1r + a_1r^2 + a_1r^3 + \cdots + a_1r^{n-2} + a_1r^{n-1}$$

If we multiply both members of this equation by r, we get

$$rS_n = a_1r + a_2r^2 + a_1r^3 + a_1r^4 + \cdots + a_1r^{n-1} + a_1r^n$$

Subtracting the second equation from the first yields the equation

$$S_n - rS_n = a_1 - a_1r^n$$

which may be solved for S_n:

$$S_n(1 - r) = a_1(1 - r^n)$$

$$S_n = a_1\left(\frac{1 - r^n}{1 - r}\right) \quad (r \neq 1)$$

In this we have a formula for finding the sum of the first n terms of a geometric progression.

EXAMPLES

1. Find the sum of 1, 2, 4, 8, 16, 32, 64, 128.

SOLUTION The terms here form a geometric progression in which $a_1 = 1$, $r = 2$, and $n = 8$; the sum can be found as follows:

$$S_n = a_1\left(\frac{1 - r^n}{1 - r}\right)$$

$$S_8 = 1\left(\frac{1 - 2^8}{1 - 2}\right)$$

$$= 1\left(\frac{-255}{-1}\right)$$

$$= 255$$

2. Find the sum of the ten terms that form a geometric progression in which the first term is 0.35 and the common ratio is 1.6.

SOLUTION Logarithms will be useful here; $a_1 = 0.35$, $r = 1.6$, $n = 10$.

$$S_n = a_1\left(\frac{1 - r^n}{1 - r}\right)$$

$$S_{10} = (0.35)\left(\frac{1 - 1.6^{10}}{1 - 1.6}\right)$$

$$\begin{cases} \log 1.6 \approx 0.2041 \\ 10 \log 1.6 \approx 2.0410 \\ 1.6^{10} \approx 109.9 \end{cases}$$

$$\approx (0.35)\left(\frac{1 - 109.9}{1 - 1.6}\right)$$

$$\approx (0.35)\left(\frac{108.9}{0.6}\right)$$

$$\begin{cases} \log 108.9 \approx \quad\quad 2.0370 \\ \log 0.35 \approx \quad\; \dfrac{9.5441 - 10}{11.5811 - 10} \;(+) \\ \\ \log 0.6 \approx \quad\; 9.7782 - 10 \\ \log S_n \approx \quad\; \dfrac{}{1.8029} \;(-) \\ \\ S_n \approx 63.51 \end{cases}$$

$$\approx 63.51$$

The elements of a geometric progression are the five we have been representing by $a_1, a_n, n, r,$ and S_n. Given any three of these elements, appropriate use of the formulas:

(1) $$a_n = a_1 r^{n-1}$$

(2) $$S_n = a_1\left(\frac{1 - r^n}{1 - r}\right) \quad (r \neq 1)$$

(3) $$S_n = \frac{a_1 - r a_n}{1 - r} \quad (r \neq 1)$$

makes it possible to determine the remaining two elements. Formula (3) is a variation of formula (2):

(2) $$S_n = a_1\left(\frac{1 - r^n}{1 - r}\right) = \frac{a_1 - a_1 r^n}{1 - r}$$

If $a_n = a_1 r^{n-1}$, which is formula (1), then

$$(r)(a_n) = (r)(a_1 r^{n-1}) = a_1 r^n.$$

So, if we replace $a_1 r^n$ in formula (2) above with its equivalent $r a_n$, we obtain

(3) $$S_n = \frac{a_1 - r a_n}{1 - r}$$

EXAMPLES

1. A geometric progression has a sum of $10\frac{1}{12}$, a first term of $\frac{1}{12}$, and a last term of $6\frac{3}{4}$. Find the remaining elements.

SOLUTION   Here $a_1 = \frac{1}{12}$, $a_n = 6\frac{3}{4}$, $S_n = 10\frac{1}{12}$, $n = ?$, $r = ?$ Using formula (3)

$$S_n = \frac{a_1 - ra_n}{1 - r}$$

to find r:

$$10\frac{1}{12} = \frac{\left(\frac{1}{12}\right) - (r)\left(6\frac{3}{4}\right)}{1 - r}$$

$$\frac{121}{12} - \frac{121}{12}r = \frac{1}{12} - \frac{27}{4}r$$

$$121 - 121r = 1 - 81r$$

$$121 - 1 = 121r - 81r$$

$$120 = 40r$$

$$3 = r$$

Using formula (1), $a_n = a_1 r^{n-1}$, to find n:

$$6\frac{3}{4} = \frac{1}{12}(3)^{n-1}$$

$$(12)\left(\frac{27}{4}\right) = 3^{n-1}$$

$$81 = 3^{n-1}$$

$$3^4 = 3^{n-1}$$

$$4 = n - 1$$

$$5 = n$$

2. The last two terms of a four-term geometric progression are $\frac{4}{175}$ and $\frac{8}{875}$. Find the remaining elements.

SOLUTION   The common ratio, r, can be found by dividing the last term by the one before it:

$$r = \frac{8}{875} \div \frac{4}{175} = \frac{\overset{2}{\cancel{8}}}{\underset{5}{\cancel{875}}} \times \frac{\cancel{175}}{\cancel{4}} = \frac{2}{5}$$

Since $n = 4$, $r = \dfrac{2}{5}$, and $a_4 = \dfrac{8}{875}$, a_1 can be found by formula (1):

$$a_n = a_1 r^{n-1}$$

$$\frac{8}{875} = a_1 \left(\frac{2}{5}\right)^{4-1}$$

$$\frac{\cancel{125}}{\cancel{8}} \times \frac{\cancel{8}}{\underset{7}{\cancel{875}}} = a_1$$

$$\frac{1}{7} = a_1$$

Using formula (2) to compute the sum:

$$S_n = a_1 \left(\frac{1 - r^n}{1 - r}\right)$$

$$S_4 = \frac{1}{7} \left(\frac{1 - \left(\dfrac{2}{5}\right)^4}{1 - \dfrac{2}{5}}\right)$$

$$= \frac{1}{7} \left(\frac{\dfrac{609}{625}}{\dfrac{3}{5}}\right)$$

$$= \frac{29}{125}$$

If we analyze formula (2) for the sum of a finite number of terms in geometric progression

$$S_n = a_1 \left(\frac{1 - r^n}{1 - r}\right)$$

we can see that when r lies between -1 and 1 (i.e., $-1 < r < 1$), the r^n term will become numerically smaller and smaller as n, the number of terms and the exponent of r, increases. For example, if $r = \dfrac{1}{3}$, then $r^2 = \dfrac{1}{9}$, $r^3 = \dfrac{1}{27}$, $r^4 = \dfrac{1}{81}$, and so on.

As the number of terms in a geometric progression increases indefinitely, the value of the r^n term for r's between -1 and 1 (or $|r| < 1$) tends to diminish toward zero. In the extreme, the formula for the sum of an *infinite* number of terms in geometric progression becomes

$$S_\infty = a_1 \left(\frac{1 - 0}{1 - r}\right) = \frac{a_1}{1 - r} \quad (|r| < 1)$$

Thus for some progressions, although the number of terms is infinite, the sum of its terms will be finite.

EXAMPLES

1. Find the sum of $\frac{1}{2}, \frac{1}{4}, \frac{1}{8}, \frac{1}{16}, \ldots$, an infinite geometric progression.

SOLUTION Here the first term is $\frac{1}{2}$, the common ratio is $\frac{1}{2}$ $\left(\text{notice that} -1 < \frac{1}{2} < 1\right)$ and the number of terms is infinite; so:

$$S_\infty = \frac{a_1}{1 - r}$$

$$S_\infty = \frac{\dfrac{1}{2}}{1 - \dfrac{1}{2}} = \frac{\dfrac{1}{2}}{\dfrac{1}{2}} = 1$$

2. Express the infinite repeating decimal $0.31616161616 \ldots$ as a fraction.

SOLUTION Here $0.31616161616 \ldots$ may be written as a sum of terms that form a geometric progression, after the first digit:

$$0.31616161616 \ldots = 0.3 + \underbrace{0.016 + 0.00016 + 0.0000016 + \cdots}_{\text{G.P.}}$$

Considering 0.016 as the first term of a geometric progression, and $r = 0.01$, then

$$S_\infty = \frac{a_1}{1 - r}$$

$$S_\infty = \frac{0.016}{1 - 0.01} = \frac{0.016}{0.99} = \frac{16}{990}$$

and

$$0.31616161616 \ldots = 0.3 + \frac{16}{990}$$

$$= \frac{3}{10} + \frac{16}{990} = \frac{297 + 16}{990} = \frac{313}{990}$$

Note: This verifies that $0.3161616 \ldots$, an unending decimal but with repetition, is a rational number; recall the discussion of such numbers on page 3.

Some of the elements of a geometric progression a_1, n, r, a_n, S_n, are given. Determine the remaining elements.

1. $n = 6, r = 2, a_n = 32$

2. $a_1 = \dfrac{1}{2}, n = 6, a_n = 16$

3. $a_1 = 128, n = 7, r = -\dfrac{1}{2}$

4. $a_1 = 1, r = 3, a_n = 243$

5. $n = 6, r = -\dfrac{1}{2}, S_n = \dfrac{21}{2}$

6. $r = \dfrac{1}{2}, a_n = \dfrac{1}{4}, S_n = \dfrac{63}{4}$

7. $a_1 = 1\dfrac{1}{2}, a_n = 96, S_n = 190\dfrac{1}{2}$

8. $a_1 = -96, r = -\dfrac{3}{2}, S_n = 399$

9. Write the seven terms that form a geometric progression if the third term is 81 and the sixth term is -3.

10. Write the six terms that form a geometric progression if the second term is $\dfrac{5}{12}$ and the fifth term is $\dfrac{2}{75}$.

11. A man plans to save \$3 during January, \$6 during February, \$12 during March, and so on, each month's savings being double that of the month before.

   (a) How much must he save during August?
   (b) How much will he have saved by the end of the year?

12. A certain type of bacteria doubles its number every 3 hr. How many times as great will their number be after 24 hr?

13. If there are n of the bacteria referred to in Exercise 12 in a culture at the beginning of a time interval, how many would there be $19\dfrac{1}{2}$ hr later?

14. For figuring depreciation on certain items, accountants sometimes use a "constant percent" or "declining balance" method, in which the value of an asset is considered to depreciate the same percent each year, based on the asset's book value of the previous year. According to this method, what would be the book value of an asset after 5 years of use if it cost \$1000 originally, and if depreciation is figured at a constant 20% a year of its previous year's book value?

   Note: 20% off is the same as 80% of.

15. A ball is dropped from a height of 27 ft, and on each rebound it comes up to a height that is two-thirds of its previous fall. How far will it have traveled by the time it reaches the top of its fifth rebound?

16. Theoretically, how far will the ball of Exercise 15 travel before it comes to rest?

Find the sums of the infinite geometric progressions.

17. $8, 4, 2, \ldots$ **18.** $27, -9, 3, \ldots$ **19.** $6, 0.6, 0.06, \ldots$

20. $\sqrt{2}, 1, \frac{1}{2}\sqrt{2}, \ldots$

Express the infinitely repeating decimals as equivalent fractions.

21. $0.7777\ldots$ **22.** $0.373737\ldots$ **23.** $0.22333\ldots$
24. $4.603111\ldots$

REVIEW

PART A

Answer True or False.

1. A sequence of numbers with a common difference between successive numbers in the sequence is an arithmetic progression.
2. It is possible to predict the value of any term in an arithmetic progression if you know the first term and the common difference.
3. Two numbers and their arithmetic mean always form an arithmetic progression.
4. $10, 1, 0.1, 0.01, 0.001$ is an example of an arithmetic progression.
5. Consecutive members of a geometric progression differ from one another by a common ratio.
6. The geometric mean between 2 and 8 is 6.
7. Any infinite set of numbers in geometric progression has an exact finite sum.

PART B

Add three additional terms to each of the sequences so that the whole sequence forms an arithmetic progression.

1. $38, 44, 50, \ldots$ **2.** $98, 51, 4, \ldots$ **3.** $\frac{5}{6}, \frac{5}{12}, 0, \ldots$

For each of the following progressions, find the term indicated in parentheses.

4. $6, 11, 16, \ldots, (18\text{th})$ **5.** $7, 9\frac{1}{2}, 12, \ldots, (21\text{st})$

6. $\frac{11}{12}, \frac{2}{3}, \frac{5}{12}, \ldots, (25\text{th})$ **7.** $\sqrt{3}, 0, -\sqrt{3}, \ldots, (23\text{rd})$

In the following, a_1 represents the first term of an arithmetic progression, a_n the general term, n the number of terms, and d the common difference.

8. $a_1 = 9, d = 2, a_n = 57, n = ?$ **9.** $a_n = 2\frac{1}{2}, n = 13, d = -\frac{1}{2}, a_1 = ?$

10. $d = \dfrac{2}{7}, n = 15, a_n = 4\dfrac{3}{7}, a_1 = ?$ **11.** $d = -0.3, a_1 = 6.1, a_n = 0.7, n = ?$

12. $n = 17, a_n = -8\dfrac{17}{20}, a_1 = \dfrac{3}{4}, d = ?$

13. Insert three arithmetic means between 5 and 29.

14. Insert three arithmetic means between 3 and -33.

15. What replacement for x will make the three-term sequence

$$2x + 1, 4x - 2, 3x + 4$$

an arithmetic progression?

Some of the elements of an arithmetic progression, a_1, a_n, n, d, and S_n, are given. Determine the remaining elements.

16. $a_1 = 7, n = 11, a_n = 47$ **17.** $n = 18, d = 3, a_n = 56$

18. $a_1 = 9, a_n = 86, S_n = 570$ **19.** $n = 25, d = -\dfrac{5}{2}, S_n = 750$

20. $d = -\dfrac{3}{4}, a_n = \dfrac{3}{4}, S_n = \dfrac{117}{2}$

21. The second term of a six-term arithmetic progression is -3, and the fifth term is -15. Find the values of the remaining elements.

22. A woman plans to save $50 this year and, in each succeeding year, $50 more than the year before. If she follows this plan, how much will she have accumulated at the end of 15 years?

23. An object falls 16 ft during the first second, 48 ft during the second second, 80 ft during the third, and so on.

    (a) How far does it fall during the sixteenth second?
    (b) How far will it have fallen after sixteen seconds?
    (c) How long will it take to drop 19,600 ft?

24. A growing city has a net increase of 200 people per week. If the population is 423,650 at the beginning of the year, what is the projected population at the end of the year? (Assume 52 weeks; treat as a progression.)

Add three more terms to each sequence so that the completed sequence forms a geometric progression.

25. $8, -16, 32, \ldots$ **26.** $7, 14, 28, \ldots$

27. $3, -\dfrac{1}{2}, \dfrac{1}{12}, \ldots$ **28.** $-7, 2.1, -0.63, \ldots$

For each of the following geometric progressions, find the term indicated in parentheses.

29. $4, 12, 36, \ldots$, (6th)   **30.** $\dfrac{1}{3}, \dfrac{2}{3}, \dfrac{4}{3}, \ldots$, (10th)

31. $\sqrt{3}, -3, 3\sqrt{3}, \ldots, (11\text{th})$ **32.** $-1, -i, 1, i, -1, \ldots, (16\text{th})$

In the following, a_1 represents the first term of a geometric progression, a_n the general term, n the number of terms, and r the common ratio.

33. $a_n = 80, a_1 = 5, r = 2, n = ?$

34. $a_n = -1875, n = 5, r = 5, a_1 = ?$

35. $a_n = -1458, r = 3, n = 7, a_1 = ?$

36. $a_n = 72.9, a_1 = 0.3, r = 3, n = ?$

37. $a_n = -64, a_1 = 2\sqrt{2}, n = 10, r = ?$

38. Insert three geometric means between 7 and 567.

39. Insert four geometric means between $-\dfrac{3}{5}$ and $7\dfrac{58}{81}$.

40. For what replacement for x will this three-term sequence, $x - 1, 2x + 1, 8x - 5$, be a geometric progression?

Some of the elements of a geometric progression, a_1, a_n, n, r, and S_n, are given. Determine the remaining elements.

41. $n = 5, r = 3, a_n = 243$ **42.** $a_1 = -4, n = 7, r = -2$

43. $n = 4, r = -\dfrac{3}{4}, S_n = 100$ **44.** $a_1 = 4096, a_n = 1, S_n = 5461$

45. Write the seven terms that form a geometric progression if the third term is 8 and the sixth term is -1.

46. A man plans to save $5 during January, $10 during February, $20 during March, and so on, each month's savings being double that of the month before.

(a) How much must he save during October?

(b) How much will he have saved by the end of the year?

47. If there are n of a certain bacteria which double their number every 4 hr, how many will there be 22 hr later?

48. A ball is dropped from a height of 128 ft, and on each rebound it comes up to a height that is half of its previous fall. How far will it have traveled by the time it reaches the top of its fifth rebound?

49. If inflation is predicted to be 10% a year over the previous year for each of the next 6 years, what might you expect to pay for an item 6 years hence that costs $2 now?

50. A machine that costs $3000 new is to be depreciated at 10% of its book value each year. What will be its book value at the end of its third year of service?

Determine the sum of the terms of the infinite geometric progression.

51. $10, 5, 2\dfrac{1}{2}, \ldots$ **52.** $8, 0.8, 0.08, \ldots$

53. $1, -\dfrac{1}{2}, \dfrac{1}{4}, -\dfrac{1}{8}, \ldots$ **54.** $-\dfrac{2}{3}, \dfrac{1}{6}, -\dfrac{1}{24}, \dfrac{1}{96}, \ldots$

Express the following infinitely repeating decimals as equivalent fractions.

55. $0.88888\ldots$ **56.** $0.326666\ldots$

57. $0.4363636\ldots$ **58.** $3.6555\ldots$

SUPPLEMENTARY UNIT: BINOMIAL THEOREM

C

1. BINOMIAL EXPANSION

The expansion of a binomial is a series with a distinct pattern among the terms. Once that pattern is known, it is possible to write directly the expansion of any binomial raised to any positive integral power—without going through the laborious detail of polynomial multiplication.

Our approach to this shorter technique is inductive. We list the expansions of $(a + b)^n$ for $n = 1, 2, 3, 4,$ and 5, and then observe consistent patterns and properties in those expansions that lead to generalizations.

$$(a + b)^1 = a + b$$
$$(a + b)^2 = a^2 + 2ab + b^2$$
$$(a + b)^3 = a^3 + 3a^2b + 3ab^2 + b^3$$
$$(a + b)^4 = a^4 + 4a^3b + 6a^2b^2 + 4ab^3 + b^4$$
$$(a + b)^5 = a^5 + 5a^4b + 10a^3b^2 + 10a^2b^3 + 5ab^4 + b^5$$

A study of these five expansions of $(a + b)$ reveals the following properties:

1. The first term in each expansion is a^n.
2. The last term in each expansion is b^n.
3. There are $n + 1$ terms in each expansion.
4. The exponents of a decrease by one with each successive term in the expansion, starting with a^n in the first term and ending with a^0 (which as a factor is unwritten, $a^0 = 1$) in the last term.
5. The exponents of b increase by one with each successive term in the expansion, starting with b^0 (which as a factor is unwritten, $b^0 = 1$) in the first term and ending with b^n in the last term.
6. The sum of the exponents for each term is n.
7. The numerical coefficient of the second term in each expansion is n.
8. After the first term, the numerical coefficient of any term is the product of the numerical coefficient and the exponent of the a factor of the previous term, divided by the number of the previous term.

Example: In the expansion of $(a + b)^5$, the second term is $5a^4b$, and the third term is $10a^3b^2$; the numerical coefficient of the third term (10) is the product of the coefficient of the previous term (5) and the exponent of the a factor of that term (4), divided by the number of the previous (second) term: 2; or $(5)(4)/2 = 10$.

9. The sequence of numerical coefficients from the first term to the middle of the expansion is the same as the sequence of numerical coefficients, only in reverse order, from the middle of the expansion to the last term. (Or, coefficients equidistant from the extremities of the expansion are equal.)

If we assume that what has been consistent for $n = 1, 2, 3, 4,$ and 5 will also be consistent for $n = 6$, the expansion of $(a + b)^6$ should have seven terms (property 3) and should be as follows:

The first term should be a^n: a^6 (property 1)

The second term should be $na^{n-1}b$: $6a^{6-1}b^{0+1} = 6a^5b$ (properties 4, 5, 7)

The third term should be $\dfrac{(6)(5)}{2} a^{5-1}b^{1+1}$: $15a^4b^2$ (properties 4, 5, 8)

The fourth term should be $\dfrac{(15)(4)}{3} a^{4-1}b^{2+1}$: $20a^3b^3$ (properties 4, 5, 8)

The fifth term should be $\dfrac{(20)(3)}{4} a^{3-1}b^{3+1}$: $15a^2b^4$ (properties 4, 5, 8)

The sixth term should be $\dfrac{(15)(2)}{5} a^{2-1}b^{4+1}$: $6ab^5$ (properties 4, 5, 8)

The seventh (last) term should be b^n: b^6 (property 2)

Thus, according to our inductive reasoning, we say

$$(a + b)^6 = a^6 + 6a^5b + 15a^4b^2 + 20a^3b^3 + 15a^2b^4 + 6ab^5 + b^6$$

That this expansion is in fact correct can be verified if we compute the product of the six factors, $(a + b)(a + b)(a + b)(a + b)(a + b)(a + b)$, in the usual way.

What we have developed intuitively here is expressed more generally in the **Binomial Theorem**.

For any positive integer n,

$$(a + b)^n = a^n + na^{n-1}b + \frac{(n)(n - 1)}{2!} a^{n-2}b^2 + \frac{(n)(n - 1)(n - 2)}{3!} a^{n-3}b^3$$

$$+ \frac{(n)(n - 1)(n - 2)(n - 3)}{4!} a^{n-4}b^4 + \cdots$$

$$+ \frac{(n)(n - 1)(n - 2) \cdots (n - r + 2)}{(r - 1)!} a^{n-r+1}b^{r-1} + \cdots$$

$$+ nab^{n-1} + b^n$$

where r is the number of the term in the expansion.

Note: $r!$ means the product $1 \times 2 \times 3 \times 4 \times 5 \times \cdots \times r$; $0! = 1! = 1$. For instance, $3! = 1 \times 2 \times 3 = 6$; $5! = 1 \times 2 \times 3 \times 4 \times 5 = 120$. The symbol $r!$ is read "r-factorial"; $5!$ is "five-factorial"; etc.

EXAMPLES

1. Write the expansion for $(a + b)^5$.

SOLUTION There will be six terms in the expansion—one more than the exponent, 5.
First term: a^5
Second term: $5a^4b$ (or $na^{n-1}b$)
Third term: The previous (2nd) term is $5\,a^4\,b$; this term is:

$$\frac{(5)(4)}{\rightarrow 2} a^{4-1}b^{1+1} = 10a^3b^2$$

Fourth term: The previous (3rd) term is $10\,a^3\,b^2$; this term is:

$$\frac{(10)(3)}{\rightarrow 3} a^{3-1}b^{2+1} = 10a^2b^3$$

Fifth term: The previous (4th) term is $10\,a^2b^3$; this term is:

$$\frac{(10)(2)}{4}\,a^{2-1}b^{3+1} = 5ab^4$$

Sixth term: The previous (5th) term is $5\,a^1b^4$; this term is:

$$\frac{(5)(1)}{5}\,a^{1-1}b^{4+1} = 1a^0b^5 = b^5$$

The expansion:

$$(a+b)^5 = a^5 + 5a^4b + 10a^3b^2 + 10a^2b^3 + 5ab^4 + b^5$$

2. Expand $(3x + 2y)^4$.

SOLUTION We use $(a+b)^4$ as a model.

$$(a+b)^4 = a^4 + 4a^3b + 6a^2b^2 + 4ab^3 + b^4$$

Then we replace a with $3x$ and b with $2y$:

$$(3x+2y)^4 = (3x)^4 + 4(3x)^3(2y) + 6(3x)^2(2y)^2 + 4(3x)(2y)^3 + (2y)^4$$

$$= 81x^4 + 4(27x^3)(2y) + 6(9x^2)(4y^2) + 4(3x)(8y^3) + 16y^4$$

$$= 81x^4 + 216x^3y + 216x^2y^2 + 96xy^3 + 16y^4$$

3. Expand $(x - 2y)^5$.

SOLUTION The model:

$$(a+b)^5 = a^5 + 5a^4b + 10a^3b^2 + 10a^2b^3 + 5ab^4 + b^5$$

Replace a with x and b with $-2y$:

$$(x-2y)^5 = (x)^5 + 5(x)^4(-2y) + 10(x)^3(-2y)^2 + 10(x)^2(-2y)^3$$
$$+ 5(x)(-2y)^4 + (-2y)^5$$

$$= x^5 + 5(x^4)(-2y) + 10(x^3)(4y^2) + 10(x^2)(-8y^3)$$
$$+ 5(x)(16y^4) + (-32y^5)$$

$$= x^5 - 10x^4y + 40x^3y^2 - 80x^2y^3 + 80xy^4 - 32y^5$$

4. Write the first four terms of the expansion $(x^2 - 2y)^{12}$.

SOLUTION The model:

$$(a+b)^{12} = a^{12} + 12a^{11}b + 66a^{10}b^2 + 220a^9b^3 + \cdots$$

The first four terms of

$$(x^2 - 2y)^{12} = (x^2)^{12} + 12(x^2)^{11}(-2y) + 66(x^2)^{10}(-2y)^2$$
$$+ 220(x^2)^9(-2y)^3 + \cdots$$
$$= x^{24} - 24x^{22}y + 264x^{20}y^2 - 1760x^{18}y^3 + \cdots$$

5. Write the expansion of $(1.02)^7$ by the binomial theorem, carrying it to four terms.

SOLUTION Let $a = 1$ and $b = 0.02$; then $(a + b) = 1 + 0.02 = 1.02$, and $(a + b)^7 = (1.02)^7$:

$$(a + b)^7 = a^7 + 7a^6b + 21a^5b^2 + 35a^4b^3 + \cdots$$

$$\left. \begin{array}{l} a = 1 \\ b = 0.02 \end{array} \right\}$$
$$= (1)^7 + 7(1)^6(0.02) + 21(1)^5(0.02)^2$$
$$+ 35(1)^4(0.02)^3 + \cdots$$

$$= 1 + (7)(1)(0.02) + (21)(1)(0.0004)$$
$$+ (35)(1)(0.000008) + \cdots$$

$$= 1 + 0.14 + 0.0084 + 0.000280 + \cdots$$

As the expansion proceeds, the increasingly higher powers of the b factor—here 0.02—makes the value of the terms progressively smaller. The sum of the first four terms is 1.14868; by five-place logarithms, $(1.02)^7$ is 1.1487.

EXERCISE C-1

Expand.

1. $(a + b)^8$ **2.** $(a - b)^9$ **3.** $(x - 2a)^5$

4. $(3a + b)^4$ **5.** $(3x - 2y)^6$ **6.** $(x^2 - y)^7$

7. $\left(\dfrac{1}{2}x - \dfrac{1}{3}y \right)^5$ **8.** $\left(\dfrac{a}{3} - b \right)^4$ **9.** $(1.03)^4$

10. $(x - \sqrt{y})^4$ **11.** $(\sqrt{a} - \sqrt{b})^3$ **12.** $\left(\dfrac{1}{\sqrt{2}} + \sqrt{2} \right)^5$

Write and simplify the first four terms in the expansions of the following.

13. $(a - 4b)^8$ **14.** $(3a + 2b)^{12}$ **15.** $(a - 4b)^{10}$

16. $(a - 2)^{17}$ **17.** $(2x + 1)^{14}$ **18.** $\left(1 + \dfrac{a}{b} \right)^{16}$

19. $(\sqrt{3} - 2)^{12}$ **20.** $(\sqrt{2} + \sqrt{3})^{18}$ **21.** $(1.03)^{10}$

22. $(2.01)^8$

2: THE rth TERM OF THE BINOMIAL EXPANSION

The general, or rth, term of the binomial expansion of $(a + b)^n$, given in the Binomial Theorem of the previous section, is

$$\frac{(n)(n - 1)(n - 2) \cdots (n - r + 2)}{(r - 1)!} a^{n-r+1}b^{r-1}$$

It would hardly be economical of time and energy to use it as a formula to determine the third term in the expansion of $(a + b)^2$, but we could. In that instance, we have

$$n = 2 \text{ (exponent)} \quad \text{and} \quad r = 3 \quad \text{(i.e., third term)}$$

So

$n - r + 2 = 2 - 3 + 2 = 1$ (last factor of the numerator of the numerical coefficient)

$r - 1 = 3 - 1 = 2$ (first factor of the denominator of the numerical coefficient, and exponent of the b factor)

$n - r + 1 = 2 - 3 + 1 = 0$ (exponent of the a factor)

By this process, then, the third term of $(a + b)^2$ would be found to be

$$\overset{2}{\overbrace{(n)(n - 1)}} \cdots \overset{1}{\overbrace{(n - r + 2)}} \underbrace{}_{2!} \, a^{\overset{0}{\overbrace{n-r+1}}}b^{\overset{2}{\overbrace{r-1}}} = \frac{(2)(1)}{2!} a^0 b^2$$

$$= 1a^0 b^2$$

$$= b^2$$

The rth-term formula is much more appropriate for instances illustrated in the following examples.

EXAMPLES

1.   Write the tenth term of $(a + b)^{41}$.

SOLUTION   In this case, $n = 41$ and $r = 10$. The first factor in the numerator of the numerical coefficient of the desired term is 41 (i.e., n); the last factor is $n - r + 2$, or $41 - 10 + 2$, or 33. The first factor in the denominator of the numerical coefficient of the desired term is $r - 1$, or $10 - 1$, or 9, which is also the exponent of the b term. The exponent of the a term is $n - r + 1$, or $41 - 10 + 1$, or 32. Thus the tenth term of the expansion $(a + b)^{41}$, given by the formula

$$\frac{(n)(n - 1)(n - 2) \cdots (n - r + 2)}{(r - 1)!} a^{n-r+1}b^{r-1}$$

is

$$\frac{(41)\cancel{(40)}(39)\cancel{(38)}(37)\cancel{(36)}\cancel{(35)}\cancel{(34)}\cancel{(33)}}{\cancel{(9)}\cancel{(8)}\cancel{(7)}\cancel{(6)}\cancel{(5)}\cancel{(4)}\cancel{(3)}\cancel{(2)}(1)}\, a^{32}b^9$$

$$(41)(13)(19)(37)(5)(17)(11)a^{32}b^9$$

or, more simply,

$$350{,}343{,}565a^{32}b^9$$

2. Write the twelfth term of the expansion $(2x - y)^{15}$.

SOLUTION Here $n = 15$, $r = 12$, and

$$n - r + 2 = 5$$
$$r - 1 = 11$$
$$n - r + 1 = 4$$

Incorporating these data into

$$\frac{(n)(n - 1)(n - 2) \,\cdots\, (n - r + 2)}{(r - 1)!}\, a^{n-r+1}b^{r-1}$$

we determine the twelfth term of $(a + b)^{15}$ to be

$$\frac{(15)\cancel{(14)}(13)\cancel{(12)}\cancel{(11)}\cancel{(10)}\cancel{(9)}\cancel{(8)}\cancel{(7)}\cancel{(6)}\cancel{(5)}}{\cancel{(11)}\cancel{(10)}\cancel{(9)}\cancel{(8)}\cancel{(7)}\cancel{(6)}\cancel{(5)}\cancel{(4)}\cancel{(3)}\cancel{(2)}(1)}\, a^4 b^{11}$$

or

$$1365a^4 b^{11}$$

Substituting $2x$ for a and $-y$ for b, we obtain the twelfth term of $(2x - y)^{15}$:

$$1365a^4 b^{11} = 1365(2x)^4(-y)^{11} = 1365(16x^4)(-y^{11})$$
$$= -21{,}840x^4 y^{11}$$

3. How much does the sixth term in the binomial expansion of $(1.02)^7$ contribute to the total value of the expansion?

SOLUTION The sixth term of $(a + b)^7$ is

$$\frac{(7)\cancel{(6)}\cancel{(5)}\cancel{(4)}\cancel{(3)}}{\cancel{(5)}\cancel{(4)}\cancel{(3)}\cancel{(2)}(1)}\, a^2 b^5 = 21a^2 b^5$$

Substituting 1 for a and 0.02 for b, we obtain the sixth term of the expansion for $(1 + 0.02)^7$:

$$21a^2b^5 = 21(1)^2(0.02)^5$$

$$= (21)(1)(0.0000000032)$$

$$= 0.0000000672$$

The sixth term of the expansion $(1 + 0.02)^7$ is 0.0000000672; the seventh and eighth terms each contribute considerably less to the total value because of their higher powers of the factor 0.02.

EXERCISE C-2

Write the term indicated in expansions.

1. Sixth term of $(a + 2)^{12}$ **2.** Fifth term of $(x - 3)^{14}$

3. Fourth term of $(2x - 3)^{10}$ **4.** Seventh term of $\left(x + \dfrac{1}{2}\right)^{12}$

5. Fifth term of $(1 + 0.03)^8$ **6.** Fifth term of $(1 - 0.02)^7$

7. Tenth term of $(a - b)^{12}$ **8.** Fourteenth term of $(3x + y)^{15}$

9. What term in the expansion of $(a + 3b)^9$ contains the factor b^3?

10. In the binomial expansion of $(0.97)^9$, or $(1 - 0.03)^9$, what is the value of the fourth term?

REVIEW

PART A

Answer True or False.

1. The first term of the expansion of $(3x + y)^{17}$ is $(3x)^{17}$.

2. There will be 17 terms in the expansion of $(3x + y)^{17}$.

3. In each term of the expansion of $(3x + y)^{17}$, the sum of the exponents attached to x and y will always be 17.

4. The product, $7 \times 6 \times 5 \times 4 \times 3 \times 2 \times 1 \times 0$, is known as 7-factorial, or 7!.

PART B

Expand.

1. $(a + b)^9$ **2.** $(x - 2a)^4$ **3.** $(2x - 3y)^6$

4. $\left(\dfrac{1}{3}a - \dfrac{1}{2}b\right)^5$ **5.** $(1.04)^4$ **6.** $(\sqrt{x} + \sqrt{y})^3$

Write and simplify the first four terms in the expansions of the following.

7. $(a - 3b)^9$ **8.** $(a + 4b)^{10}$ **9.** $(2x - 1)^{14}$

10. $(\sqrt{3} + 2)^{12}$ **11.** $(1.02)^{10}$

Write the term indicated in the expansions of the following.

12. Sixth term of $(a - 2)^{12}$ **13.** Fourth term of $(2x + 3)^{10}$

14. Fifth term of $(1 + 0.03)^9$ **15.** Tenth term of $(a + b)^{12}$

16. What term in the expansion of $(x + 2y)^8$ contains the factor y^3?

Table I. Squares, square roots, and prime factors

n	n^2	$\sqrt{n}$	Prime factors	n	n^2	$\sqrt{n}$	Prime factors
1	1	1.000		51	2,601	7.141	$3 \cdot 17$
2	4	1.414	2	52	2,704	7.211	$2^2 \cdot 13$
3	9	1.732	3	53	2,809	7.280	53
4	16	2.000	2^2	54	2,916	7.348	$2 \cdot 3^3$
5	25	2.236	5	55	3,025	7.416	$5 \cdot 11$
6	36	2.449	$2 \cdot 3$	56	3,136	7.483	$2^3 \cdot 7$
7	49	2.646	7	57	3,249	7.550	$3 \cdot 19$
8	64	2.828	2^2	58	3,364	7.616	$2 \cdot 29$
9	81	3.000	3^2	59	3,481	7.681	59
10	100	3.162	$2 \cdot 5$	60	3,600	7.746	$2^2 \cdot 3 \cdot 5$
11	121	3.317	11	61	3,721	7.810	61
12	144	3.464	$2^2 \cdot 3$	62	3,844	7.874	$2 \cdot 31$
13	169	3.606	13	63	3,969	7.937	$3^2 \cdot 7$
14	196	3.742	$2 \cdot 7$	64	4,096	8.000	2^6
15	225	3.873	$3 \cdot 5$	65	4,225	8.062	$5 \cdot 13$
16	256	4.000	2^4	66	4,356	8.124	$2 \cdot 3 \cdot 11$
17	289	4.123	17	67	4,489	8.185	67
18	324	4.243	$2 \cdot 3^2$	68	4,624	8.246	$2^2 \cdot 17$
19	361	4.359	19	69	4,761	8.307	$3 \cdot 23$
20	400	4.472	$2^2 \cdot 5$	70	4,900	8.367	$2 \cdot 5 \cdot 7$
21	441	4.583	$3 \cdot 7$	71	5,041	8.426	71
22	484	4.690	$2 \cdot 11$	72	5,184	8.485	$2^3 \cdot 3^2$
23	529	4.796	23	73	5,329	8.544	73
24	576	4.899	$2^3 \cdot 3$	74	5,476	8.602	$2 \cdot 37$
25	625	5.000	5^2	75	5,625	8.660	$3 \cdot 5^2$
26	676	5.099	$2 \cdot 13$	76	5,776	8.718	$2^2 \cdot 19$
27	729	5.196	3^3	77	5,929	8.775	$7 \cdot 11$
28	784	5.292	$2^2 \cdot 7$	78	6,084	8.832	$2 \cdot 3 \cdot 13$
29	841	5.385	29	79	6,241	8.888	79
30	900	5.477	$2 \cdot 3 \cdot 5$	80	6,400	8.944	$2^4 \cdot 5$
31	961	5.568	31	81	6,561	9.000	3^4
32	1,024	5.657	2^5	82	6,724	9.055	$2 \cdot 41$
33	1,089	5.745	$3 \cdot 11$	83	6,889	9.110	83
34	1,156	5.831	$2 \cdot 17$	84	7,056	9.165	$2^2 \cdot 3 \cdot 7$
35	1,225	5.916	$5 \cdot 7$	85	7,225	9.220	$5 \cdot 17$
36	1,296	6.000	$2^2 \cdot 3^2$	86	7,396	9.274	$2 \cdot 43$
37	1,369	6.083	37	87	7,569	9.327	$3 \cdot 29$
38	1,444	6.164	$2 \cdot 19$	88	7,744	9.381	$2^3 \cdot 11$
39	1,521	6.245	$3 \cdot 13$	89	7,921	9.434	89
40	1,600	6.325	$2^3 \cdot 5$	90	8,100	9.487	$2 \cdot 3^2 \cdot 5$
41	1,681	6.403	41	91	8,281	9.539	$7 \cdot 13$
42	1,764	6.481	$2 \cdot 3 \cdot 7$	92	8,464	9.592	$2^2 \cdot 23$
43	1,849	6.557	43	93	8,649	9.644	$3 \cdot 31$
44	1,936	6.633	$2^2 \cdot 11$	94	8,836	9.695	$2 \cdot 47$
45	2,025	6.708	$3^2 \cdot 5$	95	9,025	9.747	$5 \cdot 19$
46	2,116	6.782	$2 \cdot 23$	96	9,216	9.798	$2^5 \cdot 3$
47	2,209	6.856	47	97	9,409	9.849	97
48	2,304	6.928	$2^4 \cdot 3$	98	9,604	9.899	$2 \cdot 7^2$
49	2,401	7.000	7^2	99	9,801	9.950	$3^2 \cdot 11$
50	2,500	7.071	$2 \cdot 5^2$	100	10,000	10.000	$2^2 \cdot 5^2$

Table II.   Common logarithms

N	0	1	2	3	4	5	6	7	8	9
10	0.0000	0.0043	0.0086	0.0128	0.0170	0.0212	0.0253	0.0294	0.0334	0.0374
11	0.0414	0.0453	0.0492	0.0531	0.0569	0.0607	0.0645	0.0682	0.0719	0.0755
12	0.0792	0.0828	0.0864	0.0899	0.0934	0.0969	0.1004	0.1038	0.1072	0.1106
13	0.1139	0.1173	0.1206	0.1239	0.1271	0.1303	0.1335	0.1367	0.1399	0.1430
14	0.1461	0.1492	0.1523	0.1553	0.1584	0.1614	0.1644	0.1673	0.1703	0.1732
15	0.1761	0.1790	0.1818	0.1847	0.1875	0.1903	0.1931	0.1959	0.1987	0.2014
16	0.2041	0.2068	0.2095	0.2122	0.2148	0.2175	0.2201	0.2227	0.2253	0.2279
17	0.2304	0.2330	0.2355	0.2380	0.2405	0.2430	0.2455	0.2480	0.2504	0.2529
18	0.2553	0.2577	0.2601	0.2625	0.2648	0.2672	0.2695	0.2718	0.2742	0.2765
19	0.2788	0.2810	0.2833	0.2856	0.2878	0.2900	0.2923	0.2945	0.2967	0.2989
20	0.3010	0.3032	0.3054	0.3075	0.3096	0.3118	0.3139	0.3160	0.3181	0.3201
21	0.3222	0.3243	0.3263	0.3284	0.3304	0.3324	0.3345	0.3365	0.3385	0.3404
22	0.3424	0.3444	0.3464	0.3483	0.3502	0.3522	0.3541	0.3560	0.3579	0.3598
23	0.3617	0.3636	0.3655	0.3674	0.3692	0.3711	0.3729	0.3747	0.3766	0.3784
24	0.3802	0.3820	0.3838	0.3856	0.3874	0.3892	0.3909	0.3927	0.3945	0.3962
25	0.3979	0.3997	0.4014	0.4031	0.4048	0.4065	0.4082	0.4099	0.4116	0.4133
26	0.4150	0.4166	0.4183	0.4200	0.4216	0.4232	0.4249	0.4265	0.4281	0.4298
27	0.4314	0.4330	0.4346	0.4362	0.4378	0.4393	0.4409	0.4425	0.4440	0.4456
28	0.4472	0.4487	0.4502	0.4518	0.4533	0.4548	0.4564	0.4579	0.4594	0.4609
29	0.4624	0.4639	0.4654	0.4669	0.4683	0.4698	0.4713	0.4728	0.4742	0.4757
30	0.4771	0.4786	0.4800	0.4814	0.4829	0.4843	0.4857	0.4871	0.4886	0.4900
31	0.4914	0.4928	0.4942	0.4955	0.4969	0.4983	0.4997	0.5011	0.5024	0.5038
32	0.5051	0.5065	0.5079	0.5092	0.5105	0.5119	0.5132	0.5145	0.5159	0.5172
33	0.5185	0.5198	0.5211	0.5224	0.5237	0.5250	0.5263	0.5276	0.5289	0.5302
34	0.5315	0.5328	0.5340	0.5353	0.5366	0.5378	0.5391	0.5403	0.5416	0.5428
35	0.5441	0.5453	0.5465	0.5478	0.5490	0.5502	0.5514	0.5527	0.5539	0.5551
36	0.5563	0.5575	0.5587	0.5599	0.5611	0.5623	0.5635	0.5647	0.5658	0.5670
37	0.5682	0.5694	0.5705	0.5717	0.5729	0.5740	0.5752	0.5763	0.5775	0.5786
38	0.5798	0.5809	0.5821	0.5832	0.5843	0.5855	0.5866	0.5877	0.5888	0.5899
39	0.5911	0.5922	0.5933	0.5944	0.5955	0.5966	0.5977	0.5988	0.5999	0.6010
40	0.6021	0.6031	0.6042	0.6053	0.6064	0.6075	0.6085	0.6096	0.6107	0.6117
41	0.6128	0.6138	0.6149	0.6160	0.6170	0.6180	0.6191	0.6201	0.6212	0.6222
42	0.6232	0.6243	0.6253	0.6263	0.6274	0.6284	0.6294	0.6304	0.6314	0.6325
43	0.6335	0.6345	0.6355	0.6365	0.6375	0.6385	0.6395	0.6405	0.6415	0.6425
44	0.6435	0.6444	0.6454	0.6464	0.6474	0.6484	0.6493	0.6503	0.6513	0.6522
45	0.6532	0.6542	0.6551	0.6561	0.6571	0.6580	0.6590	0.6599	0.6609	0.6618
46	0.6628	0.6637	0.6646	0.6656	0.6665	0.6675	0.6684	0.6693	0.6702	0.6712
47	0.6721	0.6730	0.6739	0.6749	0.6758	0.6767	0.6776	0.6785	0.6794	0.6803
48	0.6812	0.6821	0.6830	0.6839	0.6848	0.6857	0.6866	0.6875	0.6884	0.6893
49	0.6902	0.6911	0.6920	0.6928	0.6937	0.6946	0.6955	0.6964	0.6972	0.6981
50	0.6990	0.6998	0.7007	0.7016	0.7024	0.7033	0.7042	0.7050	0.7059	0.7067
51	0.7076	0.7084	0.7093	0.7101	0.7110	0.7118	0.7126	0.7135	0.7143	0.7152
52	0.7160	0.7168	0.7177	0.7185	0.7193	0.7202	0.7210	0.7218	0.7226	0.7235
53	0.7243	0.7251	0.7259	0.7267	0.7275	0.7284	0.7292	0.7300	0.7308	0.7316
54	0.7324	0.7332	0.7340	0.7348	0.7356	0.7364	0.7372	0.7380	0.7388	0.7396
N	0	1	2	3	4	5	6	7	8	9

N	0	1	2	3	4	5	6	7	8	9
55	0.7404	0.7412	0.7419	0.7427	0.7435	0.7443	0.7451	0.7459	0.7466	0.7474
56	0.7482	0.7490	0.7497	0.7505	0.7513	0.7520	0.7528	0.7536	0.7543	0.7551
57	0.7559	0.7566	0.7574	0.7582	0.7589	0.7597	0.7604	0.7612	0.7619	0.7627
58	0.7634	0.7642	0.7649	0.7657	0.7664	0.7672	0.7679	0.7686	0.7694	0.7701
59	0.7709	0.7716	0.7723	0.7731	0.7738	0.7745	0.7752	0.7760	0.7767	0.7774
60	0.7782	0.7789	0.7796	0.7803	0.7810	0.7818	0.7825	0.7832	0.7839	0.7846
61	0.7853	0.7860	0.7868	0.7875	0.7882	0.7889	0.7896	0.7903	0.7910	0.7917
62	0.7924	0.7931	0.7938	0.7945	0.7952	0.7959	0.7966	0.7973	0.7980	0.7987
63	0.7993	0.8000	0.8007	0.8014	0.8021	0.8028	0.8035	0.8041	0.8048	0.8055
64	0.8062	0.8069	0.8075	0.8082	0.8089	0.8096	0.8102	0.8109	0.8116	0.8122
65	0.8129	0.8136	0.8142	0.8149	0.8156	0.8162	0.8169	0.8176	0.8182	0.8189
66	0.8195	0.8202	0.8209	0.8215	0.8222	0.8228	0.8235	0.8241	0.8248	0.8254
67	0.8261	0.8267	0.8274	0.8280	0.8287	0.8293	0.8299	0.8306	0.8312	0.8319
68	0.8325	0.8331	0.8338	0.8344	0.8351	0.8357	0.8363	0.8370	0.8376	0.8382
69	0.8388	0.8395	0.8401	0.8407	0.8414	0.8420	0.8426	0.8432	0.8439	0.8445
70	0.8451	0.8457	0.8463	0.8470	0.8476	0.8482	0.8488	0.8494	0.8500	0.8506
71	0.8513	0.8519	0.8525	0.8531	0.8537	0.8543	0.8549	0.8555	0.8561	0.8567
72	0.8573	0.8579	0.8585	0.8591	0.8597	0.8603	0.8609	0.8615	0.8621	0.8627
73	0.8633	0.8639	0.8645	0.8651	0.8657	0.8663	0.8669	0.8675	0.8681	0.8686
74	0.8692	0.8698	0.8704	0.8710	0.8716	0.8722	0.8727	0.8733	0.8739	0.8745
75	0.8751	0.8756	0.8762	0.8768	0.8774	0.8779	0.8785	0.8791	0.8797	0.8802
76	0.8808	0.8814	0.8820	0.8825	0.8831	0.8837	0.8842	0.8848	0.8854	0.8859
77	0.8865	0.8871	0.8876	0.8882	0.8887	0.8893	0.8899	0.8904	0.8910	0.8915
78	0.8921	0.8927	0.8932	0.8938	0.8943	0.8949	0.8954	0.8960	0.8965	0.8971
79	0.8976	0.8982	0.8987	0.8993	0.8998	0.9004	0.9009	0.9015	0.9020	0.9025
80	0.9031	0.9036	0.9042	0.9047	0.9053	0.9058	0.9063	0.9069	0.9074	0.9079
81	0.9085	0.9090	0.9096	0.9101	0.9106	0.9112	0.9117	0.9122	0.9128	0.9133
82	0.9138	0.9143	0.9149	0.9154	0.9159	0.9165	0.9170	0.9175	0.9180	0.9186
83	0.9191	0.9196	0.9201	0.9206	0.9212	0.9217	0.9222	0.9227	0.9232	0.9238
84	0.9243	0.9248	0.9253	0.9258	0.9263	0.9269	0.9274	0.9279	0.9284	0.9289
85	0.9294	0.9299	0.9304	0.9309	0.9315	0.9320	0.9325	0.9330	0.9335	0.9340
86	0.9345	0.9350	0.9355	0.9360	0.9365	0.9370	0.9375	0.9380	0.9385	0.9390
87	0.9395	0.9400	0.9405	0.9410	0.9415	0.9420	0.9425	0.9430	0.9435	0.9440
88	0.9445	0.9450	0.9455	0.9460	0.9465	0.9469	0.9474	0.9479	0.9484	0.9489
89	0.9494	0.9499	0.9504	0.0509	0.9513	0.9518	0.9523	0.9528	0.9533	0.9538
90	0.9542	0.9547	0.9552	0.9557	0.9562	0.9566	0.9571	0.9576	0.9581	0.9586
91	0.9590	0.9595	0.9600	0.9605	0.9609	0.9614	0.9619	0.9624	0.9628	0.9633
92	0.9638	0.9643	0.9647	0.9652	0.9657	0.9661	0.9666	0.9671	0.9675	0.9680
93	0.9685	0.9689	0.9694	0.9699	0.9703	0.9708	0.9713	0.9717	0.9722	0.9727
94	0.9731	0.9736	0.9741	0.9745	0.9750	0.9754	0.9759	0.9763	0.9768	0.9773
95	0.9777	0.9782	0.9786	0.9791	0.9795	0.9800	0.9805	0.9809	0.9814	0.9818
96	0.9823	0.9827	0.9832	0.9836	0.9841	0.9845	0.9850	0.9854	0.9859	0.9863
97	0.9868	0.9872	0.9877	0.9881	0.9886	0.9890	0.9894	0.9899	0.9903	0.9908
98	0.9912	0.9917	0.9921	0.9926	0.9930	0.9934	0.9939	0.9943	0.9948	0.9952
99	0.9956	0.9961	0.9965	0.9969	0.9974	0.9978	0.9983	0.9987	0.9991	0.9996
N	0	1	2	3	4	5	6	7	8	9

ANSWERS TO SELECTED EXERCISES

exercise 1-1

1. Re, Ra, In, P   3. Re, Ra, N
5. Re, Ra, N   7. Re, Ir, P
9. Re, Ra, In   11. Re, Ra, P
13. Re, Ir, N   15. *B*   17. *E*
19. *A*   21. *C*   23. *C*   25. T
27. F   29. T   31. T   33. F
35. T

exercise 1-2

1. 5   3. 6   5. $\frac{1}{4}$ 7. 0

9. $7\frac{2}{5}$ 11. $+19$ 13. $+12$
15. -89 17. -20 19. $+61$
21. -578 23. $+94$ 25. $+261$
27. $+\frac{4}{7}$ 29. -9 31. -28
33. -21 35. $+184$ 37. $+6$
39. -12 41. $+5$ 43. -12
45. $+19$ 47. -55 49. -72
51. $+114$ 53. -3 55. $+3$
57. -102 59. 0   61. -4
63. $+7$ 65. -9 67. $+27$

433

exercise 1-3

1. $+30$ **3.** $+120$ **5.** -70

7. -120 **9.** $+2100$ **11.** $+108$

13. -6 **15.** $+30$ **17.** 0

19. -2 **21.** -2 **23.** $+4$

25. -3 **27.** $+7$ **29.** $+\dfrac{1}{5}$

31. $+3\dfrac{1}{2}$ **33.** -1 **35.** $+2$

37. $+2$ **39.** $+\dfrac{2}{27}$

exercise 1-4

1. Commutative—multiplication

3. Associative—multiplication

5. Distributive **7.** Distributive

9. Associative—multiplication

11. Associative—addition **13.** T

15. F **17.** T **19.** F **21.** T

23. F **25.** 0 **27.** -28 **29.** 3

31. 15 **33.** -3 **35.** 8 **37.** -18

39. 5 **41.** -94 **43.** 0

45. Undefined **47.** 0

49. Undefined **51.** 0 **53.** 0

exercise 1-5

1. $+21x$ **3.** $+7p$ **5.** $-34s$

7. $-38x$ **9.** $94n-16m$ **11.** $+12f$

13. $-16x$ **15.** $-129a$

17. $14c-14x$ **19.** $-3m-17k$

21. $7r$ **23.** $-2k$ **25.** $17c+12b$

27. $34e-27d$ **29.** $\dfrac{5}{4}x$ **31.** $7xy$

33. 0 **35.** $4k-2a$ **37.** $-31s$

39. 0 **43.** 0

exercise 1-6

1. a^5 **3.** d^{13} **5.** m^{10} **7.** a^7b^7

9. x^9y^4 **11.** $-10x$ **13.** $10n^2$

15. $-8xy$ **17.** $6m^2n$ **19.** $-60a^3b^3$

21. $-3a^2b^2$ **23.** $30a^2b$ **25.** $18bx^2$

27. $27s^3t^3$ **29.** $-24m^2n^2p^2$ **31.** 0

33. -80 **35.** 729

exercise 1-7

1. 2^2 or 4 **3.** $\dfrac{1}{8}$ **5.** x^6 **7.** $\dfrac{1}{(xy)^2}$

9. a^2 **11.** $-7a$ **13.** $-4a$

15. $-5p$ **17.** -2 **19.** $\dfrac{8a}{y^2}$

21. $\dfrac{3b^3}{c}$ **23.** $\dfrac{16ab}{5x}$ **25.** $-\dfrac{2}{3y}$

27. $\dfrac{b^4}{2a^4}$ **29.** $-\dfrac{2xy}{3z}$ **31.** xy^2

exercise 1-8

1. $6xy-8xy^2$ **3.** $8xy-2y+3z$

5. $3d+7z$ **7.** $4x-2y$

9. $-3x+5y$ **11.** $3x-10y-12$

13. $x-y$ **15.** $8+4a-6b$

17. $x+6y-12$

19. $33a-34b-22c-6d$

21. $-18b-10c$ **23.** $14x-4y$

25. $12y-2x$ **27.** $m+2n$

29. $-4y-2x$ **31.** $q-p-p^2$

33. $-5x^2-3x$ **35.** $-4a-2b+c$

37. $13x+9y$ **39.** $3m-10s$

41. $6x-4$ **43.** $2x-(-6y+3t+8)$

45. $ab-(c-gk+h)$ **47.** $-13x+10y$

49. $-4x+3y$

exercise 1-9

1. $12x^2+6xy-3xz$

3. $-6a^2b+9ab^2-3a^2b^2$

5. $6a^2x-15ax^2+18ax$

7. $-6a^3c+8a^2c^2+4a^3c^2$

9. $-3p^2qr-2pq^2r^2-7pqr$

11. $10x^2-31x-14$

13. $2pm-2pn-m^2+mn$

15. $9x^3-21x^2+31x-35$

17. $-10a^2+9a^3+4a^4-3a^5$

19. $3a^6b^2-18a^5b^3+3ab^4-4a^6b^3$
$+24a^5b^4-4ab^5$

21. $12m^4+8m^3n-49m^2n^2+33mn^3-6n^4$

23. $(12x^2-13x-14)$ sq ft **25.** x^3-y^3

exercise 1-10

1. $2 - 3a^2$ **3.** $\dfrac{1}{x} - 6y^2$ **5.** $3n - 7k$

7. $7a^2 + 4ab - 1$

9. $-6xy + 11y - \dfrac{9y^3}{x}$

11. $-5a + 4b + 3c^2$

13. $-\dfrac{a}{3} + \dfrac{b^2}{a^2} + 2a^3$

15. $\dfrac{m}{2} - \dfrac{n}{4} - \dfrac{3mn^2}{8}$

17. $3x - 2$, R 0   **19.** $2x - 5$, R 0
21. $3a^2 + 3a + 2$, R 0
23. $m^2 + 4m + 1$, R 6
25. $x^2 + 2x + 1$, R 10
27. $a^3 + a^2 + 3a + 1$, R 0
29. $x^3 + x^2 + x + 2$, R 0   **31.** $2a + 3$
33. $-4x - 2$ **35.** $5a^2 + 21a + 4$
37. $a - 3$ **39.** $x - 1$, R 2   **41.** $a + 1$
43. $x^2 + 2$, R -2 **45.** $2x^5 - 1$, R -1

REVIEW—CHAPTER 1

part a

1. T
2. F (some rational numbers do not; e.g., $\frac{2}{3}$)
3. F (undefined)   **4.** T   **5.** T
6. F (one)   **7.** T   **8.** T   **9.** T
10. T

part b

1. $0, 6, -5$ **3.** All except $\sqrt{2}$ and π
5. $>$ **7.** $<$ **9.** $>$ **11.** $>$
13. $=$
15. (a) $-6; +3; -15; +6; +2$
(b) $6; 3; 15; 6; 2$

17. (a) $+19; +\dfrac{2}{3}; 0; -6; +100$

(b) $19; \dfrac{2}{3}; 0; 6; 100$ **19.** -1 **21.** $+15$
23. $+32$ **25.** $+488$ **27.** $+151$
29. $+124$ **31.** $+72$ **33.** -360

35. $+2400$ **37.** $+48$ **39.** -6
41. $+7$ **43.** undef.   **45.** -1
47. Associative—addition
49. Commutative—multiplication
51. Distributive
53. Commutative—addition   **55.** No
57. 5   **59.** -16 **61.** 107   **63.** 4
65. 0   **67.** $7x$ **69.** $31x$
71. $-139y$ **73.** $47a - 42b$
75. $1041a$ **77.** $-75m$
79. $-43a - 16b$ **81.** $5x - 2y + 7$
83. $9p - 8d$ **85.** a^{15} **87.** $-84xy$
89. $21a^3b^3$ **91.** x^6 **93.** $(xy)^2$

95. $-3x$ **97.** $-4ab$ **99.** $\dfrac{5}{4}a$

101. $44xy - 3xy^2$ **103.** $9x - 10y + 7z$
105. $8x - 16y$ **107.** $-2a + 9b$
109. $3a - 41b - 45c + 32d$
111. $-1 - 5x + 8y$
113. $5a - 7m + 5k$
115. $18x^2 + 9xy - 6x$
117. $-27a^5 + 36a^4 - 18a^3m$
119. $6a^2 - 29ax + 28x^2$
121. $6x^2 - 33xy + 2xz + 15y^2 - 10yz$
123. $a^3 + ab - ac + a^2b + b^2 + a^2c - c^2$
125. $3a^5 - 8a^4b - 2a^3b^2 + 12a^2b^3$

127. $\dfrac{1}{n} - 12mn^2$ **129.** $2x - 3y + 9xy^2$

131. $-6x^2y^2 + 3xz - 5$ **133.** $7x - 5y$

135. $4x - 9 - \dfrac{2}{2x + 7}$ **137.** $3a - 2$

139. $2x + 1$, R 2

exercise 2-1

1. $-3a + 18$ **3.** $4a^2c - 4c^3$
5. $3x^2y - 3xy^2 + 3xyz$
7. $-p^5 - p^4 + 3p^3$
9. $-8x^2y + 12xy^2 + 20x^2y^2$
11. $-20a^3 + 15a^2 - 10a - 5$
13. $5(x - 2y)$ **15.** $3a(a - 4)$
17. $17p(2q - 3)$ **19.** $m^2(a - 1)$
21. $p^2(1 + 3q)$ **23.** $2axy(7a - x)$
25. $3a(a - 1)$ **27.** $2(mn - 2m + 3n)$
29. $2xy(x^2 + 2xy + 8y^2)$

31. $a^2b^2c^3(ac + b + a^2)$

33. $3xy(3x - 2y + 1)$

35. $x^2(h^2 - 8p + 4y + 6)$

exercise 2-2

1. $x^2 + 2xy + y^2$ 3. $m^2 + 2mn + n^2$

5. $a^2 - 2ab + b^2$ 7. $x^2 + 6x + 9$

9. $9x^2 + 6x + 1$ 11. $9a^2 - 12a + 4$

13. $9a^2 - 12ab + 4b^2$

15. $64x^2 + 112xy + 49y^2$

17. $a^2x^2 - 2abxy + b^2y^2$

19. $4x^2y^2 - 4xy^3 + y^4$

21. $100 + 40 + 4 = 144$

23. $400 + 200 + 25 = 625$ 25. 2

27. 90 29. 12 31. $2ac$

33. $(x + y)(x + y)$ 35. $(x - 2)(x - 2)$

37. $16a^2 + 8a - 1$ 39. $2(3x + 2y)^2$

41. $(5x^2 + 8y^2)^2$ 43. $2(x - 3y)^2$

45. $5(9x + 4y)^2$ 47. $(ax + b)^2$

exercise 2-3

1. $m^2 - n^2$ 3. $x^2 - 9$ 5. $4x^2 - y^2$

7. $1 - n^2$ 9. $a^2b^2 - c^2$

11. $1600 - 9 = 1591$ 13. $(a - b)(a + b)$

15. $(a + 3b)(a - 3b)$

17. $(4a - 1)(4a + 1)$

19. $(3xy - 8z)(3xy + 8z)$

21. $(5p - 2q)(5p + 2q)$

23. $\left(x - \dfrac{1}{2}y\right)\left(x + \dfrac{1}{2}y\right)$

25. $(a^2 + b^2)(a + b)(a - b)$

27. $3(2m + 3n)(2m - 3n)$

29. $a^2(6b - 1)(6b + 1)$

31. $(1 + 3x)(1 - 3x)(1 + 9x^2)$

33. $a^3(ab - 1)(ab + 1)$

35. $(a - b)(a + b)(a^2 + b^2)$

exercise 2-4

1. $x^2 + 5x + 6$ 3. $3x^2 + 13x + 12$

5. $8y^2 - 34y + 35$ 7. $2x^2 + x - 28$

9. $4m^2 + 16m + 15$

11. $9x^2 - 42x + 49$ 13. $9 + 3a - 2a^2$

15. $a^2 - 5ab + 6b^2$

17. $6x^2 - 5xy - 6y^2$

19. $9x^2 + 6xy + y^2$

21. $9a^2 - 12ab + 4b^2$

23. $8d^2 - 2dg - 3g^2$

25. $acx^2 + (ad + bc)x + bd$

27. $6a^2x^2 - 11abx - 10b^2$

exercise 2-5

1. $(x + 3)(x + 2)$ 3. $(x + 3)(x - 2)$

5. $(x - 5)(x + 3)$ 7. 17

9. $(m - 3)(m - 12)$

11. $(a + 2b)(a + 7b)$

13. $(2a - 3)(2a - 1)$

15. $(6y - 5)(y + 1)$ 17. -20

19. $(3x + 5)(x + 8)$

21. $(2x - 3y)(2x - 5y)$

23. $(2x + 3)(x - 3)$ 25. $(2 - 3b)^2$

27. $(x - 2)(3x + 5)$

29. $m(m + 6n)(m + 4n)$

31. $a(18a^3 + 21a - 4)$

33. $2(5x - y)(3x + y)$

35. $a(3a - 2b)(4a + 5b)$

37. $(3x - 1)(3x + 1)(2x^2 + 1)$

39. -503 41. $(5x - 3y)(2x + y)$

exercise 2-6

1. $(a + b)(x + 3)$ 3. $(d + 1)(x - c)$

5. $g(1 + h)(x - 3)$ 7. $(x + 3)(y - 5)$

9. $(a + 1)(a - 1)(x + d)$

11. $(t - s)(b + 2a)$

13. $(a + 2b)(5c - d)$

15. $(k - m)(x^2 + y^2)$

17. $[(x + 1) + 1]^2 = (x + 2)^2$

19. $3(2x + 7)$

21. $(2x + 2y - 3)(x + y + 1)$

23. $(2a + 2b + 3)(a + b - 3)$

25. $5(2x - 1)$ 27. $(4a - 3b)(2a + b)$

29. $(a + 3b)(1 - a + 3b)$

31. $(a^2 - 2)(2a + 5)$ 33. $(p - q)(r - s)$

35. $(a - b - 4)(a - b + 3)$

37. $(27x - 31y)(27x + 31y)$

39. $(2a + 9)(4a + 1)$

41. $(4p^2 + 9q^2)(2p + 3q)(2p - 3q)$

43. $(a + 3b + 2)(a + 3b - 2)$

45. $2(x)(x + 3)(x - 2)(x + 2)$
47. $(2x - 3y - 4)(2x - 3y + 4)$
49. $(2a - 2b - 3)(a - b + 4)$

exercise 2-7

1. $(x - y)(x^2 + xy + y^2)$
3. $(x + 2)(x^2 - 2x + 4)$
5. $(3 - y)(9 + 3y + y^2)$
7. $(2r + 3t)(4r^2 - 6rt + 9t^2)$
9. $(4ab - c)(16a^2b^2 + 4abc + c^2)$
11. $(x^2 - y)(x^4 + x^2y + y^2)$
13. $(3x^3 - y^2)(9x^6 + 3x^3y^2 + y^4)$
15. $8t^3(2t^6 - 1)(4t^{12} + 2t^6 + 1)$
17. $(a + 2)(a^2 - 5a + 13)$
19. $(2x)(x^2 + 27)$

exercise 2-8

1. $5(2t - 1)$
3. $(12m + 6n - 7)(2m + n + 1)$
5. $(6x + 2y - 5)(6x + 2y + 1)$
7. $16(a - 1)$ **9.** $(8m - 5n)(2m + n)$
11. $(c^3 - 5)(3c + 5)$
13. $(4s + 3t)(1 - 4s + 3t)$
15. $(2m - n)(2m + n)(4m^2 + n^2)$
17. $(xy - 3z)(xy + 3z)(x^2y^2 + 9z^2)$
19. $(4a^2b - 1)(4a^2b + 1)$
21. $8m(m + 5)$
23. $(16x^2 + 9y^2)(4x - 3y)(4x + 3y)$
25. $(3x + 2y + 4)(3x + 2y - 4)$

REVIEW—CHAPTER 2

part a

1. T **2.** T
3. F [e.g., $(x - y)(x + y) = x^2 - y^2$]
4. T **5.** F (signs will differ)
6. F (used often) **7.** T
8. F ($b^2 - 4ac$ must be the *square* of an integer)
9. T
10. F [$x^3 + y^3 = (x + y)(x^2 - xy + y^2)$]

part b

1. $-4m - 28$
3. $-4k^2m + 12km^2 - 8kmn$
5. $-21a^2b - 28ab^2 + 35a^2b^2$
7. $7(2a - 3b)$ **9.** $18y(3x - 5)$
11. $m(7m^2 + 5m - 9)$
13. $8ab(a + 2b)(a + b)$
15. $m^3(6a^2 + 8b - 3n^2 - 1)$
17. $x^2 + 10xy + 25y^2$
19. $64 + 16b + b^2$
21. $49x^2 - 70x + 25$
23. $16a^2 + 72ab + 81b^2$
25. $9a^2c^2 + 12abc + 4b^2$ **27.** 961
29. 60 **31.** $6a^2$ **33.** $(x + 5)(x + 5)$
35. $(3m - n)^2$
37. Not square trinomial
39. $(a^2m + 2y^2)^2$ **41.** $a^2 - 25$
43. $1 - y^2$ **45.** 396
47. $(6t + 1)(6t - 1)$
49. $(11a - 9b)(11a + 9b)$
51. $m^5(1 + mn)(1 - mn)$
53. $3t^2 + 19t + 28$ **55.** $12a^2 + a - 88$
57. $64a^2 - 48a + 9$
59. $20m^2n^2 - 3mn - 9$
61. $6x^2 + xy - 2y^2$
63. $10a^2b^2 - 3ab - 1$
65. $(t - 5)(t - 4)$ **67.** -4
69. $(3m - 4)(5m - 2)$
71. $3(3 - a)(1 - 2a)$
73. $x(x - 7)(x - 3)$
75. $4(5t - 4s)(2t + 3s)$
77. $(4x - 1)(x - 8)$ **79.** 56
81. $(a - 1)(x + y)$ **83.** $(m + 7)(n - 5)$
85. $(a - y)(2b + d)$
87. $(3a - b)(s^2 + t^2)$ **89.** $5(2m + 1)$
91. $(x + y - 2)(x + y - 1)$
93. $(15x - y)(x - 4y)$
95. $(2k - 3a)(1 - 2k - 3a)$
97. $a^2(a^2 + 4)(a - 2)(a + 2)$
99. $(m - 3)(3m + 1)$
101. $(2x - 5y + 3)(2x - 5y - 3)$
103. $(x^2 - 3)(x + 2)$
105. $(y + x)(y^2 - xy + x^2)$
107. $2(3x - y)(9x^2 + 3xy + y^2)$
109. $(x + y^3)(x^2 - xy^3 + y^6)$
111. $-2b(3a^2 + b^2)$

exercise 3-1

1. $=$ **3.** $=$ **5.** $\neq$ **7.** $\neq$

9. $=$ **11.** $\dfrac{3a}{5a-2}$ **13.** $\dfrac{b}{5}$

15. $\dfrac{2+3b}{x+3b}$ **17.** $\dfrac{1}{xy-3}$ **19.** $\dfrac{x-3}{x-2}$

21. $\dfrac{2x-5}{x-7}$ **23.** $\dfrac{x-y}{a-5}$

25. $\dfrac{x(2x-3)}{a-b}$ **27.** $\dfrac{x+y-5}{x}$

29. $\dfrac{a-b}{x+y}$

exercise 3-2

1. $+$ **3.** $-$ **5.** $+$ **7.** $+$ **9.** $-$

11. $+$ **13.** -1 **15.** $-a$ **17.** $\dfrac{1}{p}$

19. $\dfrac{3}{4y}$ **21.** $-x$ **23.** $-\dfrac{1}{5+3x}$

25. $-\dfrac{x+2}{x+1}$ **27.** -1

29. $\dfrac{x^2(3+x)}{1-y}$ **31.** $-\dfrac{x^2+xy+y^2}{x+y}$

exercise 3-3

1. $\dfrac{2(x^2-2)}{x-2}$ **3.** $\dfrac{1}{x}$ **5.** $\dfrac{3(2x-3)}{xy-2}$

7. $\dfrac{2(a-3)}{(a+2)(2a-3)}$ **9.** $\dfrac{3y^2}{5a}$

11. $2-3x$ **13.** $(m+n)^2$ **15.** $\dfrac{x+3}{x-1}$

17. $\dfrac{a-2}{3}$ **19.** $\dfrac{6x}{x-3}$ **21.** 1

23. $\dfrac{1}{3a-b}$ **25.** $a-2b$

27. $\dfrac{(d-3c)(a+b-3)}{4a^2+2ab+b^2}$

exercise 3-4

1. $\dfrac{2x^3}{9a^2}$ **3.** $\dfrac{b^2}{6c^2}$ **5.** $6xy^5$

7. $\dfrac{1}{(x+2)(x-3)}$ **9.** $(p-3q)(p-q)$

11. $\dfrac{x+3}{x+2}$ **13.** $\dfrac{x+4}{3(x-2)}$

15. $\dfrac{(x+2)^2}{(x+4)(x^2-3)}$ **17.** $\dfrac{m+9}{2(m-4)}$

19. $\dfrac{x-2}{2x-5}$ **21.** $\dfrac{(a+b)^2}{a-3b}$

23. $\dfrac{2a+5b}{(a^2+b^2)(a+b)}$ **25.** $\dfrac{x-3}{x-2}$

27. $\dfrac{6(a-b-4)}{a-2b}$ **29.** $-\dfrac{1}{3}$

31. $\dfrac{(a-b)^2}{3ab^2(a+b)}$ **33.** $\dfrac{3x}{2x+3y}$

exercise 3-5

1. $3(3x-1)$ **3.** $x-y$

5. $(x-3)(2x-1)$

7. $(x-2)(x+3)(x-3)$

9. $(m-1)^2(2m+3)$

11. $(a-b)(a^2+b^2)(a+b)$

13. LCD: $(x-2)(x+3)$; $\dfrac{x(x+3)}{\text{LCD}}; \dfrac{x+1}{\text{LCD}}$

15. LCD: $(x)(2x-1)(2x+3)$;

$$\dfrac{(x-1)(2x+3)}{\text{LCD}}; \dfrac{x(x-2)}{\text{LCD}}$$

17. LCD: $(x-2)(2x+1)(x-3)$;

$$\dfrac{(x-4)(x-3)}{\text{LCD}}; \dfrac{(3x-1)(2x+1)}{\text{LCD}}$$

19. LCD: $(a+b)(2a-3)(b+2)$;

$$\dfrac{(a-b)(b+2)}{\text{LCD}}; \dfrac{(b-1)(a+b)}{\text{LCD}}$$

21. LCD: $(2x-3)^2(3x+2)(x-1)$;

$$\dfrac{(x+1)(2x-3)(x-1)}{\text{LCD}};$$

$$\dfrac{(x+2)(3x+2)(x-1)}{\text{LCD}}; \dfrac{(x+3)(2x-3)^2}{\text{LCD}}$$

23. LCD: $(2x-3)(3x-1)(4x-5)(3x+4)$;

$$\dfrac{(2x-1)(3x+4)(4x-5)}{\text{LCD}};$$

$$\dfrac{(3x+4)(2x-3)(4x-5)}{\text{LCD}};$$

$$\dfrac{(3x-7)(2x-3)(3x+4)}{\text{LCD}}$$

exercise 3-6

1. $\dfrac{x^2 + 3}{x}$ **3.** $\dfrac{2b + 3a}{ab}$ **5.** $\dfrac{7x + 6y}{72}$

7. $\dfrac{6a^2 - ab - 18b^2}{36a^2 b^2}$ **9.** $\dfrac{13t - 6r}{rt}$

11. $\dfrac{3 - xy + 2x^2}{x}$ **13.** $\dfrac{4x}{2 - 3x}$

15. $\dfrac{x^2 - 5x + 17}{x - 7}$ **17.** $\dfrac{4x^2}{2x - 5y}$

19. $\dfrac{5x^2 - 12x + 17}{x^2 - x - 12}$ **21.** $\dfrac{7}{a - b}$

23. $\dfrac{x + 4}{x^2 - 4}$ **25.** $\dfrac{16}{(4 + a)(5 - a)}$

27. $\dfrac{d^2 - cd - 3c^2}{d^2 - 4c^2}$ **29.** $\dfrac{5}{(x - 3)(x + 2)}$

31. $\dfrac{x^2 - x - 15}{x^2 - 5x + 6}$

33. $\dfrac{x^2 + 5x + 2}{(2x + 3)(x - 3)(3x - 2)}$

35. $\dfrac{x^2 - 20x - 11}{(2 - x)(1 + x)(2 + x)}$

exercise 3-7

1. $\dfrac{2a}{b}$ **3.** $\dfrac{xy}{x + y}$ **5.** $\dfrac{2a^2}{4 - a^2}$

7. $\dfrac{2}{x(x - 2)}$ **9.** $\dfrac{a + b}{2b}$ **11.** $\dfrac{2 - x}{3x + 2}$

13. $\dfrac{2x + 3}{x + 5}$ **15.** $\dfrac{(x - y)^2}{3(x + y)^2}$

17. $\dfrac{b - a}{a^2 + b^2}$ **19.** $\dfrac{-3a(a + b^2)}{b^2}$

21. $\dfrac{3 - x}{x + 12}$ **23.** $\dfrac{a^2 - 2a + 1}{2a^2 - a}$

25. $\dfrac{a^2 + a - 2}{a^2 + a}$

REVIEW—CHAPTER 3

part a

1. T   **2.** F [if $(x)(n) = (y)(m)$]

3. T

4. F (nonnegative when a or x represent negative numbers)

5. T   **6.** T   **7.** T

8. F ($x - a$ and $a - x$ are negatives of one another)

9. T   **10.** F ($\frac{1}{2} \div xy$)

part b

1. $+$ **3.** $-$

5. Already in lowest terms   **7.** $\dfrac{2x - 5}{2x + 1}$

9. -1 **11.** $\dfrac{1}{a}$ **13.** $-(x + 3)$

15. $\dfrac{6 - x}{x}$ **17.** $-(c + d)^2$

19. $\dfrac{7x - 1}{(5x + 2)(2x + 3)}$ **21.** $\dfrac{a + 2}{a - 1}$

23. $\dfrac{-x}{1 + x^2}$ **25.** $\dfrac{m - n}{m + n}$ **27.** $\dfrac{3}{a + 2b}$

29. $a^2 - b^2$ **31.** $\dfrac{(2a - 7)(3a + 2)}{(3a - 2)(2a + 7)}$

33. $\dfrac{pa}{a + 2}$ **35.** $\dfrac{(b + 1)(n + 2)}{(m + 2)(a - 1)}$

37. $\dfrac{4 - m}{3 - m}$ **39.** $\dfrac{3 - a}{a - 4}$

41. $\dfrac{6ay - 6x + 2by - 3xy}{6xy}$

43. $\dfrac{3xy - 3y^2 - 5x - 3y}{y(x^2 - y^2)}$

45. $\dfrac{-2a^2 + 2a + 11}{2a + 6}$ **47.** $\dfrac{y - x - y^2}{xy - 2y^2}$

49. $\dfrac{11a^4 - 22a^2 + 6}{(a^4 - 4)(2a^2 - 3)}$ **51.** $\dfrac{3a^2 + 5a}{1 - a^2}$

53. $\dfrac{2y - x}{3}$ **55.** -1

57. $\dfrac{(a - 3)(a + 1)^2}{(a - 2)(a^2 + a - 1)}$

59. $\dfrac{3a - 19}{7a^2 + 24a + 2}$

exercise 4-1

1. 10   **3.** 4   **5.** 4   **7.** -2

9. 7   **11.** 6   **13.** -4 **15.** -4

17. $\dfrac{3}{25}$ **19.** 32   **21.** $\dfrac{1}{2}$ **23.** 4

25. 2　27. $\dfrac{4}{3}$　29. -2　31. 2

33. $2\dfrac{1}{3}$　35. $-\dfrac{1}{9}$　37. -2　39. 0

41. 2　43. -7　45. -24

47. -70　49. -4　51. 0　53. $1\dfrac{2}{3}$

55. 2　57. -5　59. No; identities

exercise 4-2

1. $\dfrac{b}{5}$　3. $d - 6$　5. $t - 3$

7. $\dfrac{a - 5}{a + 3}$　9. 1　11. $4a + 3b$

13. $\dfrac{3a + 4b}{2}$　15. $2b$　17. $6a$

19. a　21. $2a$　23. $p - 2$

25. $a - 3$　27. $a^2 + ab + b^2$

29. $b = y - mx; x = \dfrac{y - b}{m}; m = \dfrac{y - b}{x}; 4$

31. $n = \dfrac{2S}{a + l}; a = \dfrac{2S - nl}{n}; l = \dfrac{2S - na}{n}; 10$

exercise 4-3

1. $\dfrac{1}{12}$　3. 2　5. 1　7. $\dfrac{1}{2}$

9. No solution　11. 0　13. -2

15. 2　17. No solution　19. 3

21. $\dfrac{6 + 2m}{1 - m}$　23. -7　25. $-\dfrac{16}{7}$

27. 3

exercise 4-4

1. 3　3. -3

5. 6　7. -2

9. 0　11. $x < 5$　13. $p \geq 1$

15. $x \leq -3$　17. $m \geq 12$

19. $x > -5$　21. $a \geq -1$　23. $k > 2$

25. $p \leq 7$　27. $m \geq -8$　29. $x > 2$

31. $x \leq -70$　33. $x \leq 3$　35. $x \geq 3$

exercise 4-5

1. $\{+8, -8\}$　3. $\{2, 4\}$　5. $\{1, 7\}$

7. $\{2, -6\}$　9. $\{-2, -4\}$

11. $\left\{4\dfrac{1}{3}, -1\right\}$　13. $\left\{1\dfrac{1}{2}\right\}$

15. $\left\{-1, -1\dfrac{2}{5}\right\}$　17. $\{0, 12\}$

19. $\left\{-3, -3\dfrac{2}{3}\right\}$　21. $\{1, 4\}$

23. Impossible　25. $\left\{1\dfrac{2}{9}, 2\dfrac{1}{9}\right\}$

27. $\left\{1\dfrac{7}{16}, 1\dfrac{9}{16}\right\}$　29. $\left\{-1\dfrac{1}{5}, \dfrac{2}{5}\right\}$

exercise 4-6

1. $\{x < -4 \text{ or } x > 4\}$

3. $\{m \leq -3 \text{ or } m \geq 3\}$

5. $\{x < -4 \text{ or } x > 10\}$

7. $\{a \leq -10 \text{ or } a \geq 6\}$

9. $\{p \leq -10 \text{ or } p \geq -6\}$

11. $\left\{p \leq -\dfrac{2}{3} \text{ or } p \geq 2\right\}$

13. $\left\{x < \dfrac{4}{5} \text{ or } x > \dfrac{4}{5}\right\}$

15. $\{-4 < x < 4\}$

17. $\{-2 \leq x \leq 2\}$

19. $\{-1 < a < 7\}$

21. $\{1 < x < 3\}$

23. $\{-5 \le m \le -1\}$

25. $\left\{-3\frac{1}{2} < x < -\frac{1}{2}\right\}$

27. $\{-2 \le x \le 0\}$

29. $\{2 < x < 4\}$

31. $\left\{\frac{1}{3} \le x \le 1\right\}$

33. $\{-1 < x < 7\}$

35. $\{-2 \le x \le 3\}$

37. $\left\{x < \frac{1}{3} \text{ or } x > \frac{7}{15}\right\}$

39. $\left\{x \le -\frac{2}{3} \text{ or } x \ge -\frac{1}{3}\right\}$

REVIEW—CHAPTER 4

part a

1. T **2.** T
3. F (equivalent equations have same solution set)
4. T
5. F (e.g., $x + 1 = x$ has no solution)
6. T **7.** T **8.** T
9. F ($-x > 6$ and $x < -6$ are equivalent)
10. T

part b

1. $\frac{4}{25}$ **3.** -8 **5.** No solution

7. -2 **9.** 24 **11.** $\frac{8}{3}$ **13.** 0

15. Identity **17.** 3 **19.** -15

21. m **23.** $\frac{3}{7}a$ **25.** $-\frac{4}{3}c$

27. $a - 3$ **29.** $a - b$ **31.** Identity

33. $l = \dfrac{2S - na}{n}$; 33 **35.** -9 **37.** $\frac{1}{2}$

39. -12 **41.** $1\frac{1}{2}$ **43.** No solution

45. 5 **47.** $x < 7$ **49.** $t < -9$
51. $x \le -5$ **53.** $p > 18$ **55.** $x < 3$

57. $x \le -\frac{2}{3}$ **59.** $m > 3$

61. $x \le -3$ **63.** $y < -2$ **65.** $\{6, 8\}$

67. $\left\{\frac{1}{3}\right\}$ **69.** $\{0, 5\}$ **71.** $\left\{\frac{1}{4}, 1\frac{1}{4}\right\}$

73. $\{m < -8 \text{ or } m > 8\}$

75. $\{-4 \le x \le 9\}$ **77.** $\left\{-1 \le x \le 2\frac{1}{3}\right\}$

79. $\left\{x < -4\frac{2}{3} \text{ or } x > 0\right\}$

81. $\{-9 < m < 3\}$

exercise 5-1

1. 12 ft; 24 ft **3.** 18 in.

5. $8\frac{1}{2}$ in. by 17 in. **7.** 15 ft

9. 18 in. **11.** 8 in. **13.** $2

15. 94 mi

exercise 5-2

1. 30 lb @ 50¢; 15 lb @ 80¢

3. 4 g @ 65¢; 6 g @ 40¢ **5.** 6 kg

7. $4\frac{1}{2}$ lb **9.** 6 oz **11.** 8 gal cream

13. $\frac{1}{4}$ liter **15.** 2 qt

17. 8 nickels, 14 quarters

19. 7 dimes, 5 quarters

21. 220 end, 440 side

exercise 5-3

1. $3000 @ 4%; $4000 @ 7%   3. $7200

5. $3750   7. $4000 @ 6%; $3000 @ 8%

9. $4\frac{1}{2}\%$ and $2\frac{1}{4}\%$ 11. $11,000

13. $1800   15. 9.18%

exercise 5-4

1. 18 mph   3. 24 mph   5. 57 mph

7. 8 min   9. 8.7 yd   11. 125 mi

13. 60 mph   15. $35\frac{5}{11}$ min

17. 5 min

exercise 5-5

1. 1 hr 12 min   3. 18 min   5. 36 hr

7. 4 hr   9. $66\frac{2}{3}$ min   11. 6 hr

13. $10\frac{4}{5}$ min   15. 120 copies/min

17. 5 min

exercise 5-6

1. 12, 13, 14   3. 16, 18, 20

5. 37 or 73   7. 2 and 4   9. 417

11. 21, 24, 27, 30   13. -2, 1, 4, 7

15. Any three consecutive even integers

17. 9, 11, 13, 14

exercise 5-7

1. {(9, 5), (8, 4), (7, 3), ...}

3. $\left\{n < -\frac{2}{3}\right\}$ 5. {n > 6}

7. {1, 3, 5}, {3, 5, 7}, {5, 7, 9}

9. Less than 10

11. (a) Less than or equal to 8 m (b) No

13. Greater than 40 mph

15. More than 5 ft 4 in.

17. Greater than 9.57%

REVIEW—CHAPTER 5

1. 30 in. × 44 in.   3. 41°, 60°, 79°

5. Legs; 16 in.; base: 12 in.   7. 3 kg

9. 10%: 27 cc; 4%: 54 cc   11. 3 gal

13. $40,000   15. $11,000

17. $10,000   19. $4\frac{2}{7}$ lb   21. 156 mi

23. $\frac{3}{8}$ mi   25. $12\frac{3}{11}$ min   27. $5\frac{5}{6}$ hr

29. Will not overflow   31. $6\frac{2}{3}$ hr

33. -1, 3, 7   35. 52

37. -8, -1, $+6$, $+13$, $+20$

39. 6 yr; 12 yr   41. $3.40

43. 7 and 10; 8 and 11   45. 6 or less

47. More than $2\frac{1}{2}$ hr

49. More than $11\frac{1}{3}\%$

exercise 6-1

1. 8   3. -9 5. -27 7. -64

9. $-\frac{8}{27}$ 11. 0.0081   13. a^8

15. x^9 17. x^3 19. $\frac{b^2}{a}$ 21. $\frac{bc}{a}$

23. $-b^4$ 25. 64   27. 108

29. $-a^{15}b^{12}$ 31. $81a^8b^4c^4$ 33. -8

35. $\frac{a^4}{16}$ 37. 2^{m+n} 39. 7^{mn}

41. $\frac{1}{3^{xy}}$ 43. $\frac{x^{12}}{y^{20}}$ 45. $\frac{81}{v^8}$

47. $\frac{3^{3n}}{8} = \frac{27^n}{8}$ 49. x^{3n} 51. $\frac{1}{x^n}$

53. $9a^2 + 6ax + x^2$ 55. $3x^{2n}$

57. $a^{2r}b^{3r}$ 59. $a^3b^2c^2$ 61. $\frac{ax}{y^4}$

63. $\frac{a^4b^8}{81c^4}$

exercise 6-2

1. 1   3. m^5 5. 1   7. 1

9. 1   11. 36   13. $\frac{1}{x^{-2}}$

15. $-n^{-4}$ 17. $\frac{1}{8}$ 19. $\frac{1}{6}$ 21. $\frac{1}{32}$

23. $-\frac{1}{4}$ 25. $9\frac{1}{9}$ 27. 64 29. $\frac{b^3}{a}$

31. $1 - \frac{1}{a^2b^2}$ 33. $\frac{a(a+b^2)}{b^3}$

35. $\frac{1}{27x^3}$ 37. x^4 39. $\frac{y^2}{x^2}$ 41. $\frac{x^6}{64}$

43. a^2b^2 45. $a^4x^{12}y^5$ 47. $\frac{x^4}{y^6}$

49. $\frac{3(y^2+x^2)}{x^2y^2}$ 51. $\frac{a}{b^3}$ 53. $\frac{x^2y}{x+y}$

55. $\frac{y^2}{x^2+y^2}$ 57. $-\frac{x^2y^2(x-y)}{x+y}$

59. $\frac{a^3b^3}{2a^2-3b^2}$ 61. $\frac{1}{x^2}-\frac{1}{y^2}$

63. $\frac{4}{a^2}+\frac{4}{ab}+\frac{1}{b^2}$

exercise 6-3

1. 3.6×10^4 3. 6.28×10^4

5. 3.7×10^{-5} 7. 6.21×10^{-5}

9. 1.4×10 11. 6.1×10^{-1}

13. 7×10^{-1} 15. 1.3×10^0

17. 4500 19. $602,000$ 21. 0.067

23. 0.00000671 25. 12×10^4 or $120,000$

27. 2×10^6 or $2,000,000$

29. 108×10^0 or 108

31. 1.26×10^4 or $12,600$

33. 7.5×10^{-7} or 0.00000075

35. 1×10^{12} or $1,000,000,000,000$ (1 trillion)

exercise 6-4

1. $+2$ 3. $+2$ 5. -2 7. $+\frac{1}{2}$

9. None 11. $+\frac{2}{5}$ 13. $-\frac{2}{3}$

15. -2 17. $3^{1/2}$ 19. $11^{1/4}$

21. $2^{3/2}$ 23. $\frac{1}{2}$ 25. $\frac{1}{2}$ 27. $2^{1/2}$

29. $p^{1/2}$ 31. $t^{1/10}$ 33. $p^{2/3}$

35. $y^{2/n}$ 37. $p^{-(s/r)}$ 39. $(c-6)^{1/2}$

41. $\frac{c^{3/4}}{b^{1/4}}$ 43. $\sqrt[3]{3}$ 45. $\frac{1}{2}$ 47. $\sqrt[5]{36}$

49. $\sqrt[5]{a}$ 51. $\sqrt[8]{a^5}$ 53. $\sqrt{3x}$

55. $\sqrt[3]{\frac{3}{2}}$ 57. $\frac{1}{4\sqrt{6}}$ 59. $\sqrt{\frac{2}{3}}$

61. 1 63. $\sqrt{b-c}$ 65. $\sqrt[3]{(x+y)^2}$

67. $\sqrt[4n]{s^8t}$ 69. 0 71. $\sqrt[6]{x}$ 73. a^3

75. $\sqrt[5]{a^3}$ 77. $\sqrt[12]{m^8n}$ 79. $\frac{4}{25}$

81. $\frac{1}{x+y}$ 83. $\frac{1}{16}$ 85. $\frac{x^2\sqrt{x}}{8\sqrt[4]{8y}}$

87. $\sqrt[10]{s}$ 89. $\sqrt[24]{x}$

exercise 6-5

1. $2\sqrt{2}$ 3. $2\sqrt{21}$ 5. 5

7. $-3\sqrt[3]{3}$ 9. $\frac{3}{5}$ 11. $2\sqrt{23}$

13. $x\sqrt[3]{x}$ 15. $a\sqrt{3}$ 17. $\frac{1}{x}$

19. $x\sqrt[5]{x^2}$ 21. $\sqrt{3a}$ 23. $\sqrt[3]{2m}$

25. $\sqrt[3]{3m^2}$ 27. $\sqrt[3]{x^2+y^2}$

29. $3x\sqrt[3]{xy^2}$ 31. $3xyz^2\sqrt{2y}$

33. $6x\sqrt{3x}$ 35. $\frac{1}{4}\sqrt{x}$ 37. $2\sqrt[3]{m^2}$

39. $\frac{6x}{y^2}$ 41. $\frac{2y^2}{\sqrt[3]{x^2z}}$ 43. $\frac{1}{(a+b)^2}$

45. $\frac{\sqrt{xy+1}}{\sqrt{y}}$ 47. $2x+3$

49. In simplest terms 51. $\frac{1}{\sqrt{x+a}}$

exercise 6-6

1. $5\sqrt{a}$ 3. $3\sqrt{5}$ 5. $7\sqrt[3]{3}$

7. $\sqrt{2}$ 9. $\sqrt[3]{6}$ 11. $-2x\sqrt{6}$

13. $6\sqrt[3]{2}-3\sqrt{2}$ 15. $7\sqrt{2}-\sqrt[3]{2}$

17. $3-x\sqrt{6}$

19. $(a-b+1)\sqrt[3]{(a-b)^2}$

21. $\frac{1}{12}a\sqrt{b}+\frac{1}{6}\sqrt{ab}$

23. $(a-a^2b-b^2)\sqrt{b}$ 25. 0

27. $\frac{2+\sqrt{a}}{\sqrt{b}}$ 29. $\frac{\sqrt{y}-y\sqrt{2}}{\sqrt{x}}$

exercise 6-7

1. $\sqrt[4]{9}$ 3. $\sqrt[6]{16x^2y^2}$ 5. $\sqrt[3]{64}$

7. $\sqrt[6]{(x^2-y^2)^3}$ 9. $2a$ 11. $3a\sqrt{2a}$

13. $\sqrt[6]{72}$ 15. $\sqrt[6]{16x^2y^4}$

17. $2ab\sqrt[6]{27a^5b^2}$ 19. $3-2\sqrt{15}+3\sqrt{2}$

21. $x-\sqrt{x}$ 23. $a\sqrt{2}+\sqrt{3a}$

25. $\sqrt[6]{a^5}-\sqrt[3]{a^2}$ 27. $19-8\sqrt{3}$

29. -1 31. 2 33. $57-12\sqrt{15}$

35. $a-b$ 37. $-4-3\sqrt{15}$

39. $6+5\sqrt{3a}+3a$

41. $4x+4\sqrt{xy}-15y$

43. $2x-4\sqrt{6xy}+9y$

45. $18a-4\sqrt{6ab}-4b$

47. $\sqrt[3]{x^2}-2\sqrt[3]{xy}+\sqrt[3]{y^2}$

49. $a+19+8\sqrt{a+3}$

51. $2a-2\sqrt{a^2-b^2}$ 53. $7-5\sqrt{2}$

55. $21+14\sqrt{3}$

exercise 6-8

1. $\frac{1}{3}\sqrt{3}$ 3. $\frac{3x\sqrt{5}}{5}$ 5. $\frac{3\sqrt[3]{2}}{2}$

7. $12\sqrt[3]{x^2}$ 9. $\frac{\sqrt{7m}}{m}$ 11. $\frac{\sqrt[3]{-6abc^2}}{2bc^3}$

13. $\sqrt{3}+\sqrt{2}$ 15. $-(2+\sqrt{6})$

17. $\frac{x+x\sqrt{x}}{1-x}$ 19. $-5-2\sqrt{6}$

21. $-1-\frac{1}{2}\sqrt{6}$ 23. $1-\frac{1}{3}\sqrt{6}$

25. $\frac{-1}{5-2\sqrt{6}}$ 27. $\frac{1}{3+\sqrt{6}}$

exercise 6-9

1. $2\sqrt{2}$ 3. $\sqrt{2}-2\sqrt{5}$

5. $\frac{2}{21}\sqrt{7}-\frac{1}{6}\sqrt{6}+\frac{1}{21}\sqrt{14}$

7. $2\sqrt[6]{3}+\sqrt[3]{18}$ 9. $\frac{1}{3}(\sqrt{10}-1)$

11. $4\frac{4}{7}+\frac{3}{7}\sqrt{15}$ 13. $\frac{8\sqrt{6}+11\sqrt{3}}{3}$

15. $\frac{9-6\sqrt{a}+a}{9-a}$ 17. -1

19. $\frac{a-5\sqrt{ab}+6b}{a-9b}$

21. $\frac{6m-13\sqrt{mn}+6n}{4m-9n}$

23. $\frac{x\sqrt{2}-2\sqrt{xy}+3\sqrt{2xy}-6y}{x-9y}$

exercise 6-10

1. $+1$ 3. $+1$ 5. i 7. $+1$

9. -1 11. $-i$ 13. -3

15. $i-1$ 17. $(\sqrt{3}-2)i-3$

19. $4i-2$ 21. $5-2i$

23. $6-5i$ 25. $12+i$ 27. $\frac{9}{8}$

29. $18+i$ 31. $1-21i$ 33. $3-2i$

35. $9-46i$ 37. $91+91i$

39. $\frac{10}{17}+\frac{11}{17}i$ 41. 1 43. $-\frac{3}{5}+\frac{2}{5}i$

45. $\frac{5+11i}{6}$

REVIEW—CHAPTER 6

part a

1. F $(3.264^0=1)$
2. F $(6.23\times10^{-2}=0.0623)$ 3. T
4. T
5. F $[(-3)^4=+81;\ -3^4=-81]$
6. F $(\sqrt[3]{n^2}=n^{2/3})$ 7. T
8. F (not incorrect, but preferred form)
9. T
10. F $(4+7i$ is conjugate of $4-7i)$

part b

1. 81 3. (-0.027) 5. $\left(-\frac{8}{125}\right)$

7. $\frac{1}{m^2}$ 9. $-b^{23}$ 11. $\frac{27}{8}$

13. $-\dfrac{243p^5}{q^{20}}$ **15.** a^{5n-4} **17.** $a^{2a}b^{am}$

19. $\dfrac{b^8c^6}{x^4y^4}$ **21.** 1 **23.** 1 **25.** x^{-2}

27. $\dfrac{1}{m^{-3}}$ **29.** $-n^{-3}$ **31.** $\dfrac{4y^2}{x^2}$

33. a^{1+m} **35.** $\dfrac{b^3}{a^{12}c^6}$

37. $\dfrac{yz - az + ay}{ayz}$ **39.** $\dfrac{1}{s^5t^5}$

41. $\dfrac{x^4y^4}{y^4 - 2y^2x^2 + x^4}$ **43.** $-ab(a + 2b)$

45. 2.8×10^5 **47.** 3.21×10^{-4}

49. 3×10^0 **51.** $63{,}200$ **53.** 0.0034

55. 0.4 **57.** 3.6×10^8 **59.** 4×10^1

61. 100 **63.** None   **65.** $-\dfrac{2}{3}$

67. $-y^{3/5}$ **69.** $s^{2/r}t^{4/r}$ **71.** $\dfrac{a^{p/5}}{b^{q/5}}$

73. $3\sqrt[3]{a}$ **75.** $\sqrt[2p]{a^4b}$ **77.** 0

79. $\dfrac{\sqrt[6]{a^5}}{a}$ **81.** $\dfrac{\sqrt[12]{a^5b^6}}{b}$ **83.** 9

85. x^4y^3 **87.** $\dfrac{b^3c^2}{a^8}$ **89.** $2a\sqrt{2ab}$

91. $\sqrt[m]{6a^2b}$ **93.** $5xy^3\sqrt{2y}$

95. $2\sqrt[3]{-2}$ or $-2\sqrt[3]{2}$ **97.** $\dfrac{x}{m}\sqrt[3]{amx}$

99. $\dfrac{3}{x}\sqrt[3]{\dfrac{y^2}{z^2}}$ **101.** $\dfrac{m\sqrt{m}}{4n^3}$ **103.** $\sqrt[6]{27}$

105. $\sqrt[12]{16x^8y^{-4}}$ **107.** $\dfrac{3x\sqrt{p}}{p}$

109. $\dfrac{2\sqrt[3]{b}}{b}$ **111.** $\dfrac{3\sqrt{2} + 3\sqrt{7}}{-5}$

113. $\dfrac{\sqrt{x} + x}{1 - x}$ **115.** $\dfrac{\sqrt[6]{a^5} + \sqrt[6]{a^2b^3}}{a - b}$

117. $\dfrac{8b + 10\sqrt{ab} + 3a}{4b - a}$ **119.** $9\sqrt{3}$

121. $2\sqrt[3]{6}$ **123.** $\dfrac{7\sqrt{15}}{225}$

125. $(x - 2)\sqrt{3x}$ **127.** $\dfrac{8x\sqrt{xy}}{15}$

129. $4\sqrt{2}$ **131.** $3x\sqrt[4]{3x}$

133. $3m^2n\sqrt[6]{16m^5n^3}$ **135.** $25\sqrt{2} - 40$

137. $3\sqrt{3} - 2\sqrt{15} - 3\sqrt{5} + 10$

139. $2a + 1 + 2\sqrt{a^2 + a - 6}$

141. $39 - 16\sqrt{5}$

143. $\dfrac{2}{3}\sqrt{6} - \dfrac{2}{3}\sqrt{3} + 1$

145. $\sqrt[3]{18} - 2\sqrt[6]{3}$ **147.** $\dfrac{11\sqrt{6} - 24}{25}$

149. $\dfrac{1}{21}(3\sqrt{2} - 4\sqrt{3} + 6\sqrt{6} - 9)$

151. $\dfrac{x\sqrt{6} + y\sqrt{6} - 5\sqrt{xy}}{2x - 3y}$ **153.** -3

155. $3 + 8i$ **157.** $2 + i$

159. $-1 - 8i$ **161.** $\dfrac{1}{2} - 2i\sqrt{2}$

163. $-11 - 23i$ **165.** $-\dfrac{13 + 9i}{20}$

167. $\dfrac{3 - 5i\sqrt{3}}{21}$ **169.** $\dfrac{2}{25} + \dfrac{29}{25}i$

exercise 7-1

1. $\{4, 2\}$ **3.** $\{-4, -6\}$ **5.** $\{3, 8\}$

7. $\{-4, -9\}$ **9.** $\{7\}$ **11.** $\{5, -5\}$

13. $\{9, -9\}$ **15.** $\{0, 1\}$ **17.** $\left\{0, \dfrac{2}{3}\right\}$

19. $\{7, -1\}$ **21.** $\left\{-2\dfrac{1}{3}, -5\right\}$

23. $\left\{\dfrac{3}{5}, -\dfrac{2}{3}\right\}$ **25.** $\{5a, -3a\}$

27. $\left\{\dfrac{a}{3}, \dfrac{a}{2}\right\}$ **29.** $\left\{\dfrac{2}{3a}, -\dfrac{3}{a}\right\}$

31. $\{0, 4a\}$

exercise 7-2

1. $\{3, -3\}$ **3.** $\left\{\dfrac{1}{2}, -\dfrac{1}{2}\right\}$

5. $\{\sqrt{3}, -\sqrt{3}\}$ **7.** $\{i, -i\}$

9. $\{\pm 1\}$ **11.** $\{\pm\sqrt{6}\}$ **13.** $\{\pm i\}$

15. $\{\pm 2i\}$ **17.** $\{4, -2\}$ **19.** $\{2, -8\}$

21. $\{2 \pm \sqrt{5}\}$ **23.** $\{4 \pm 2i\}$

25. $\{2, -1\}$ **27.** $\{5, 1\}$

29. $\{-4 \pm 2\sqrt{2}\}$ **31.** $\dfrac{1}{2}(1 \pm \sqrt{3})$

33. $\left\{1\dfrac{1}{2} \pm i\right\}$

exercise 7-3

1. 9   **3.** 16   **5.** $\dfrac{49}{4}$ **7.** $\dfrac{121}{4}$

9. $\dfrac{1}{16}$ **11.** $\dfrac{9}{64}$ **13.** $\{4, 2\}$

15. $\{5, -3\}$ **17.** $\{1, -6\}$

19. $\{6, -3\}$ **21.** $\left\{3, \dfrac{1}{2}\right\}$

23. $\left\{2, -\dfrac{2}{3}\right\}$ **25.** $\{3 \pm \sqrt{3}\}$

27. $\{3 \pm \sqrt{7}\}$ **29.** $\{-4 \pm 2\sqrt{2}\}$

31. $\{-1 \pm 2i\}$ **33.** $\{2 \pm 2i\}$

35. $\left\{-\dfrac{1}{3} \pm \dfrac{\sqrt{3}}{3}\right\}$ **37.** $\{1 \pm i\sqrt{2}\}$

39. $\left\{\dfrac{2 \pm i}{3}\right\}$

exercise 7-4

1. $\{3, 2\}$ **3.** $\{3\}$ **5.** $\left\{1\dfrac{1}{2}, 1\dfrac{1}{3}\right\}$

7. $\left\{\dfrac{-3 + \sqrt{5}}{4}, \dfrac{-3 - \sqrt{5}}{4}\right\}$

9. $\{1 + 2i, 1 - 2i\}$ **11.** $\left\{-\dfrac{1}{3} \pm \dfrac{1}{3}\sqrt{3}\right\}$

13. $\left\{\dfrac{1 \pm \sqrt{5}}{6}\right\}$ **15.** $\left\{1 \pm \dfrac{1}{4}\sqrt{6}\right\}$

17. $\left\{\dfrac{-3 \pm \sqrt{9 + 8k}}{4}\right\}$ **19.** $\left\{\pm \dfrac{3}{2}a\right\}$

21. $\{a \pm \sqrt{a^2 + c}\}$ **23.** $\left\{\dfrac{8 \pm i\sqrt{2}}{3}\right\}$

25. $\left\{\dfrac{2 \pm i\sqrt{2}}{3}\right\}$

exercise 7-5

1. $\{2, 3\}$ **3.** $\left\{\dfrac{1}{2}, -2\right\}$ **5.** $\{3\}$

7. $\{-1, 3\}$ **9.** $\{\pm 3i\}$

11. $\left\{\dfrac{-1 \pm \sqrt{5}}{2}\right\}$ **13.** $\left\{\pm \dfrac{1}{2}i\right\}$

15. $\left\{-\dfrac{2}{3}, 1\right\}$ **17.** $\left\{\dfrac{1}{2}, -1\right\}$

19. $\left\{\dfrac{-1 \pm i\sqrt{19}}{2}\right\}$ **21.** $\left\{\dfrac{8 \pm i\sqrt{2}}{3}\right\}$

23. $\{2\}$ **25.** $\{-1\}$

exercise 7-6

1. 7 and 9; -7 and -9

3. 2 and 7; -2 and -7

5. -2 and 7   **7.** 6 in. and 8 in.

9. 4 ft   **11.** 1 in.   **13.** $n = 7$

15. 2 mph   **17.** 30 mph; 50 mph

19. 8   **21.** 6 hr; 9 hr   **23.** \$1144.90

25. 20%

exercise 7-7

1. $\{33\}$ **3.** No solution   **5.** $\{2\}$

7. $\{5\}$ **9.** $\{2\}$ **11.** $\{3, 7\}$

13. $\{1\}$ **15.** $\{6, 2\}$ **17.** $\left\{\dfrac{9}{4}\right\}$

19. $\{23\}$ **21.** $\{5, -5\}$

exercise 7-8

1. $\{-2 < x < 2\}$

3. $\{x < -1 \text{ or } x > 1\}$

5. $\{-\sqrt{3} < m < \sqrt{3}\}$ **7.** $\{2 \le x \le 4\}$

9. $\{n < 3 \text{ or } n > 5\}$ **11.** $\{-3 < x < 2\}$

13. $\{r \le -4 \text{ or } r \ge 1\}$

15. $\{m < -3 \text{ or } m > -2\}$

17. $\{-6 < s < -2\}$

19. $\left\{-\dfrac{1}{2} \le x \le 4\right\}$

21. $\left\{x < 1\dfrac{1}{2} \text{ or } x > 5\right\}$

23. $\left\{-\dfrac{2}{3} < y < \dfrac{1}{2}\right\}$

25. $\left\{n < -1\dfrac{2}{3} \text{ or } n > -\dfrac{1}{2}\right\}$

27. $\left\{\dfrac{5 - \sqrt{5}}{2} \le x \le \dfrac{5 + \sqrt{5}}{2}\right\}$

29. $\left\{m \le \dfrac{-3 - \sqrt{3}}{2} \text{ or } m \ge \dfrac{-3 + \sqrt{3}}{2}\right\}$

31. $\left\{t < \dfrac{-3 - \sqrt{5}}{4} \text{ or } t > \dfrac{-3 + \sqrt{5}}{4}\right\}$

33. $\left\{x < \dfrac{1}{2} \text{ or } x > \dfrac{1}{2}\right\}$ **35.** $\{\ \}$

37. $\left\{0 < x < 2\dfrac{1}{2}\right\}$

39. $\left\{k < -\dfrac{1}{3} \text{ or } k > 0\right\}$

41. $\{-2 < x < 2\}$

43. $\{x < -7 \text{ or } x \geq 7\}$

45. $\{-3 < x < 0\}$

47. $\{m < -3 \text{ or } m \geq 2\}$

49. $\{0 < x \leq 1\}$ **51.** $\{x < -3 \text{ or } x \geq 1\}$

53. $\{-5 \leq x < -2\}$

REVIEW—CHAPTER 7

part a

1. T   **2.** T   **3.** T

4. F (doing so may result in loss of a solution)

5. T   **6.** F (all)

7. F (not always; e.g., 0 denominators)

8. T   **9.** T   **10.** T

part b

1. $\{3, 10\}$ **3.** $\{6, -3\}$ **5.** $\left\{\dfrac{3}{4}, -\dfrac{1}{3}\right\}$

7. $\left\{\dfrac{a}{5}, \dfrac{2a}{3}\right\}$ **9.** $\left\{0, \dfrac{2}{3}a\right\}$ **11.** $\{\pm 3\}$

13. $\{\pm 2i\}$ **15.** $\left\{\pm 1\dfrac{1}{2}\right\}$ **17.** $\{2, -6\}$

19. $\{3 + \sqrt{3}, 3 - \sqrt{3}\}$ **21.** $\left\{\dfrac{1}{3}, -1\right\}$

23. $36a^2$ **25.** $\dfrac{1}{100}$ **27.** $\dfrac{1}{4}a^2b^2$

29. $\{2 \pm \sqrt{5}\}$ **31.** $\{0, 3\}$

33. $\left\{\dfrac{1}{4}, -3\right\}$ **35.** $\left\{\dfrac{7 \pm \sqrt{41}}{2}\right\}$

37. $\left\{-2, \dfrac{7}{4}\right\}$ **39.** $\{-3 \pm \sqrt{22}\}$

41. $\{3 \pm i\}$ **43.** $\dfrac{1 \pm \sqrt{1 + 5a}}{2}$

45. $\{-a \pm 3\sqrt{a^2 + 1}\}$ **47.** $\left\{\pm\dfrac{2}{3}\sqrt{3}\right\}$

49. $\{1 \pm \sqrt{2}\}$ **51.** $\left\{\dfrac{3 \pm \sqrt{21}}{2}\right\}$

53. $\left\{0, 3\dfrac{1}{2}\right\}$ **55.** $\{3, 2\}$ **57.** 12 ft

59. -5 **61.** 4 mph

63. Up: 2 mph; down: 3 mph

65. $3 - \sqrt{5}$, or 0.764 in., approx.

67. $\{7\}$ **69.** $\{4, 1\}$ **71.** 10

73. $\{-2, -1\}$ **75.** $\{x < -5 \text{ or } x > 3\}$

77. $\left\{1\dfrac{1}{2} < x < 4\right\}$

79. $\{x \leq -1 - \sqrt{6} \text{ or } x \geq -1 + \sqrt{6}\}$

81. No solution   **83.** No solution

85. $\{x < -5 \text{ or } x > 5\}$

87. $\left\{a \leq \dfrac{1}{2} \text{ or } a > 3\right\}$

89. $\left\{-1\dfrac{1}{2} < y \leq 4\right\}$ **91.** $\{(4, 6), (2, 4)\}$

93. $\{(4, 7), (-4, -7)\}$

95. Between 1 in. and 3 in.

exercise 8-1

1. $(1, 2), (2, 0)$

3. $(1, -3), (2, -1), (4, 3)$ **5.** $(-4, -1)$

7. $(1, -2), (2, 4), (-1, -2)$

9. $8; -2; -4$ **11.** $-\dfrac{1}{5}; 2; 1\dfrac{1}{5}$

exercise 8-2

1–9.

11. $A: (3, 1); B: (-4, 2); C: \left(-2, -3\dfrac{1}{2}\right);$

$D: \left(2\dfrac{1}{2}, 0\right); E: \left(2\dfrac{1}{2}, -3\dfrac{1}{2}\right)$

13. Square   **15.** $(-3, 4), (2, 4), (2, -1)$

exercise 8-3

1. 9   **3.** 9   **5.** 3   **7.** 7
9. $\sqrt{26}$ **11.** 8   **13.** $4\sqrt{2}$
15. $2\sqrt{29}$ **17.** 5   **19.** $14\sqrt{2}$
21. $\sqrt{58}$ **23.** $AB = AC = \sqrt{221}$
25. $\sqrt{13}, 2\sqrt{13}, \sqrt{65}$; right triangle

exercise 8-4

13. Parallelism   **15.** $x = -8$

exercise 8-5

1. x intercept: $(4, 0)$; y intercept: $(0, 3)$

3. $(1, 0); (0, -5)$ **5.** $\left(-\dfrac{7}{2}, 0\right); \left(0, \dfrac{7}{3}\right)$

7. $(0, 0); (0, 0)$ **9.** $\left(-1\dfrac{1}{2}, 0\right); (0, -1)$

11. $\left(\dfrac{5}{3}, 0\right)$; no y intercept

13. $(2, 0)$ and $(-2, 0); (0, 4)$
15. $(1, 0)$ and $(-1, 0); (0, -1)$
17. $(-2, 0)$ and $(4, 0); (0, 8)$
19. $(-2, 0); (0, 2)$ and $(0, -1)$

exercise 8-6

1. -1 **3.** $-\dfrac{5}{3}$ **5.** $3\dfrac{1}{2}$ **7.** 0

9. $-4\dfrac{1}{2}$ **11.** Undefined   **13.** 2

15. $\dfrac{1}{2}$ **17.** $-\dfrac{2}{5}$ **19.** $\dfrac{2}{3}$

exercise 8-7

1. $3x - y - 3 = 0$ **3.** $3x + y - 7 = 0$
5. $x - 2y - 5 = 0$ **7.** $x + 3y + 2 = 0$
9. $3x + 4y + 8 = 0$ **11.** $x + y - 5 = 0$
13. $x - y - 4 = 0$ **15.** $3x - y + 2 = 0$
17. $2x + y + 3 = 0$ **19.** $y + 2 = 0$
21. $3x - 5y + 30 = 0$

23. $3x + 2y - 2 = 0$
25. $3x - y - 8 = 0$ **27.** $y - 4 = 0$
29. $3x - 2y + 7 = 0$

exercise 8-8

1.

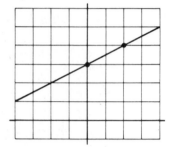

5.

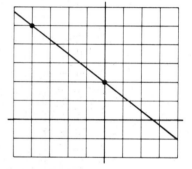

7.

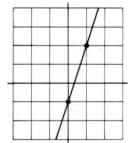

13.

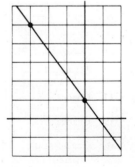

17.

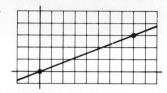

exercise 8-9

1. $(3, 0)$ **3.** $\left(\dfrac{1}{2}, 0\right)$ **5.** $(-3, 0)$

7. $(0, 0)$ **9.** $(3, 0)$

11.

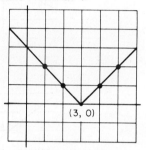

$(3, 0)$

13.

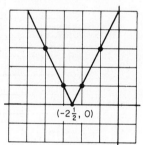

$\left(-2\dfrac{1}{2}, 0\right)$

15.

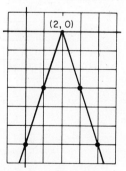

$(2, 0)$

17. Change at $\left(\dfrac{1}{2}, 0\right)$

19. Change at $(-2, 0)$

21. Change at $\left(-\dfrac{3}{4}, 0\right)$

23. Change at $(1, 0)$
25. Change at $(0, -3)$
27. Change at $(2, -2)$

exercise 8-10

1.

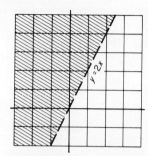

9.

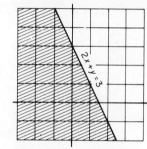

15.

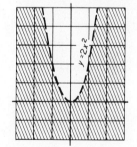

REVIEW—CHAPTER 8

part a

1. F (same pair, not same *ordered* pair)
2. T **3.** F (same ordinate) **4.** T
5. T **6.** T **7.** F (undefined)
8. F$\left(m = \dfrac{3}{2}; \text{intercept}, \left(0, -\dfrac{1}{2}\right)\right)$ **9.** T
10. T

part b

1. $(3, 4), (1, -2)$ 3. $(2, 0), (7, 3)$

5. $6\frac{1}{2}; -3; \frac{1}{4}$

7. (a) Parallelogram; (b) $\left(1, 1\frac{1}{2}\right)$; (c) $2\sqrt{2}; 5$

9. $(-3, 4), (3, -4)$ 11. $(-4, 4), (4, -4)$

13. 9 15. $2\sqrt{5}$ 17. 6 19. 5

31. $(3, 0); (0, -2)$ 33. $(0, 0); (0, 0)$

35. $(0, -4); \left(2\frac{2}{3}, 0\right)$

37. $(0, -4); (2, 0), (-2, 0)$

39. $(3, 0); (0, 3), (0, -1)$ 41. $1\frac{1}{2}$

43. $-\frac{3}{4}$ 45. Undefined

47. $3x - y - 2 = 0$

49. $3x - 2y + 1 = 0$

51. $3x + 2y - 3 = 0$

53. $3x - y + 2 = 0$ 55. $x + y = 0$

57. $3x + 4y + 20 = 0$ 59. $2x - y = 0$

61. $2x + 3y - 1 = 0$

63. $m = \frac{1}{3}; (0, -2)$ 65. $m = -2; (0, 1)$

67. $m = \frac{1}{2}; \left(0, -1\frac{1}{2}\right)$ 69. $m = \frac{2}{3}; (0, 0)$

71. Vertex $\left(\frac{2}{3}, 0\right)$; opens up

73. Vertex $(-2, 0)$; opens down

75. Vertex $(0, -1)$; opens up

77. Vertex $(-1, -3)$; opens down

exercise 9-1

1. Yes; single unique solution.

3. No; dependent   5. No; inconsistent

7. Yes; single unique solution

9. Yes; single unique solution   11. $(2, 3)$

13. $(-2, 3)$ 15. $\left(4, -\frac{1}{2}\right)$

17. $\left(-2\frac{1}{2}, 1\right)$ 19. $\left(-3\frac{1}{2}, -4\frac{1}{2}\right)$

21. $(-2, 0)$ 23. $\left(-3, -1\frac{1}{2}\right)$

exercise 9-2

1. $(2, 3)$ 3. $\left(-\frac{1}{2}, 3\frac{1}{2}\right)$ 5. $(5, 1)$

7. $(-2, -3)$ 9. $(7, 1)$ 11. $(0, -3)$

13. $\left(1, \frac{1}{2}\right)$ 15. $(1, -2)$ 17. $(-6, 2)$

19. $(2, -1)$

exercise 9-3

1. $(6, 1)$ 3. $(5, -1)$ 5. $(2, 1)$

7. $(-2, -3)$ 9. $(3, -2)$

11. $\left(\frac{3}{4}, -2\right)$ 13. $(-4, 6)$

15. Dependent   17. $(-22, -19)$

19. $\left(\frac{5}{7}, -\frac{1}{3}\right)$ 21. Dependent

23. Inconsistent   25. $\left(6, 2\frac{1}{2}\right)$

27. Inconsistent

exercise 9-4

1. $(2)(4) - (3)(1)$

3. $(2)(-4) - (-1)(3)$ 5. -11

7. -30 9. 15 11. 0 13. 0

15. 15 17. $-x$ 19. $-ab$

exercise 9-5

1. $(2, 3)$ 3. $\left(-\frac{1}{2}, 3\frac{1}{2}\right)$ 5. $(5, 1)$

7. $(-2, -3)$ 9. $(7, 1)$ 11. $(0, -3)$

13. Inconsistent   15. $(-3, -7)$

17. Dependent   19. $\left(2, \frac{1}{2}\right)$

exercise 9-6

1. $(1, 2, 3)$ 3. $(2, -2, -1)$

5. $(1, 2, -1)$ 7. $(3, 2, 1)$

9. $(0, 0, -4)$ 11. Dependent

13. $(-7, 8, 1)$ 15. $(1, 2, 3)$

17. $\left(\frac{1}{2}, \frac{1}{3}, -\frac{1}{2}\right)$ 19. $(4, 3, 2, 1)$

exercise 9-7

1. $2\frac{1}{2}, 3\frac{1}{2}$ 3. 7, 11 5. 7, −2

7. 4 nickels, 3 dimes 9. 23 ft by 15 ft

11. $1.40 first 3 min; 35¢ each add. min.

13. Pete: 12 days; Al: 6 days

15. Plane: 175 mph; wind 50 mph

17. 145 res.; 55 gen. 19. 8 m, 10 m, 12 m

21. A = 6 g, B = 4 g, C = 13 g

exercise 9-8

15. $\left(-4\frac{1}{2}, 3\right), (1, 3), \left(1, -\frac{2}{3}\right)$

17. $\begin{cases} x \le y \\ y \le 6 \\ x \ge 3 \end{cases}$

REVIEW—CHAPTER 9

part a

1. T 2. T 3. F (inconsistent)

4. T

5. F (The solution set of the resulting equation includes the solution of the system.)

6. T 7. T

8. F (Some systems will be dependent, some inconsistent.)

9. T 10. T

part b

1. Independent 3. Dependent

5. Independent 7. Inconsistent

9. Independent 11. $\left(3\frac{1}{2}, -2\right)$

13. Dependent 15. Inconsistent

17. (−2, −1) 19. (6, 8) 21. (−3, 0)

23. $\left(-\frac{1}{3}, 3\right)$ 25. (3, −4)

27. (1, 2, −1) 29. (−3, 4, 5)

31. (0, 4, −1) 33. 6 35. 29

37. 2 + mn 49. $4\frac{1}{2}, 7\frac{1}{2}$

51. 35 children, 125 adults

53. 7 of $8, 6 of $10 55. 40°, 60°, 80°

59.

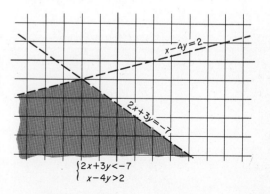

$\begin{cases} 2x+3y < -7 \\ x-4y > 2 \end{cases}$

exercise 10-1

1. $\log_2 8 = 3$ 3. $\log_3 243 = 5$

5. $\log_{16} 2 = \frac{1}{4}$ 7. $\log_7 \frac{1}{49} = -2$

9. $\log_3 0.33 = -1$ 11. $\log_{10} \frac{1}{100} = -2$

13. $\log_8 \frac{1}{2} = -\frac{1}{3}$ 15. $4^3 = 64$

17. $0.1^3 = 0.001$ 19. $8^{0.33} = 2$

21. $10^{0.301} = 2$ 23. $10^{1.230} = 17$

25. $p = 2^3$ 27. $L = 4$ 29. $a = 2$

31. $n = 25$ 33. $a = 36$ 35. $L = \frac{2}{3}$

37. $n = 125$ 39. $a = 0.8$

41. $n = \frac{1}{121}$ 43. $b = 3$ 45. $x = \frac{1}{16}$

47. $a = 16$ 49. $b = 4$

exercise 10-2

1. 4×10^3 3. 1.836×10^3

5. 3×10^6 7. 8.3×10^{-1}

9. 4×10^{-4} 11. 6.325×10^{-4}

13. (a) p = 1.4; (b) p = 2.4; (c) p = −0.6; (d) p = −1.6; (e) p = 3.4

15. 0.8657 17. 0.9375 19. 0.9031

21. 1.7803 23. 3.4771

25. −1 + 0.8000 or −0.2

27. −3 + 0.8 or −2.2

exercise 10-3

1. 2.5065   3. 1.7993   5. 1.6232
7. 9.5682 − 10   9. 3.6990
11. 7.7160 − 10   13. 2.0000
15. 9.9996 − 10   17. 3.4829
19. 5.2095   21. 6.5051 − 10
23. 0.7042

exercise 10-4

1. 26.0   3. 340   5. 0.0399
7. 74,300   9. 0.674   11. 168
13. 0.0626   15. 207,000   17. 0.474
19. 7970   21. 54.9
23. 0.0000000227 or 2.27×10^{-8}

exercise 10-5

1. 0.6651   3. 1.8714   5. 2.7188
7. 7.8064 − 10   9. 5.6281
11. 9.9899 − 10   13. 4.144
15. 813.6   17. 48,360   19. 72.52
21. 0.001701   23. 860,600

exercise 10-6

[*Note*: Results may differ slightly if interpolation is not used.]

1. 504   3. 520   5. 0.09045
7. 0.01762   9. 0.172   11. 245.9
13. − 26.96   15. 1.015×10^8
17. 24.56   19. 252.2   21. 8
23. 1.631   25. 0.3282   27. 1.148
29. 0.2614   31. 0.9915   33. − 18.87
35. 1.159   37. 637.3
39. 4.27×10^{-5}

exercise 10-7

[*Note:* Results may differ slightly if interpolation is not used.]

1. 300,800   3. 3.324×10^6
5. 0.001815   7. 2.524×10^{-7}
9. 422.7   11. 9.506×10^{-5}
13. 2.331×10^{-9} 15. 2.291   17. 4.729

19. 7.06×10^{-25} 21. 27   23. 3.7
25. 8   27. 7.984   29. 0.3958
31. 1.335   33. 0.5947   35. 0.2153
37. 0.8212   39. − 1.341   41. 0.9064
43. 0.02272   45. $1360   47. $1373
49. $767,300,000

exercise 10-8

[*Note*: Results may differ slightly if interpolation is not used.]

1. 3   3. − 3   5. 4   7. $\frac{1}{3}$
9. $-\frac{1}{3}$ 11. 9   13. $1, \frac{1}{2}$
15. 6, − 1   17. 2   19. 0.5
21. 0.5821   23. − 6.3   25. 1.649
27. − 9.318   29. 6.689   31. 0.8695
33. 1.108   35. − 10.31   37. 7 yrs
39. 3420 yrs

exercise 10-9

[*Note*: Results may differ slightly if interpolation is not used.]

1. 5.0687   3. 3.8382   5. 0.6258
7. − 0.2970   9. − 4.9818
11. − 2.7030   13. 3.4780
15. − 2.8402

REVIEW—CHAPTER 10

part a

1. T   2. T   3. T   4. T
5. T   6. T   7. T
8. F (0.308 has three significant digits; 0.038 has two)
9. T   10. F ($x = \log b \div \log a$)

part b

[*Note*: Results may differ slightly if interpolation is not used.]

1. $\log_{0.5} 0.25 = 2$ 3. $\log_2 64 = 6$
5. $\log_{64} 16 = 0.67$ 7. $9^2 = 81$

9. $16^{1/4} = 2$ 11. $10^{0.415} = 2.6$
13. $L = 2$ 15. $n = 4096$
17. $n = \dfrac{1}{512}$ 19. 3.864×10^3
21. 4.0×10^7 23. 7.89×10^{-4}
25. $p = 0.5740$ 27. $p = 0.7067$
29. $p = -0.3372$ 31. 2.9360
33. $4.0000 - 10$ 35. 4.5024
37. 2.4524 39. $4.5311 - 10$
41. 7.37 43. 130 45. 1.06×10^{-13}
47. 8.212 49. 74.98
51. 6.314×10^{-14} 53. 297.6
55. 5.325×10^{-4} 57. 0.06413
59. 3.425×10^{-7} 61. 69.9
63. 0.8628 65. 0.05721
67. 0.003564 69. 1.040×10^7
71. 7.063×10^{-18} 73. 263.4
75. $-10{,}640$ 77. 2.892
79. 0.5756 81. 0.2802 83. 3.941
85. 5 87. 3 89. $-\dfrac{1}{5}$
91. $1, -\dfrac{1}{2}$ 93. 2 95. 0.6637
97. -1.649 99. -0.2248
101. 0.3021 103. 3.6642
105. -3.6665 107. 0.4511
109. -2.3617 111. $\$160{,}900$
113. 12.3% 115. $155{,}300$
117. 9.44 hr

exercise A-1

1. $\{-6, 0, 9, 15\}$ 3. $\{11, 5, 2, -10\}$
5. $\left\{-\dfrac{2}{5}, -\dfrac{2}{3}, -1, -2\right\}$ 7. $\{4, 2, 5, 8\}$
9. $\left\{0, -\dfrac{1}{2}\sqrt{2}, \dfrac{1}{2}\sqrt{2}, \dfrac{3}{10}\sqrt{10}\right\}$
11. $y = f(x)$ 13. $y = f(x)$
15. $y = f(x)$ 17. $y \neq f(x)$
19. $y \neq f(x)$ 21. $y = f(x)$
23. All real numbers 25. $y \geq -1$
27. $y \geq 0$ 29. $y > -1$

exercise A-2

1. 4 3. 2 5. 1 7. $-\dfrac{1}{2}$

9. $2p^2 - 1$ 11. $2a^2 - 12a + 17$
13. $-\dfrac{1}{8}; 0$ 15. 1; undefined
17. $5; 1$ 19. 1; undefined 21. 3
23. 7 25. $\dfrac{5}{3}$ 27. 3 29. $4x + 5$
31. 3 33. $x < 3$
35. $A = f(x) = 8x$
37. $P = f(x) = 16 + 2x$
39. $A = f(r) = \pi r^2$ 41. $A = f(c) = \dfrac{c^2}{4\pi}$
43. 5 45. 24.

REVIEW—UNIT A

part a

1. T 2. F ("f of x") 3. T 4. T

part b

1. $21; 1$ 3. $3; 15$
5. Undefined; $\pm\sqrt{5}$ 7. $1\dfrac{3}{5}; 0$
9. $a - 2; \dfrac{a^2 + 2a}{a + 2}$ 11. -10
13. -1 15. $4p^2 - 12p + 5$
17. $x \geq \sqrt{3}$
19. All real numbers except 1
21. $x \leq -\dfrac{1}{2}, x \geq 3$ 23. $y \neq f(x)$
25. $y = f(x)$ 27. $y = f(x)$
29. $A = f(x) = 5x$ 31. 6

exercise B-1

1. $24, 27, 30$ 3. $1, -4, -9$
5. $81, 243, 729$ 7. $-16, 32, -64$
9. $4, 5, 6, 7$ 11. $-2, 1, 4, 7$
13. $3, 0, -3, -6$ 15. $2, 8, 18, 32$
17. $3, 2, \dfrac{5}{3}, \dfrac{3}{2}$ 19. $-\dfrac{1}{3}, 0, \dfrac{1}{9}, \dfrac{1}{6}$
21. 20 23. 24 25. 6 27. -58
29. $2\dfrac{2}{15}$

exercise B-2

1. 37, 43, 49 **3.** $-50, -102, -154$

5. $-\dfrac{1}{12}, -\dfrac{1}{3}, -\dfrac{7}{12}$ **7.** 119 **9.** 54

11. $-1\dfrac{3}{4}$ **13.** $-16\sqrt{2}$ **15.** 13

17. 9 **19.** $\dfrac{3}{4}$ **21.** 17 **23.** $-\dfrac{5}{7}$

25. 9, 15, 21 **27.** $-4, -15, -26$

29. 2

exercise B-3

1. $d = 3, S_n = 282$ **3.** $a_1 = 7, S_n = 160$
5. $n = 17, d = 4$ **7.** $a_1 = 13, a_n = -92$
9. $a_1 = -7, n = 15$
11. $a_1 = 5, a_n = 23, d = 3, S_n = 98$
13. \$6750

15. (a) 624 ft; (b) 6400 ft; (c) $37\dfrac{1}{2}$ sec

exercise B-4

1. 54, 162, 486 **3.** $-27, 81, -243$
5. $0.024, -0.0048, 0.00096$ **7.** 243
9. 256 **11.** $32\sqrt{2}$ **13.** 1 **15.** 5
17. -2 **19.** 3 **21.** 0.1
23. $-\sqrt{3}$ **25.** $\pm 12, 36, \pm 108$

27. $1, -1\dfrac{1}{2}, 2\dfrac{1}{4}, -3\dfrac{3}{8}$ **29.** $\dfrac{1}{3}$ and 7

exercise B-5

1. $a_2 = 1, S_n = 63$ **3.** $a_n = 2, S_n = 86$

5. $a_1 = 16, a_n = -\dfrac{1}{2}$ **7.** $n = 7, r = 2$

9. $729, -243, 81, -27, 9, -3, 1$
11. (a) \$384; (b) \$12,285

13. $64\sqrt{2}\,n$ or $91n$ (approx.) **15.** $117\dfrac{2}{9}$ ft

17. 16 **19.** $6\dfrac{2}{3}$ **21.** $\dfrac{7}{9}$ **23.** $\dfrac{67}{300}$

REVIEW—UNIT B

part a

1. T **2.** T **3.** T
4. F (geometric) **5.** T **6.** F (4)
7. F (only when $-1 < r < 1$)

part b

1. 56, 62, 68 **3.** $-\dfrac{1}{12}, -\dfrac{5}{6}, -1\dfrac{1}{4}$

5. 57 **7.** $-21\sqrt{3}$ **9.** $8\dfrac{1}{2}$

11. 19 **13.** 11, 17, 23 **15.** 3
17. $a_1 = 5, S_n = 549$
19. $a_1 = 60, a_n = 0$
21. $a_1 = 1, d = -4, a_n = -19, S_n = -54$
23. (a) 496 ft; (b) 4096 ft; (c) 35 sec
25. $-64, 128, -256$

27. $-\dfrac{1}{72}, \dfrac{1}{432}, -\dfrac{1}{2592}$ **29.** 972

31. $243\sqrt{3}$ **33.** 5 **35.** -2

37. $-\sqrt{2}$ **39.** $1, -1\dfrac{2}{3}, 2\dfrac{7}{9}, -4\dfrac{17}{27}$

41. $a_1 = 3, S_n = 363$
43. $a_1 = 256, a_n = -108$

45. $32, -16, 8, -4, 2, -1, \dfrac{1}{2}$

47. $32\sqrt{2}\,n$ or $45n$ (approx.) **49.** \$3.22

51. 20 **53.** $\dfrac{2}{3}$ **55.** $\dfrac{8}{9}$ **57.** $\dfrac{12}{275}$

exercise C-1

1. $a^8 + 8a^7b + 28a^6b^2 + 56a^5b^3 + 70a^4b^4 + 56a^3b^5 + 28a^2b^6 + 8ab^7 + b^8$
3. $x^5 - 10ax^4 + 40a^2x^3 - 80a^3x^2 + 80a^4x - 32a^5$
5. $729x^6 - 2916x^5y + 4860x^4y^2 - 4320x^3y^3 + 2160x^2y^4 - 576xy^5 + 64y^6$

7. $\dfrac{1}{32} x^5 - \dfrac{5}{48} x^4 y + \dfrac{5}{36} x^3 y^2 - \dfrac{5}{54} x^2 y^3$
$+ \dfrac{5}{162} xy^4 - \dfrac{1}{243} y^5$

9. $1 + 0.12 + 0.0054 + 0.000108$
$+ 0.00000081$

11. $a\sqrt{a} - 3a\sqrt{b} + 3b\sqrt{a} - b\sqrt{b}$

13. $a^8 - 32a^7 b + 448a^6 b^2 - 3584a^5 b^3 + \cdots$

15. $a^{10} - 40a^9 b + 720a^8 b^2 - 7680a^7 b^3 + \cdots$

17. $16{,}384x^{14} + 114{,}688x^{13} + 372{,}736x^{12}$
$+ 745{,}472x^{11} + \cdots$

19. $729 - 5832\sqrt{3} + 64{,}152 - 142{,}560\sqrt{3}$
$+ \cdots$

21. $1 + 0.3 + 0.0405 + 0.00324 + \cdots$

exercise C-2

1. $25{,}344a^7$ **3.** $-414{,}720x^7$

5. 0.0000567 **7.** $-220a^3 b^9$

9. $2268a^6 b^3$

REVIEW—UNIT C

part a

1. T   **2.** F (18 terms)   **3.** T

4. F (does not include 0)

part b

1. $a^9 + 9a^8 b + 36a^7 b^2 + 84a^6 b^3$
$+ 126a^5 b^4 + 126a^4 b^5 + 84a^3 b^6 + 36a^2 b^7$
$+ 9ab^8 + b^9$

3. $64x^6 - 576x^5 y + 2160x^4 y^2 - 4320x^3 y^3$
$+ 4860x^2 y^4 - 2916xy^5 + 729y^6$

5. $1 + 0.16 + 0.0096 + 0.000256$
$+ 0.00000256 = 1.1698586$

7. $a^9 - 27a^8 b + 324a^7 b^2 - 2268a^6 b^3 + \cdots$

9. $16{,}384x^{14} - 114{,}688x^{13} + 372{,}736x^{12}$
$- 745{,}472x^{11} + \cdots$

11. $1 + 0.2 + 0.018 + 0.00096 = 1.21896$

13. $414{,}720x^7$ **15.** $220a^3 b^9$

INDEX